Manufacturing Planning and Control

Manufacturing Planning and Control

Patrik Jonsson
and
Stig-Arne Mattsson

McGraw-Hill
Higher Education

London Boston Burr Ridge, IL Dubuque, IA Madison, WI New York San Francisco
St. Louis Bangkok Bogotá Caracas Kuala Lumpur Lisbon Madrid Mexico City
Milan Montreal New Delhi Santiago Seoul Singapore Sydney Taipei Toronto

Manufacturing Planning and Control.
Patrik Jonsson and Stig-Arne Mattsson
ISBN-13 978-0-07-711739-9
ISBN-10 0-07-711739-5

McGraw-Hill
Higher Education

Published by McGraw-Hill Education
Shoppenhangers Road
Maidenhead
Berkshire
SL6 2QL
Telephone: 44 (0) 1628 502 500
Fax: 44 (0) 1628 770 224
Website: *www.mcgraw-hill.co.uk*

British Library Cataloguing in Publication Data
A catalogue record for this book is available from the British Library

Library of Congress Cataloging in Publication Data
The Library of Congress data for this book has been applied for from the Library of Congress

Acquisitions Editor: Rachel Gear
Development Editor: Jennifer Rotherham
Marketing Manager: Mark Barratt
Production Editor: Louise Caswell

Text Design by Ken Vail Design
Cover design by SCW
Printed and bound by CPI Group (UK) Ltd, Croydon, CR0 4YY

ISBN-13 978-0-07-711739-9
ISBN-10 0-07-711739-5

The *McGraw·Hill* Companies

Brief Table of Contents

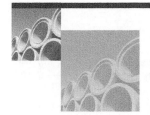

Detailed Table of Contents

Guided Tour

Learning Objectives

At the beginning of each chapter, you are provided with a list of learning objectives to help you identify what you should have learnt after reading the chapter.

Key Concepts and Glossary

The key terms are highlighted the first time they are introduced, alerting you to the core concepts and techniques in each chapter. A full explanation is contained in the glossary at the end of the book.

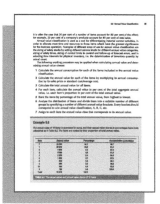

Problems

Some of the chapters feature problems to help explain concepts. Full solutions to the problems are provided in Appendix A.

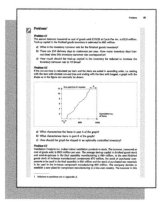

Examples

Boxed examples show how a particular concept or idea is used in practice.

Figures and Tables

Each chapter provides a number of figures, illustrations and tables to help you visualise the examples. Descriptive captions summarise important concepts and explain the relevance of the illustration.

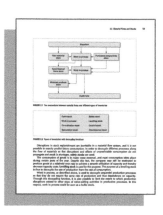

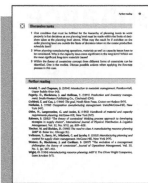

Discussion Tasks

This feature encourages you to review and apply the knowledge you have acquired from each chapter.

Summary

This briefly reviews and reinforces the main topics that you will have covered in each chapter to ensure that you have acquired a solid understanding of the key topics. Use it as a quick reference to check you have understood the chapter.

Further Reading

Found at the end of each chapter, these provide you with the opportunity to research the subject further.

Cases and Discussion Questions

This book includes cases at the end of many of the chapters to illustrate current practice and key concepts defined and described in the book. Each case is followed by a set of related questions to help you critically apply your understanding and further develop some of the topics introduced to you.

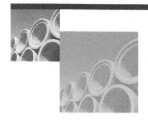

Technology to enhance learning and teaching

*Visit **www.mcgraw-hill.co.uk/textbooks/jonsson** today*

Online Learning Centre (OLC)

Resources for students include:

- *Glossary*
- *Weblinks*
- *Learning Objectives*

Also available for lecturers:

- *Instructor's Manual*
- *PowerPoint slides*
- *Answers to discussion questions*

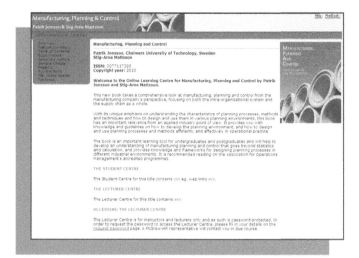

Custom Publishing Solutions: Let us help make our **content** your **solution**

At McGraw-Hill Education our aim is to help lecturers to find the most suitable content for their needs delivered to their students in the most appropriate way. Our **custom publishing solutions** offer the ideal combination of content delivered in the way which best suits lecturer and students.

Our custom publishing programme offers lecturers the opportunity to select just the chapters or sections of material they wish to deliver to their students from a database called Primis at www.primisonline.com.

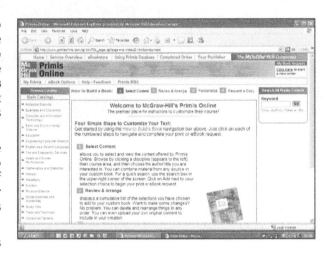

Primis contains over two million pages of content from:

- textbooks
- professional books
- case books – Harvard Articles, Insead, Ivey, Darden, Thunderbird and BusinessWeek
- Taking Sides – debate materials

Across the following imprints:

- McGraw-Hill Education
- Open University Press
- Harvard Business School Press
- US and European material

There is also the option to include additional material authored by lecturers in the custom product – this does not necessarily have to be in English.

We will take care of everything from start to finish in the process of developing and delivering a custom product to ensure that lecturers and students receive exactly the material needed in the most suitable way.

With a **Custom Publishing Solution**, students enjoy the best selection of material deemed to be the most suitable for learning everything they need for their courses – something of real value to support their learning. Teachers are able to use exactly the material they want, in the way they want, to support their teaching on the course.

Please contact *your local McGraw-Hill representative* with any questions or alternatively contact Warren Eels **e:** *warren_eels@mcgraw-hill.com*.

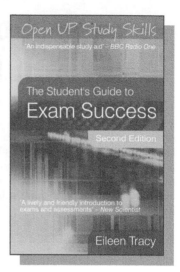

Acknowledgements

Our thanks go to the following reviewers for their comments at various stages in the text's development:

Jalal Ashayeri, University of Tilburg
John Bicheno, Cardiff University
Derek Ford, University of Cambridge
Willem van Gronendaal, University of Tilburg
Chandra Lalwani, University of Hull
Morag Mallins, Cardiff University
Henny van Ooijen, Eindhoven
Mikael Rönnqvist, Norwegian School of Economics and Business Administration
Martin Rudberg, Linköping University
Willem Selen, Middle East Technical University
Anders Thorstenson, The University of Aarhus

Every effort has been made to to trace and acknowledge ownership of copyright and to clear permission for material reproduced in this book. The publishers will be pleased to make suitable arrangements to clear permission with any copyright holders whom it has not been possible to contact.

Authors' Acknowledgements
We would like to acknowledge Peter Olsson at Chalmers University of Technology for writing some of the end of chapter problems and Anna Fredriksson, Mats Johansson and Linea Kjellsdotter (all at Chalmers University of Technology) for their feedback on the manuscript. All companies allowing us to use their material in case studies and illustrations also need to be thanked. We are also grateful to all reviewers who have provided valuable feedback and suggestions on semi-finished book chapters, and to the editorial team at McGraw-Hill – Rachel Gear, Karen Harlow and Jennifer Rotherham – who have supported us in a great way during the entire development process of the book.

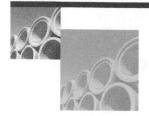

About the Authors

Patrik Jonsson is Professor of Operations and Supply Chain Management at Chalmers University of Technology. He holds a PhD in production management from Lund University and is CFPIM and CSCP certified through APICS. He has during the last 15 years co-ordinated and taught numerous operations and supply chain management courses at Växjö University and Chalmers University of Technology and in executive and further education programmes for people working in industry, e.g. the CPIM and CSCP certification programmes of APICS. His research, which mainly has dealt with manufacturing planning and control, information systems, supplier structures and relations and performance management, has been published in journals such as *Journal of Operations Management, International Journal of Operations and Production Management, International Journal of Production Research, International Journal of Production Economics, Supply Chain Management: An International Journal* and *International Journal of Physical Distribution and Logistics Management.*

Stig-Arne Mattsson has some 30 years of experience in operations management, supply chain management and information systems, from national as well as international companies. Parallel to his industrial career he has during 12 years been adjunct Professor in Supply Chain Management, first at Växjö University and later at Lund University. He is the former president of the logistics association PLAN, Swedish Production and Inventory Management Society, an affiliate to APICS. He has also for several years been a member of the board of ELA, the European Logistics Association, in Brussels. Stig-Arne has a licentiate degree in Production Management and is certified CFPIM by APICS and certified Eslog by ELA. He has taught numerous courses for practitioners, e.g. within the CPIM certification programmes of APICS. He has also written a number of textbooks and papers published in journals such as *International Journal of Production Research, International Journal of Operations and Production Management* and *International Journal of Physical Distribution and Logistics Management.* He is a frequent speaker at conferences and seminars.

Preface

The purpose of this book is to convey a complete description of the subject area of manufacturing planning and control, and to supplement traditional ways to look at different planning processes, methods and theories in terms of their characteristics and applicability, both on a conceptual and methodological level and with respect to the planning environment in which they will be applied.

The book is written for university courses in technical and business programmes, for the APICS (the Association for Operations Management) CPIM certification programme, and for people in industry working with manufacturing planning and control. Focus is on the design and use of manufacturing planning and control processes and methods. The aim has been to take perspectives and cover issues which are of important practical relevance for manufacturing companies when designing and using approaches, processes and methods for manufacturing planning and control. Therefore, some issues of more academic interest are not emphasised and covered to the same extent.

The book is divided into four parts and 18 chapters. Parts 1 and 2 describe the starting points and preconditions for manufacturing planning and control. Parts 3 and 4 cover the design and use of manufacturing planning and control processes and methods on different hierarchical levels, starting with forecasting and long-term planning and ending with materials planning, execution and control. Each chapter starts with learning objectives and ends with key terms lists and discussion tasks. Definitions of the key concepts are available in the Glossary. Chapters including quantitative planning methods also end with quantitative problems with full solutions in the Appendix. In order to give practical illustrations, cases with discussion questions are included in the 11 chapters focusing on planning processes and methods. All cases are based on real-life companies but some are made anonymous.

APICS has defined much of the vocabulary and concepts in manufacturing planning and control. The ambition has been to use the APICS language when applicable. Detailed instructions of how the book supports and can be used in the APICS CPIM certification programme is available on the OLC.

PART 1
Starting Points of Manufacturing Planning and Control

The first part of the book contains Chapters 1–3 and covers the starting points of manufacturing planning and control, including definitions, objectives, perspectives and structures. Chapter 1 defines the subject field and explains how it mainly concerns management of production and the flow of materials in a manufacturing company. It also gives a retrospective and introduces related concepts and philosophies. Chapter 2 describes in more detail what is meant by production and material flows. It also provides an overall categorisation of manufacturing companies, partly related to production conditions and partly to material flow conditions. The appropriateness of manufacturing planning and control strategies depends on the unique situation, the context. To be able to describe different problem contexts it is necessary to characterise and classify the conditions that exist in a company. The purpose of the categorisation presented in the chapter is thus to enable the positioning of an individual company from a manufacturing planning and control perspective. The planning of production and material flows is normally carried out in a hierarchical structure of planning processes, each of which has different time horizons and degrees of detail. It is within this framework that the planning methodology presented in later chapters of the book is used. It may thus be described as a frame of reference for the planning of production and material flows in manufacturing companies. This planning structure is introduced in Chapter 3.

Part Contents

Introduction and Background

For some considerable time there has been much discussion on the value of approaches, methods and information technology (IT) support used in the planning and control of material flows and production. The theme has been taken up at many conferences and seminars, and numerous articles in newspapers, magazines and journals have treated the subject from different viewpoints. Much attention has been paid to the subject by logistics and production personnel in industry. Many people have questioned investments made, and claim that results achieved in terms of reducing tied-up capital, for example, have been extremely modest.

Developing and introducing manufacturing planning and control software systems is thought by many to be complicated, expensive and time-consuming, not least during their initial introduction. In some companies their failure is almost immediate. In a number of cases, installation projects have even been terminated. The American manufacturing planning and control pioneer, George Plossl, in an article concerning an **enterprise resource planning (ERP)** system, paraphrased Churchill as follows: "Never in the history of materials and production management has so much been promised and expected, and so little actually obtained." This still holds true for several companies.

A simplistic approach has evolved after all the difficulties that have been encountered. People say that Western industry has taken the wrong path, and that the use of so many complex planning methods and planning systems – which cannot be made to work properly anyhow – should be abandoned. Simple, easily understood and more user-friendly approaches and tools should be introduced to replace them. There have also been examples of companies that have gone back to using planning methods scrapped decades ago because they were considered inadequate.

However, other conclusions have also been drawn from the sometimes less successful attempts to bring more efficiency to companies' manufacturing planning and control. Planning methods and planning systems can never be made to operate efficiently if the personnel that use them do not understand their characteristics and how they must be used. This is true whether they are built on simple or complex theoretical relationships. Neither can they function efficiently if they are used in the wrong context and under conditions they were not intended for, in which they are unable to prove their worth. A minimum of basic knowledge and understanding of the subject area and of the systems and methods that can be used to control

3

material flows and production is always a first condition for achieving results that can be reasonably expected.

1.1 The Manufacturing Company

The manufacturing company and its environmental relationships may be considered as flows of materials, money and information, as shown in Figure 1.1. The large, solid arrows indicate flows of material, narrow solid arrows indicate flows of information and dotted arrows indicate monetary flows.

The raw materials, components and other semi-finished items required for manufacturing a company's products are sourced from external suppliers. This sourcing takes place principally in the form of deliveries to stock for later usage in production or in the form of deliveries for direct usage. Flows of information initiate the flows of these materials.

The processing of the materials procured takes place during production. This processing, and value-adding activity generally, is carried out in a number of steps and results in manufactured parts and semi-finished items. The manufactured items may be delivered to stock for later usage or direct for use in end-product manufacturing/final assembly operations. It is controlled by the flow of information. The company's final products are manufactured either to stock for future deliveries to customers or directly to customer orders. These activities are also controlled by the flow of information.

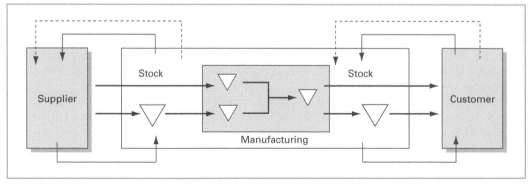

FIGURE 1.1 The manufacturing company as a flow of materials, information and money. Solid arrows indicate flows of material, narrow solid arrows indicate flows of information and dotted arrows indicate monetary flow.

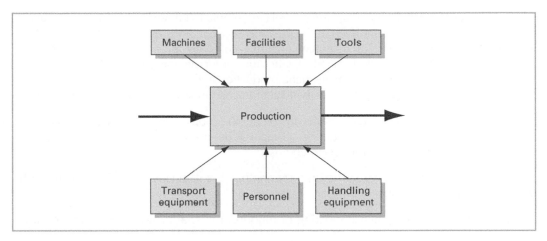

FIGURE 1.2 Production resources for value adding in the flow of materials

Somewhat simplified, we can state that in every supply chain there is a flow of materials that starts at the company's suppliers and ends with its customers. There is a monetary flow in the opposite direction. Customers pay for products delivered and the company in turn pays suppliers for the delivery of raw materials and various components. The differences in value of both these monetary flows consist of direct production costs and contributions to other costs and profits in the business. Monetary flows are also initiated by flows of information.

Flows of information that are connected with material flows in supply chains are two-way, both between the company and its suppliers and between the company and its customers. Information flows from customers to the company and from the company to suppliers are of the demand type, whereas flows of information in the opposite direction are of the supply type.

In general terms, trading companies may be described in the above fashion from the flow perspective. One particular characteristic of manufacturing companies is the value adding that takes place during production. This value adding is achieved with the help of a number of different types of production resources, as shown in Figure 1.2. Flows of resources are used to achieve value-adding processes. Thus, a manufacturing company can be characterised by a number of material flows, monetary flows, resource flows and information flows.

1.2　The Concept of Manufacturing Planning and Control

This book is about the management of material flows and production processes in manufacturing companies. The area consists of a materials management part and a production management part which can each be defined in the following ways:

Materials management, referring to planning, control and follow-up of material flows from suppliers to customers, e.g. decisions about what and when to produce, deliver and replenish stocks.

Production management, referring to planning, control and follow-up of the use of resources for production, e.g. decisions about access to appropriate capacity in order to efficiently produce, deliver and replenish stocks.

The concept of materials management is related to material flows and production management to resource flows, as discussed in the previous section. Since the material flow processes and the value-adding processes in a manufacturing company are intimately connected with each other, both terms are often joined together into one common concept. It is traditionally called production and inventory management, production and materials management, or manufacturing planning and control. The term manufacturing planning and control is used in this book.

Manufacturing planning and control is one part of the subject area of logistics, often defined as planning, development, co-ordination, organisation, management and control of material flows from raw materials suppliers to end-users.

Using such a definition of logistics as a starting point, a company's logistics system can be divided into three sub-systems:

- A material supply system consisting of inbound material flows to the company from suppliers
- A production system consisting of the flow of materials in the company during value-adding processes
- A distribution system consisting of the outbound flow of materials from the company to customers.

The main emphasis in this book is on the logistics of the production system. However, it will not be limited to this sub-system alone. The three above sub-systems are closely integrated with each other in most companies and constitute a supply chain. This implies that the entire flow of materials is considered but, in contrast to the total concept of logistics, it is considered from the perspective of production.

Manufacturing planning and control decisions are made on different time horizons and on different degrees of detail. Decisions that are six months or a year ahead in the future, for example, could be about capacity investments in a plant or securing the quality and volume of inbound material flows from suppliers. More short-term and detailed decisions concern, for example, what products to make each week, what and when to purchase individual items and the actual execution of production of unique customer-specific products. The book covers both the materials and capacity perspectives of manufacturing planning and control, and approaches and methods to planning with different time horizons and levels of detail.

1.3 A Retrospective

Manufacturing planning and control is a relatively recent subject area. In the professional arena, it does not go back for much more than some 50 years. The theory began seriously to emerge during the 1950s, and it was then that the existence of specialized managerial posts for planning of production operations started to be seen on a larger scale.

The roots of the subject are older than this, though. The first parts lists showing the bill of material for a product were used for planning purposes as early as the turn of the last century. In 1913, **Harris** derived the square root **formula** for calculating economic order quantities which is still widely used today. The next large step in development was in 1934 when **Wilson** developed the theory of the reorder point (see **Wilson's formula** in Chapter 12) and published his results in the *Harvard Business Review*. Wilson combined Harris's order quantity calculation with his own re-order point theory and produced a complete inventory management concept.

He also claimed that if his methods were introduced, companies could reduce their tied-up capital by 15 per cent and the number of shortages by 20 per cent. But neither Harris's nor Wilson's method[1] was used to any great extent until the 1950s.

The first specialized posts within the area of production management were the so-called expediters, or order hunters, who were introduced at the beginning of the 1940s. Their task was to keep track of and prioritise materials needed for fabrication and assembly. Planning posts in a more comprehensive form did not arrive until the 1950s. In comparison with the situation today, both in terms of concepts and organisation, the areas of material management and production management were kept separate. This can be observed in organisation charts and other documents from that time. In general terms, this dual approach and form of organisation continued throughout the 1960s. During the 1970s, a more holistic view on manufacturing planning and control concepts was introduced. This development left its mark on organisation charts and the naming of posts at companies.

Until the 1950s, production was based largely on customer orders. Acceptable delivery times to customers were so long that most of the manufacturing and purchasing could be carried out within this time frame. Manufactured quantities in fabrication and assembly were usually equivalent to the direct needs of each customer order, and it was generally possible to apply a make to order strategy to manufactured parts and semi-finished items. As a result of these conditions, relatively simple methods and planning tools could be used.

Conditions for planning changed during the 1950s. Acceptable delivery times to customers became shorter at the same time as companies' manufacturing lead times increased as product complexity grew. The increased lead times were also a result of rationalization measures in production at the time, with the purpose of reducing manufacturing costs. The prevalent perspective in companies was production oriented, leading to low manufacturing costs being prioritised over throughput times and delivery performance. This all resulted in an increased dependency on forecasts, not least for managing the flow of raw materials and semi-finished items.

Another significant change as a result of this striving to cut manufacturing costs was that set-up times for machines and production plants, and consequentally ordering costs, increased dramatically. For reasons of capacity utilization and costs it was no longer acceptable to manufacture only quantities required to cover individual customer orders on each manufacturing occasion. Companies started instead to manufacture for future known or expected needs; the make to stock strategy had commenced.

To be able to manage make to stock manufacturing, access to new supporting tools was required and the methods already developed for calculating economic order quantities and management of re-order points started to be used on a wider scale. The methodology was also refined. During the Second World War, a new area of knowledge called operations research had been developed by the British and American military powers. The latter half of the three decades after the war was the heyday of operations research theory, including its industrial applications. Inventory management theory was one of its main areas, and a multitude of books were published describing mathematical methods and algorithms for determining re-order points and order quantities. A negligible amount of this theoretical development was put into practice.

In many companies that introduced the new materials management methods, problems were certainly solved in a reasonable way, for example concerning the determination of due dates and quantities when releasing new manufacturing orders. However, new problems were also created as a result of using the methods, which in many cases were more difficult to handle

---

1   Harris, F. (1913) "How many parts to make at once", *Factory*, Vol. 10, No. 2, pp. 135–136; Wilson, R. (1934) "A scientific routine for stock control", *Harvard Business Review*, Vol. 13, pp. 116–128.

than the old ones. Tied-up capital in inventory grew, and the number of shortages became embarrassingly large, causing disruptions in production, productivity losses and low delivery-performance to customers.

Out of the ashes of many failures resulting from the uncritical introduction of systems like re-order point systems, new planning methods arose which were almost exclusively developed by practical men in companies. Examples of such methods are gross requirements calculations, cyclic planning and re-order point systems supplemented with shortage list procedures. These methods were dominant during the whole of the 1960s and the first half of the 1970s.

It was realised at the beginning of the 1960s that the theoretical assumptions underlying re-order point methods were insufficient for managing inventories of materials in manufacturing companies. A significant step in development was taken when Joseph Orlicky clarified the difference between independent and dependent demand. Another great pioneer in this area, Oliver Wight, wrote an article in 1968 entitled "To order-point or not to order-point".[2] This article became the foundation of the next generation of material planning systems.

Computer-based systems had started to be used for manufacturing planning and control around 1960. In particular, the advent of disk memory created conditions for the application of more powerful methods. **Material requirements planning (MRP)** was developed and started to be applied at the beginning of the 1960s. This time it was primarily computer companies, with IBM in the lead, which provided methods development. Material requirements planning had a great impact and in USA there was somewhat of a crusade for a decade for the introduction of MRP systems.

The introduction of MRP was a great success for many companies, although there were a number of less successful implementations. By the end of the 1970s companies were beginning to realise that MRP alone could not provide the improvements in performance that had been expected. The MRP systems were supplemented with extended master planning functions, more consideration was given to the issue of capacity, and improved reporting from manufacturing was introduced. New concepts and a new generation system, the so-called **manufacturing resource planning (MRP II)** systems, were developed during the 1980s. The concept was further developed during the 1990s and the term **enterprise resource planning (ERP)** is now widely used. An ERP system is a database and a collection of software programs which provide and process information required for administrative management and control of activities in a company. It includes a database which is shared between all program functions and which can provide the necessary information for all business processes. Using this system's architecture, every user has online access to all information needed to carry out planning, and personnel can work in the same software system without problems with different interfaces – which is often the case if several different functional systems are involved. Enterprise resource planning systems are divided into modules, e.g. customer order, forecasting, master planning, material planning, production activity control and procurement modules. In later versions of ERP systems, the basic modules have been supplemented with modules which support enhanced performance measurements, alternative production strategies such as make to order and just-in-time, integration with other planning and execution systems, tools for co-ordinated planning, electronic commerce, etc. The ERP system has changed from being a more or less closed system which focused on planning and optimising one company's activities to an open system aimed at creating value through integration with processes in other companies. The new generation of ERP systems are sometimes called ERP II or extended ERP.

---

2   Wight, O. (1968) "To order point or not to order point", *Production and Inventory Management Journal*, Vol. 9, No. 3, pp. 13–18.

This description of history so far has been mainly related to materials management. A corresponding development has taken place in production management and specifically within the areas of capacity planning and production scheduling. The advent of computers was the starting point of new methods development and it became, for example, possible to schedule manufacturing orders to available capacity.

There is an interesting difference between the development of capacity planning and production scheduling methods in the USA and in Europe. In the USA, production planning and scheduling methods and systems for capacity planning had a very little significance during the 1950s and 1960s. Companies concentrated more on materials management, whereas in Europe the situation was the reverse. The differences have been explained in terms of repercussions after the Second World War. European industry had been largely destroyed during the war and for many years the greatest problem was obtaining and efficiently utilising available capacity. Methods that favoured these purposes were given priority and the dominant approach was that of production. This was especially the case in Germany, which became the leader in the development and application of systems for scheduling to finite capacity.

The opposite situation was prevalent in the USA. American industry was intact and the problem was not of obtaining capacity but of achieving flows of materials that were as effective as possible, and materials management systems were given priority for this reason. Another explanation is the American attitude to labour. With a more widespread use of hiring and firing at short notice, the need for capacity planning and other production management methods was reduced.

In the early 1980s Eli Goldratt released the optimized production technology (OPT) software. It was a scheduling software with somewhat different logic compared with the traditional software. It focused on identifying and fully utilising constraints in the material flow and subordinated the entire production system to these constraints. The basic principle of the OPT software was discussed in the industry novel, *The Goal*,[3] and has later been developed into a philosophy about constraint based planning, scheduling and management. This philosophy is often called theory of constraints (TOC).

Introducing and applying computerised production and materials management software systems is no trivial task. The difficulties of succeeding in this have also increased when moving from a production-oriented strategy to an increasingly customer and market-oriented approach to how manufacturing companies should carry out operations. Larger demands have been put on flexibility, shorter delivery times and more product variants. New approaches and methods for how manufacturing planning and control should be executed in order to meet these requirements have been developed since the 1980s. This development has been largely inspired by Japanese production philosophy, especially the Toyota production system (TPS). Approaches such as **just-in-time** and **lean production** started to be used. These approaches can be characterised by a striving towards eliminating waste such as excessive production resources, overproduction, excessive inventories and unnecessary capital investment, simplifying and improving conditions for planning and control, and promoting a flexible production organization. These approaches have sometimes also been called pull or synchronised manufacturing strategies because they focus on pulling material through the production system in a continuous material flow, based on the customer demand. The fundamental idea of lean production, namely waste elimination and simplification, has had a strong impact on manufacturing companies since the 1990s and on their conditions and prerequisites for manufacturing planning and control.

---

3   Goldratt, E. and Cox, J. (1984) *The goal*. North River Press, Croton-on-Hudson (NY).

The term APS software was introduced in the 1990s and has been used since then. The "APS" stands for **advanced planning and scheduling**. Other terms used include advanced planning and optimization (APO), advanced planning systems (APS) and **supply chain planning (SCP)**, the latter since several of the systems modules are aimed at supporting planning of material flows in the whole supply chain and not only in one plant. There are many factors which distinguish APS from the traditional ERP system, but APS can be seen as a natural development of ERP and incorporates knowledge and development from logistics, planning, mathematics, IT, computer hardware and so on. Some software vendors integrate APS functionalities in their ERP systems while others distinguish between ERP and APS systems. Advanced planning and scheduling allows very frequent replanning to enable adaptation to changes in planning conditions. They take into consideration priorities and capacities, with the aim of identifying possible and feasible plans. The decision engine is based on advanced mathematics and logical algorithms. Optimisation tools can be included as part of the planning tools with the aim of identifying the best possible plans. Advanced planning and scheduling supports planning that considers business limitations in planning processes (such as capacity, material flows and business goals). They support scenario simulation and "what if" analyses. In the planning process, consideration is given to stock levels and utilisation of resources in processes external to the company, and plans are generated which result in best total efficiency.

In recent years the production and logistics concepts have been widened considerably. The term supply chain management (SCM) is used more extensively in order to reflect the fact that the production and logistics systems involve more than one individual company. Competition is viewed from the perspective of the network of different manufacturing sites, often involving supplier and customer companies. The significance of information flow in creating effective logistics has been increasingly recognised. There is now a clear focus on sharing information within and between companies with the help of well-developed partnership relations between players in the logistics system and the use of information technology to simplify and automate the flow of information. To quickly satisfy changing customer demands is another important characteristic of the twenty-first century. Strategies building on long-term relationships between functions, processes and firms, and allowing for mass customisation and postponement are therefore developed.

Historically, work to achieve more efficient flows of materials and production conditions has been influenced by rather narrow-minded approaches. Production specialists, production technicians, purchasers, planners and IT personnel have all worked separately on the development of their own function-oriented and specialised solutions. Co-operation over the boundaries between the various functions has been limited.

The significance of collaboration with production managers, designers, purchasers, suppliers and customers is simply a result of the fact that these categories largely create the conditions for different planning and control systems being able to function satisfactorily. It is also through these categories of specialists that it is possible to avoid over-complex and obscure systems and methods of manufacturing planning and control.

1.4 Approaches to Manufacturing Planning and Control

What lessons can be learned and what conclusions can we draw from historical developments in this area? One obvious deduction is that planning and control methods were essentially developed by practical people – those who had worked in or close to production activities at companies. Very few of the theories developed at universities and colleges have been put into practice, for different reasons. Development has taken place in two different worlds. It took about 15 years from the first installation of a system for materials requirements planning until theories of material requirement calculations were even mentioned to any significant extent in academic textbooks and journals. The dissemination of knowledge about different planning methods has taken place largely through trade organizations, ERP conferences and software packages for production and materials management. The choice of planning and control method has not always been given high priority, and those methods found in computerised systems have often been used simply because they were available, without any particular evaluation or considered selections being made.

From a historical perspective strategic decisions regarding reorientation of a company's activities on the basis of changing requirements and conditions on the market have seldom included consideration of the methods to use for managing production and flow of materials and the consequences of these methods. Measures taken in these areas have often been after the fact. It is only a slight exaggeration to claim that the forming and development of corporate strategies has generally only included companies' marketing and product development aspects. The consequences for production layout and management of operative activities have been little considered and problems which have arisen as a result of changes in strategy have been solved as they occur. Strategic decisions about, for example, increasing numbers of product variants and shorter delivery lead times to customers can have considerable consequences for the ability of manufacturing planning and control methods to function efficiently. In most situations it is desirable that both production issues and problems related to methodology and approaches to manufacturing planning and control are considered at the stage of developing new corporate strategies.

The simple approach described in the introduction to this chapter is often far too stereotyped. The statement that systems and methods used should be as simple as possible can hardly be refuted. It is an entirely rational and obvious idea. But the problems involved are often complex and cannot be solved efficiently using simple methods; the problems must first be simplified if they are to be solved using simple methods. At least, this is the case if high efficiency and impact on profits is to be achieved.

Simplifying the problems is, in this context, the same as creating the best possible planning environment. The planning environment refers to the conditions that characterise products, production and material flows, and which constitute the basic prerequisites for planning. In order to successfully develop and introduce new manufacturing planning and control methods, the planning environment should be simplified as much as possible. Designers, production managers, purchasers and so on have an important role to play here. Inter-departmental measures of this type have become increasingly common over the years. One of the main points of lean production is, for example, to simplify the planning problems.

In order to achieve efficient manufacturing planning and control, collaboration alone is not sufficient. The parties involved must also have a reasonably uniform and attuned picture of what is most important for the company to achieve. The traditional conflicts between, for example, design and production and between production and sales must be minimised. Some common differences in viewpoints between marketing departments and production departments are shown in Figure 1.3.

In addition to having a common and well-anchored perception of what is most important for the company to achieve, these objectives must also comply with the company's overall strategic direction and support the company's competitiveness on the market.

The above arguments can be summarised in the following four statements. This summary constitutes the fundamental approach to manufacturing planning and control on which this book is based.

Area	Marketing department	Production department
Capacity	Why don't we ever have enough capacity?	Why can we never get accurate forecasts?
Delivery time	We must be flexible. Our lead times are much too long.	We need more realistic lead times.
Inventory	Why do we always have wrong items in stock?	We can't have everything in stock.
New products	New products are a prerequisite for our competitiveness.	All those unnecessary engineering changes makes planning impossible.
Product variants	Our customers demand a broad product assortment.	With our product assortment we have to manufacture to small batches.
Cost	We can't compete with the product costs we have today.	You can't make whatever a customer wants with these prices.

FIGURE 1.3 Differences in viewpoints between a marketing department and a production department

1 *Alignment with corporate strategies*: a company's corporate strategies should include its manufacturing planning and control, both with regard to the opportunities it can contribute with in creating a successful and competitive company, and with respect to the limitations it may bring to bear on the choice of strategies for products, production and marketing.

2 *Goal orientation*: in the development of the manufacturing planning and control functions, those objectives which support the company's business mission and competitive strength should be identified and prioritised. The goals should also be made known and anchored with relevant personnel at the company.

3 *Planning environment simplification*: before management systems and planning methods are developed and introduced, the planning environment should be simplified and rationalised as far as this is financially justifiable. Work with simplification should be primarily carried out on the fulfilment of prioritised goals and in compliance with overall strategies.

4 *Situation-specific selection of systems and planning methods*: management systems and planning methods should be developed and selected on the basis of the existing planning environment, and should be aimed at supporting prioritised company goals. As far as possible, all personnel in the relevant competence areas in the company should be involved in the development work.

The cornerstones of this approach for the development of manufacturing planning and control processes are consequently alignments with corporate strategies, goal orientation, the simplification process and the situation-dependent selection of systems and planning methods. The main emphasis in this book is on selecting and using manufacturing planning and control systems and methods in existing situations, and not on how to simplify and rationalise the planning environment. However, manufacturing planning and control in several different planning situations is covered.

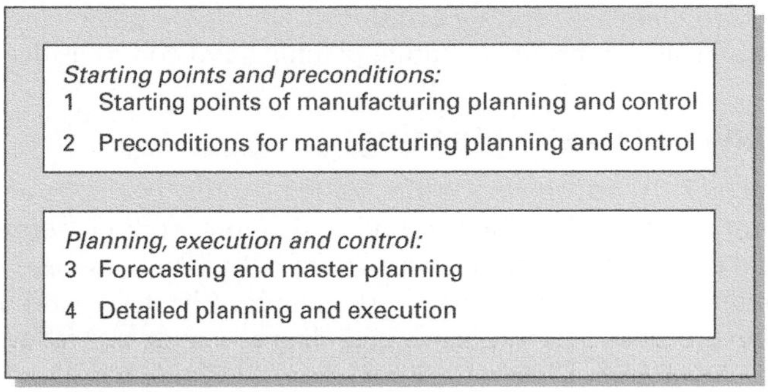

Starting points and preconditions:
1 Starting points of manufacturing planning and control

2 Preconditions for manufacturing planning and control

Planning, execution and control:
3 Forecasting and master planning

4 Detailed planning and execution

FIGURE 1.4 The four parts of the book

1.5 The Purpose and Structure of the Book

The purpose of this book is to convey a complete description of the subject area of manufacturing planning and control, and to supplement traditional ways to look at different planning processes, methods and theories in terms of their characteristics and applicability, on both conceptual and methodological levels and with respect to the planning environment in which they will be applied. The book is divided into four parts, as shown in Figure 1.4. Parts 1 and 2 describe the starting points and preconditions for manufacturing planning and control, while Parts 3 and 4 cover manufacturing planning and control strategies and methods on different hierarchical levels.

Part 1: Starting points of manufacturing planning and control

The first part consists of Chapters 1–3 and concerns the starting points of manufacturing planning and control. As a background to the various planning environments that exist, Chapter 2 describes the general characteristics of a manufacturing company. This description is made on the basis of what production consists of and how production may be considered from the logistics point of view. Different types of material flow related to the selection and application of different planning methods are also included. Planning of material flows and production normally takes place in a hierarchical structure of planning processes, each of which has different time perspectives and different levels of detail. It is within the framework of such a structure that the methods of application and planning treated in Parts 3 and 4 of the book are used. Chapter 3 presents this planning structure. The chapter is also a frame of reference for the planning of material flows and production in manufacturing companies.

Part 2: Preconditions for manufacturing planning and control

The second part contains Chapters 4–6 and describes the preconditions for manufacturing planning and control. The objectives of manufacturing planning and control and its effect on a company's performance are discussed in Chapter 4. Different types of basic data for production and materials management and how this basic data is stored, maintained and used is described in Chapter 5. There are four main types of basic data treated: item data, bill-of-material data, operation data and work centre data. Chapter 6 takes up important planning parameters and planning variables for manufacturing planning and control, such as ordering costs, holding costs, service levels, annual volume values and lead times. The parameters and variables are defined and their significance for manufacturing planning and control is discussed.

Part 3: Forecasting and master planning

The third part (Chapters 7–10) covers forecasting, customer order processes and master planning. The aim is to present the theoretical background of different work processes and methods, to describe the different characteristics of the methods from the viewpoint of applicability, and the planning environments for which they are most suitable. Chapter 7 describes the design of methods, systems and processes for forecasting, including execution and follow-up. The customer order process is described from a logistics perspective in Chapter 8. Special attention is given to the structure of the process in different types of companies, such as make to stock, assemble to order and engineer to order. Chapters 9 and 10 contain work processes and methods for sales and operations planning, and master production scheduling respectively. Sales and operations planning is a process which is carried out in gradual steps with a number of leading personnel from different units involved, and is aimed at achieving a balance between supply and demand so that the company's overall business goals are supported. This process is described in Chapter 9. The master production scheduling process is described in a similar way in Chapter 10. Its function, design and relationships with sales and operations planning are also clarified.

Part 4: Detailed planning and execution

The fourth part (Chapters 11–18) treats detailed planning and execution. Chapter 11 presents the fundamental starting points for material planning and the order processes and planning processes which initiate flows of materials. The most commonly occurring material planning methods are described and compared. Lot-sizing methods and various mechanisms for buffering against uncertainties are central parts of materials management. Chapter 12 treats lot sizing and Chapter 13 describes safety stocks. Chapter 14 takes up capacity planning at different levels of planning. Different action variables for achieving a balance between available capacity and capacity requirements in the medium- and short-term perspectives are reviewed. Finally, a summary is made of a number of commonly applied methods for capacity planning. Execution and control in pull environments are described in Chapter 15. Here, it is described how kanban and material planning without orders can be used as means of execution. Chapter 16 covers execution and control in traditional environments. The most commonly used methods for managing order release, material availability check and priority control are described. The procurement process for purchased items is described in a similar way in Chapter 17. The book ends with Chapter 18. Here, inventory accounting including the reporting of inbound deliveries and withdrawals from stock as well as physical inventory counting are described.

🔑 Key concepts

Advanced planning and scheduling (APS) system 10

Enterprise resource planning (ERP) 3, 8

Harris formula 6

Just-in-time 9

Lean production 9

Manufacturing resource planning (MRP II) 8

Material requirements planning (MRP) 8

Supply chain planning (SCP) 10

Wilson's formula 6

Discussion tasks

1 The chapter describes different viewpoints between the marketing and production functions. But what different viewpoints may exist between the production and finance functions?

2 One fundamental approach to manufacturing planning and control is to simplify the environment before methods and systems are selected and implemented. Another is to select situation-specific methods and systems on the basis of the existing planning environment. Discuss why companies normally have to work with both approaches.

3 Enterprise resource planning (ERP) and advanced planning and scheduling (APS) systems are mentioned in the chapter. What other information technologies may have major impacts on the manufacturing planning and control?

Further reading

Brown, S. (1996) *Strategic issues in manufacturing.* Prentice-Hall, Englewood Cliffs (NJ).

Fogarty, D., Blackstone, J. and Hoffman, T. (1991) *Production and inventory management.* South-Western Publishing Co., Cincinnati (OH).

Gilbert, J. and Schonberger, R. (1983) "Inventory-based production control systems – a historical analysis", *Production and Inventory Management,* Vol. 24, No. 2, pp. 1–13.

Goldratt, E. and Cox, J. (1984) *The goal.* North River Press, Croton-on-Hudson (NY).

Harris, F. (1913) "How many parts to make at once", *Factory,* Vol. 10, No. 2, pp. 135–136.

Mabert, V. (2006) "The early road to material requirements planning", *Journal of Operations Management,* Vol. 25, pp. 346–356.

Shapiro, B. (1977) "Can marketing and manufacturing coexist?", *Harvard Business Review,* September, pp. 104–114.

Watson, K., Blackstone, J. and Gardiner, S. (2007) "The evolution of a management philosophy: the theory of constraints", *Journal of Operations Management,* Vol. 25, pp. 387–402.

Wight, O. (1968) "To order point or not to order point", *Production and Inventory Management Journal,* Vol. 9, No. 3, pp. 13–18.

Wilson, R. (1934) "A scientific routine for stock control", *Harvard Business Review,* Vol. 13, pp. 116–128.

CHAPTER 02

The Manufacturing Company

In the previous chapter manufacturing planning and control was defined and it was stated that it mainly concerns the management of production and the flow of materials in the manufacturing company. The first two sections of this chapter describe in more detail what is meant by production and the flow of materials.

In solving logistics problems it is important to apply a context-oriented approach. Somewhat simplified, this means that methods and approaches should be considered and evaluated on the basis of the unique situation that is current when the methods and approaches are to be used. To be able to describe different problem contexts with such an aim, it is also necessary to characterise and classify the conditions that exist in the company. More detailed and characterising descriptions of various contexts are given in later chapters. This chapter provides only an overall categorisation of manufacturing companies, partly related to production conditions and partly to material flow conditions. The purpose of this categorisation is to enable the positioning of an individual company from a manufacturing planning and control perspective.

LEARNING OBJECTIVES

After studying this chapter, you should be able to:

- Describe different types of transformations conducted in production systems
- Explain the roles of different stock types that occur in material flows
- Compare and explain the similarities and differences between the customer order point and customer order decoupling point
- Explain the relationship between the customer order decoupling point and manufacturing strategies
- Describe and compare the characteristics of various manufacturing processes
- Describe different convergent and divergent material flow types in manufacturing

2.1 The Concept of Production

Production in general terms means a process in which goods and services are created through a combination of materials, work and capital. In line with this way of looking at production, it includes such wide-ranging phenomena as the manufacturing of heat exchangers, the transportation of milk, medical examinations and bank transactions. Clearly, such a definition of the concept of production is impractical in this context. In addition, since this book is about planning and control in manufacturing companies, production will be treated in a narrower sense and will only include processes in which goods are created. This delimitation should not be interpreted as meaning that a clear distinction is possible between production of goods and production of services. Since consumption is the overall goal for all production, goods produced must be distributed in some way to allow consumption. Production of goods, or manufacturing, is thus in most cases not relevant unless it is combined with the production of services.

The connection between manufacturing and production of services is very clear in a number of contexts. For example, in a company that makes to order the engineering of products is completely dependent on services which are produced directly to customers' requirements. The importance of services of different types as a complement to physical products has increased considerably in recent years. To achieve efficient logistics solutions, this production of services must obviously be controlled and co-ordinated with the manufacturing of goods. The concept of production in this book focuses on manufacturing of goods, but the production of different types of peripheral services which are included or required for the manufacturing of goods is also taken into consideration.

The manufacture of goods can be said to consist of a sequence of operations or value-adding activities. Raw materials are transformed during this process from a given to a desired state. In principle, this transformation can take place in five different ways as described below, and in many cases these ways take place in combination with each other.

1 *Transformation through division*, mainly with one item as input and several different items as output from the manufacturing system. Examples of this type of transformation are sawing timber into different formats from logs, or producing petrol and diesel from crude oil.

2 *Transformation through combination*, with several items as input and one item as output. The manufacture of machines and chemical products are examples in this category.

3 *Transformation through separation*, where the form of the input item is changed through the removal of material, such as in the manufacture of shafts by turning on a lathe.

4 *Transformation through shaping*, where the shape of the input item is changed through re-forming the material mass. Examples of this type of transformation process are rolling of steel profiles from billets and moulding of plastic items.

5 *Transformation through adoption of properties*, meaning that the properties of the input item are changed without changing its form. Heat treatment and surface treatment are examples in this category.

2.2 Material Flows and Stocks

Flow generally means the movement of a material or an immaterial entity. In the context of logistics it is more specifically a question of the movement of raw materials or materials in different stages of refinement. From the perspective of the individual company, these flows start with the company's suppliers and end with its customers. Ideal flows are continuous movements of materials, including their refinement, from suppliers to customers. For various reasons such ideal flows are impossible to achieve in practice. The speed of flow varies in different parts of the chain. In addition, flows are more often intermittent than continuous and there are interruptions between the value-adding operations involved. Different sub-flows are decoupled from each other, partly due to discontinuity and partly to avoid the propagation of inevitable disruptions in one sub-flow to other sub-flows. The primary function of inventory is to achieve this decoupling of flows. Inventories of various kinds may thus be seen as an integrated part of the total material flow system. The situation may be compared with ponds and lakes, which serve as reservoirs of water for a river in which the masses of water vary with the time of year and whose total water system is used to generate electricity.

With respect to the position of a stock in the flow of materials, we can talk of inventories of raw material and purchased components, inventories of semi-finished manufactured items, work in process and inventories of finished goods. *Inventories of raw materials, purchased components* and *semi-finished manufactured items* are intended for consumption in production. The primary function of these types of inventories is to decouple inbound flows of material and production processes from each other. Through keeping inventory, materials requirements in production can also be satisfied without delays.

Work in process (WIP) refers to goods which are under refinement in or between a sequence of value-adding resources. Keeping work in process means that different steps in production are decoupled from each other, which among other things enables different rates of production in different parts of a production system. Work in process also arises since production processes take time to execute.

Finished goods inventories consist of stocks of goods that are completed and ready for sale. Their primary function is to decouple the production process from the processes of sales and distribution. Deliveries to customers without delays are also enabled through finished goods inventories. The connections between the various types of inventories are illustrated in Figure 2.1.

Inventory refers to materials that are "stationary" and not moving along the flow. To a certain extent this is also true of work in process, or materials which are between sequential production resources pending further refinement. However, materials are also tied up in the flow itself due to the time it takes to achieve refinement and movement. In this respect we can distinguish between two types of inventories: work in process as described above and *in-transit inventory*. The size of these inventories is completely dependent on the speed of flow, or the time it takes to refine or transport materials. Generally it is only materials in external transport that are described as in-transit inventory. Materials in internal transport and handling are usually included in work in process.

With respect to the type of decoupling function, another division of inventory types may be made, as shown in Figure 2.2. **Cycle stock** means that part of an inventory which arises due to inbound deliveries taking place at a different speed and in larger quantities than consumption. Decoupling between stock replenishment and consumption is motivated by the fact that every order, transport and inbound delivery is associated with quantity-independent ordering costs. The larger the quantities ordered, the lower will be ordering costs per item unit.

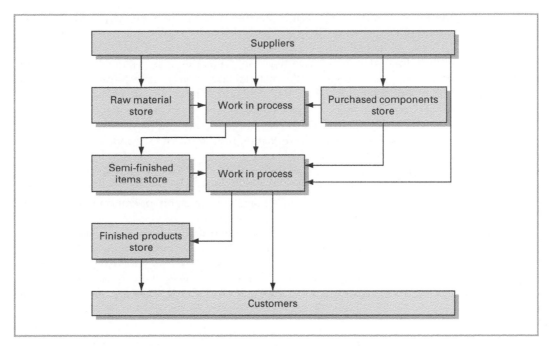

FIGURE 2.1 The connections between material flows and different types of inventories

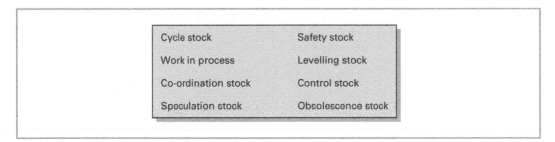

FIGURE 2.2 Types of inventories with decoupling functions

Disruptions in stock replenishment are inevitable in a material flow system, and it is not possible to exactly predict future consumption. In order to decouple different processes along the flow of materials so that disruptions and effects of unpredictable consumption do not propagate and result in shortages, **safety stocks** are used.

The consumption of goods is in many cases seasonal, and most consumption takes place during certain parts of the year. Despite this fact, the company may still be motivated to produce goods at a relatively even rate to achieve a smooth utilization of capacity and thereby decrease capacity costs. **Levelling stock** is used for this purpose. The purpose of a levelling stock is thus to decouple the rate of production from the rate of consumption.

Work in process, as described above, is used to decouple sequential production processes so that they do not require the same rate of production and thus dependence on capacity. Through this decoupling function, it is also possible to limit the extent to which production disruptions extend to other steps of value-adding activities in production processes. In this respect, work in process could be seen as a buffer stock.

Controlling the flows of materials in a manufacturing company is no easy task, especially when complex products are involved. There are many demands on planning methods and working procedures to achieve the necessary co-ordination of different sub-flows towards a finished product. As a result of these difficulties, two types of inventories arise. The first, **co-ordination stock**, is deliberately created to couple parallel material flows and achieve co-ordination effects. A co-ordination stock arises, for example, when jointly ordering several items even if they have somewhat different requirements dates. Two examples are the simultaneous inbound delivery of several items from the same supplier to decrease transport costs, and the simultaneous ordering in manufacture to decrease set-up costs in production resources. The second type of inventory arises accidentally and may be called **control stock** since its size is related to imperfections in control. In this case, too, it is an issue of certain items "waiting" for others to achieve simultaneous availability. A common example of this type of inventory is when items are required at the same time for final assembly but must wait in stock for the assembly operation to start, due to another component being delayed or to shortages in stock.

One special type of decoupling is the case of **speculation stock**, which is a type of cycle stock in which the stock replenishment is completely decoupled from expected short-term consumption. The motivation for building up a speculation stock is expected price increases in the future.

Finally, **obsolescence stock** is included in the list. Obsolescence stock consists of goods for which there is no expected consumption. In principle, the stock has to be written off through scrapping.

2.3 Manufacturing Strategies

There are different ways of classifying manufacturing companies with respect to the character of operations carried out. From the perspective of manufacturing planning and control, one important classification is to categorise companies according to the extent to which their operations are customer-order initiated. This is the equivalent of categorising the degree of integration between the company's production functions and customer orders received.

Categorisation of different company types with respect to the degree of customer order initiation may be carried out by means of the **customer order point** and the **customer order decoupling point**, also termed the order penetration point. The customer order point refers to the position in a product's bill of material from which the product has customer order specific appearance and characteristics. Expressed in another way, materials supply and value added in production is general up to this point, and not predetermined for delivery to any specific customer. The delivery lead time to a customer must be at least as long as the time it takes to complete the manufacturing operation from this point onwards.

A concept closely related to the customer order point is the customer order decoupling point, which is defined as the position in the bill of material from which material supply and value-added activities are customer-order initiated, i.e. are only executed as the result of a customer order received. The customer order decoupling point thus represents the point in the bill of material at which material planning is not dependent on forecasts. Before the decoupling point, the determination of delivery dates and order quantities is dependent on forecasts of future material requirements.

There is a close relation between the customer order point and the customer order decoupling point. In most contexts they are, in fact, at the same place. The customer order decoupling point may lie before the customer order point, but never after. When the decoupling point is earlier, material supply and value added can be customer order managed, despite the fact that it concerns standard items which are used in many different products and whose consumption

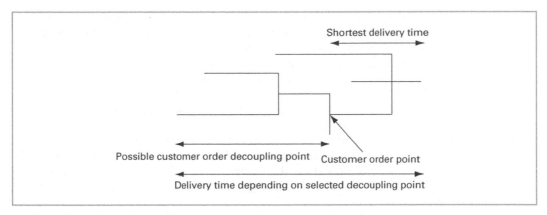

FIGURE 2.3 The connections between customer order point and the customer order decoupling point
Each horizontal line illustrates an item. The length of the lines illustrates lead times

is not specific to any particular customer order. The options for allowing the decoupling point to be before the customer order point are dependent on how large the ordering costs are. In the case of small ordering costs, it may be acceptable to procure material and manufacture those quantities for which there is a direct customer order requirement. The position of the decoupling point is also dependent on the accumulated lead time, from remaining material supply and value added up to finished product, being shorter than the delivery time required by customers. The relationship between the two concepts is illustrated in Figure 2.3.

Five different types of manufacturing strategies related to the customer order decoupling point are described below in order from highest to lowest degree of customer order integration (see Figure 2.4).

The company type of **engineer to order (ETO)** means that the company's products are to a greater or lesser degree engineered to customer specifications. Engineering work, process planning, manufacturing preparations, procurement of materials and manufacturing are to a large extent carried out and controlled in terms of time and quantities by customer orders received. The customer order decoupling point lies at a very low level in the bill of material of the product.

Make to order (MTO) is similar to the previous type, but products are generally engineered and prepared for manufacture before customer orders are received. A large part of materials procurement and manufacturing of parts and semi-finished items are carried out independently

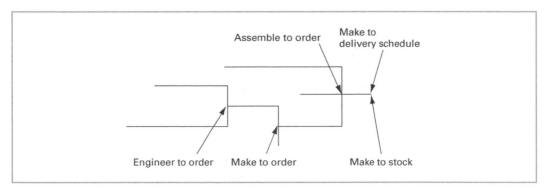

FIGURE 2.4 Position of customer order decoupling point for different types of company
Each horizontal line illustrates an item. The length of the lines illustrates lead times

of specific customer orders. Certain fabrication and all assembly/final manufacturing are performed directly to customer orders. Materials from external suppliers may also be procured to a certain extent to customer orders. The degree of integration between production and customer orders is thus less for this type of company than the previous types, and the customer order decoupling point lies higher up in the product's bill of material.

Assemble to order (ATO) means in principle that all raw materials and purchased components are procured and all fabrication of manufactured items takes place without connection to specific customer orders. The final form of the products and their properties is achieved through determining variants during assembly/final manufacture in conjunction with customer orders. Apart from the final determination of variants, the products only consist of completely standardised items that are engineered by the company itself. The degree of integration is relatively low and the customer order decoupling point lies at the bill-of-material level just below the final product.

The lowest degree of integration is in the remaining two types of company, **make to delivery schedule** and **make to stock (MTS)**. For both these company types, the products are completely known and specified on customer order receipt, i.e. the customer order decoupling point lies after the final level in the bill of material. In the case of make to stock, the products are completely standardised and kept in stock pending customer orders. A make to delivery schedule may also be entirely standardised products without any connection to specific customers, although it may also involve customer-specific products that are delivered to individual customers. This case is common for subcontractors in the automobile industry and other repetitive industries. Instead of customer orders initiating manufacturing activities, it is the delivery schedules, forecasts or stock levels that cause manufacturing to be initiated in both of these company types.

The categorisation of different types of company with respect to the degree of integration between manufacturing and customer orders is shown in Figure 2.5. The characteristics stated in the table refer to common and typical conditions for each type of company. Deviations may occur naturally for individual companies. Integration with customer means in this context the extent to which a customer via its orders influences the final form and performance of the product and the manufacturing and materials management operation.

Characteristic	Engineer to order	Make to order	Assemble to order	Make to schedule	Make to stock
Delivery time	Long	Average	Short	Short	Very short
Manufactured volumes	Very small	Small	Average	Very large	Large
Product variation	Very high	High	High	Low	Low
Base for planning	Customer order	Forecast/cust.order	Forecast/cust.order	Delivery	Forecast
Integration with customer	High	Average	Little	No	No
Number of customer orders	Very few	Few	Average	Few	Average

FIGURE 2.5 Characteristics of different company types

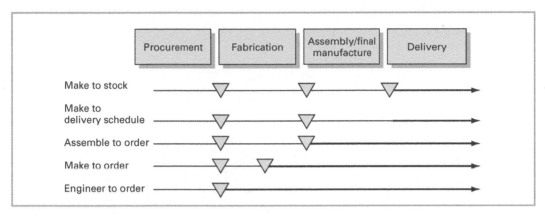

FIGURE 2.6 Stock points at different types of company

The types of companies presented above can also be categorised from a material flow perspective, and in particular with respect to where in the flow of materials the **stock point** lies for each type of company. The general conditions are illustrated in Figure 2.6, in which a triangle represents a stock point and a thick whole line represents customer-order initiated value-adding activities and transportation to customers.

The division into different types of company from the material flow perspective is based on the degree of integration between the company's manufacturing and the company's customer order. From the production and materials management viewpoint, there is also another interesting difference between the five types: the extent to which products are known and defined when the customer order is received, i.e. the difference in what is called degree of information at order receipt. The change in the degree of information from order receipt to delivery for the five different types of company is illustrated in Figure 2.7.

Make to stock and make to delivery schedule mean that standard products are delivered to the customer from stock or production according to plan respectively. The products are completely defined and known at the time of order and in this case we can refer to a 100 per

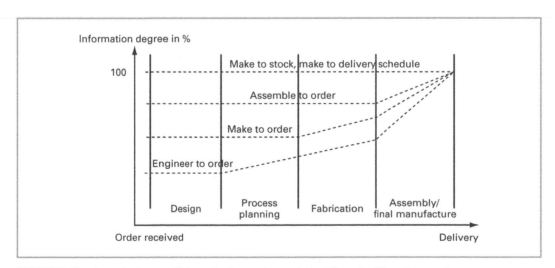

FIGURE 2.7 The change in degree of information from order receipt to delivery for different types of company

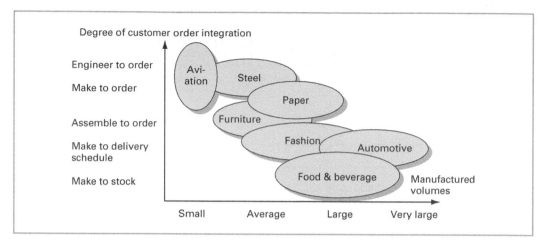

FIGURE 2.8 The relationship between production volumes and customer-order initiated manufacturing and material flow

cent degree of information. At the opposite type of company, engineer to order, it is a question of customer-order oriented products where knowledge of the product is low at the time of order. What is known is often based on some form of quotation, which specifies in general terms what will be delivered. As engineering work and process planning progress, the degree of information successively increases and when the product is finished and ready for delivery the degree is 100 per cent. The company types assemble to order and make to order lie somewhere between these two extremes. In both cases, the products are generally known to a large extent, but not the final and customer order specific variant of the customer order product.

The varying degree of information about products for the different types of company bring about different conditions for production and materials management and make various methods more or less applicable. It is obvious that with limited access to detailed information about what is to be manufactured, it is not possible to use the same approach to planning as in situations with complete and detailed information. For those planning situations which arise in most customer-oriented cases, it is very important to use methods that work well without access to detailed information and which make possible the successive utilisation of increased knowledge and information on products that is obtained as the engineering process progresses and as process planning for manufacturing is carried out.

As illustrated in Figure 2.5, production volumes in a company normally vary with the extent to which activities are controlled by customer orders. The relationship between production volumes and degree of customer order control can be shown even more clearly with the aid of a matrix, as in Figure 2.8.

The different ovals in Figure 2.8 illustrate typical areas within which companies belonging to various industries are normally active. It can be used as a basis for positioning an individual company with respect to production volume and degree of integration between customer orders and production.

2.4 Different Manufacturing Processes

Companies may also be categorised with respect to the basic type of manufacturing process which is carried out. A division may then be made by the extent of the flow orientation. Flow-oriented manufacturing processes in this context means the extent to which existing production resources are organised according to the bill of material of the products and the

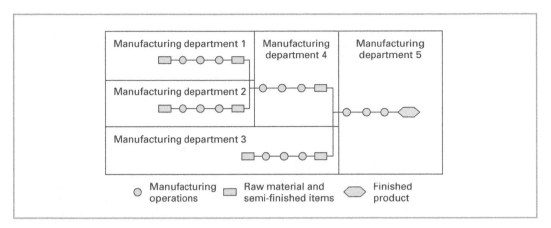

FIGURE 2.9 Illustration of an ideal-flow oriented process

manufacturing sequences. In an ideal-flow oriented manufacturing process the organization of production resources in a factory coincides with the bill of materials and manufacturing sequences of manufactured products. The manufacture of every product would also be continuous and without interruption for the manufacture of other products using the same manufacturing resources. An ideal-flow oriented process is illustrated in Figure 2.9.

In practice, however, real production processes deviate to some extent from this ideal model. The process type which deviates most of all will be called the **project process**. This type of process is characterised by basically no product flow at all; instead, production resources are organised around the product as it evolves.

Another type of process is the **jobbing or job shop process**. This process type is characterised by production resources being organised by manufacturing function, and the flow of materials during the process of production is adapted to this organisation and production layout. The shaping of the production process is, in other words, not product specific but is more general and therefore appropriate for the manufacture of different types of semi-finished items and end products. We can say that material flows must be adapted to the conditions of the production process instead of vice versa.

A higher rate of flow orientation is obtained with line processes. A linear manufacturing process is characterised by manufacturing resources being organised around the manufacturing sequences to a greater or lesser degree. For this type of manufacturing process it can be said that the organisation of production resources is subordinate to the opportunities for achieving rational material flows. The design of the manufacturing process is therefore more product specific than is the case for jobbing processes. Two main types of line manufacturing processes can be distinguished: intermittent line processes and continuous line processes.

Intermittent line process means a line process in which entire batch quantities of manufactured items are completed in one manufacturing step before they are transported to the next step. The movement of batches along the line can be paced or unpaced and it can be manually or automatically initiated. This type of process is therefore characterised by a higher degree of flow orientation than jobbing processes, but flows are interrupted at each step of manufacturing.

Continuous line process is a linear process in which manufacturing generally takes place unit by unit in all stages. In discrete manufacturing, this means that every unit of a manufactured item is handed over to the next manufacturing step as it is produced, and not after a whole batch is produced. Such process layouts put large demands on the use of lines which are specifically designed for each product in question. Continuous line processes are therefore more product specific than is the case for intermittent linear processes.

Characteristic	Project process	Job shop process	Intermittent line process	Continuous line process
Product variation	Very high	High	Average	Low
Customer order size	Very little	Little	Average	Large
Capital intensity	Very low	Low	Average	High
Flexibility	Very high	High	Average	Low
Capital Work in process	High	High	Average	Low
Capital Finished products	None	Low	Average	Low

FIGURE 2.10 Characteristics of different company types

The term continuous linear process is used here irrespective of whether it refers to process manufacturing or discrete manufacturing. In the case of discrete manufacturing, the process type is sometimes called repetitive line process.

A categorisation of the different process types is shown in Figure 2.10. The characteristics in the figure refer to common and typical conditions for each type. Exceptions may of course exist in individual companies.

In the same way as the classification by degree of integration between production and customer orders, there are certain relationships between the type of the manufacturing process and production volumes. We talk about the product/process matrix for positioning manufacturing companies' activities in this respect. Based on a division into different process types shown above, a matrix is obtained with an appearance shown in Figure 2.11.

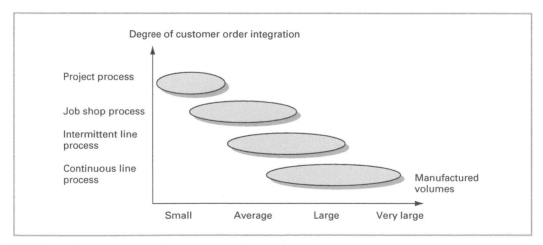

FIGURE 2.11 Product/process matrix for manufacturing companies

The ovals in Figure 2.11 show the areas of process choice and production volumes which manufacturing companies usually fall into. Examples of types of company in each oval are shipyards, manufacturers of tool machines, manufacturers of cookers and sugar refineries from top left to bottom right in the diagram.

2.5 Material Flow Types

The above method of classification applies primarily to manufacturing conditions. However, companies can also be classified with respect to types of material flows. We can then distinguish between so-called V, A, T, X and I types of company. All types have a close connection with the different transformation forms for manufacturing as described in Section 2.1. Several types of flow can occur in one company at the same time, and in many cases in combination with each other. The different material flow types are illustrated in Figure 2.12. Raw materials, purchased components and semi-finished manufactured items are represented in the figure by circles, while finished products are represented by squares.

V types are characterised by divergent material flows and occur in companies where manufacturing takes place in the form of transformation by division. The dominating characteristic for this flow structure is the occurrence of divergence points where some raw material is transformed into various end products. The number of end products may be large or very large in relation to the number of raw materials. Abattoirs and sawmills are examples of typical companies with divergent material flows.

In contrast to the V type, A, T and X types are examples of converging material flows. All three represent transformation through combination. The principal difference between them is in the number of end products in relation to the number of raw materials. In the *A type* of material flow, there are a number of convergence points where several incorporated materials are combined into single output semi-finished items and, finally, into end products. This type is characterised by a small number of end products in relation to the number of raw materials and purchased components. Complex products that are manufactured in small volumes with few variants, such as aircraft engines, are examples of A type products.

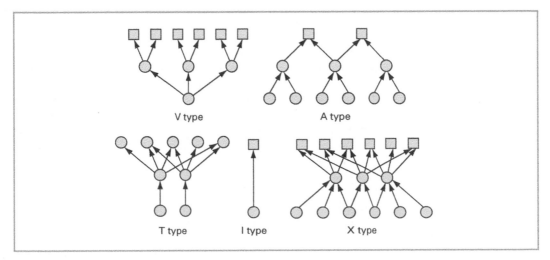

FIGURE 2.12 Different material flow types

T type has fewer convergence points than A type, and the convergence points are concentrated at the finished product level to a larger extent than is the case for the A type. In other respects, the T type is characterised by an extremely large number of finished products in comparison with the number of raw materials and purchase components. Product variants are created through the combination of different purchased components and manufactured semi-finished items. Most of these items are used in many products, i.e. they have a bigger commonality of use than is the case for items in A and X types of products. Paint is one example of a product with T type material flow. A small number of basic colours are mixed in the retail store to a large number of customer-specific colours.

X type also has several convergence points. The number of convergence points is greatest immediately under the finished product level. One characteristic of this type is that a large number of raw materials and purchased components are combined into a limited number of manufactured semi-finished items, often in some form of standardised modules. These semi-finished items/modules together with a number of purchased components can then be combined into an extremely large number of different finished products. A typical product which has X type material flows is a car, for which a large number of customer-specific variants can be built from a limited number of relatively complex modules and purchased components.

I type material flow corresponds to transformation through separation, shaping and adaptation of properties. This type is characterised above all by the fact that it has no real divergence or convergence points. In principle, one or a few starting materials result in a finished product. In terms of materials management, the I type of material flow is considerably easier to manage than the other types. The manufacturing of glasses by glass blowing is an example of a process with this type of material flow.

In practice, some divergent flows may occur in combination with A, T, X and I types. For example, there may be different types of by-product obtained during a chemical process in the manufacturing of an A type product.

2.6 Summary

The unique context that every company operates in and its characteristics create the prerequisites for manufacturing planning and control. Basic characteristics of manufacturing companies have been described in this chapter in terms of types of material flows and stocks, in terms of manufacturing strategies and the degree of customer-order initiated manufacturing, and in terms of different types of production processes.

Manufacturing is categorised in five types of transformation from raw materials. Material flow is described in terms of stock points and flows between these stock points. Every stock point represents a form of decoupling point for inbound and outbound flows of material. Eight stock types were identified with respect to various decoupling functions. Diverging and converging material flows of material are distinguished, e.g. V types, A types and X types. Manufacturing strategies and the degree of customer order control was categorized in terms of engineer-to-order, make-to-order, assemble-to-order, make-to-delivery-schedule or make-to-stock manufacturing. It was explained in what ways the conditions and principles of manufacturing planning and control are influenced by the degree of customer order integration. Manufacturing processes were divided into different types on the basis of degree of flow orientation, from the project process being the least flow oriented and the continuous line process being the most flow oriented.

The characteristic properties of a manufacturing company as described in this chapter are part of an overall categorisation related to the management of production and material flow.

Key concepts

Assemble to order (ATO) 22	Levelling stock 19
Continuous line process 25	Make to delivery schedule 22
Control stock 20	Make to order (MTO) 21
Co-ordination stock 20	Make to stock (MTS) 22
Customer order decoupling point 20	Obsolescence stock 20
Customer order point 20	Project process 25
Cycle stock 18	Safety stock 19
Engineer to order (ETO) 21	Speculation stock 20
Intermittent line process 25	Stock point 23
Jobbing or job shop process 25	Work in process (WIP) 18

Discussion tasks

1 Different kinds of inventories have been identified. Discuss why it is important to distinguish between them.

2 Choosing an assemble-to-order strategy instead of a make-to-stock strategy may be an appropriate way to manage in a situation with lots of product variants if the products are designed to allow it. Which are the two main prerequisites to be able to do so?

3 Discuss why and in what way the job shop process is more flexible than line processes.

Further reading

Gopal, C. and Cahill, G. (1992) *Logistics in manufacturing.* Business One Irwin, Homewood (IL).

Hays, R., Wheelright, S. and Clark, K. (1988) *Dynamic manufacturing.* Free Press, New York (NY).

Hill, T. (2000) *Manufacturing strategy – text & cases.* Palgrave, New York (NY).

Mather, H. (1988) *Competitive manufacturing.* Prentice-Hall, Englewood-Cliffs (NJ).

Schmenner, R. (1993) *Plant and service tours in operations management.* Macmillan, New York (NY).

Slack, N., Chambers, S. and Johnston, R. (2007) *Operations management.* Prentice-Hall, London.

Wemmerlöv, U. (1984) "Assemble-to-order manufacturing: implications for materials management", *Journal of Operations Management,* Vol. 4, No. 2, pp. 347–368.

03

Approaches in Manufacturing Planning and Control

The definition of manufacturing planning and control in the introductory chapter puts it firmly at the centre of a manufacturing company's control systems. To clarify what manufacturing planning and control means from the general level of management, this chapter treats the three generally applied levels of control (strategic, tactical and operative) which usually can be identified in a manufacturing company. Typical issues related to each of these levels of control in manufacturing planning and control are also included.

Manufacturing planning and control is normally carried out in a hierarchical structure of planning processes, each of which has different time horizons and degrees of detail. It is within this framework that the planning methodology presented later in the book is considered. The planning structure used is introduced in this chapter. It may be described as a frame of reference for the planning of material flows and production in manufacturing companies. Different approaches to capacity considerations when planning material flows are also included in the chapter, e.g. **theory of constraints (TOC)**.

LEARNING OBJECTIVES

After studying this chapter, you should be able to:

- Describe how planning of material flows and production can be conducted in a hierarchical structure of functions and processes
- Explain the relationships between different planning levels
- Explain the balance between the materials and capacity perspectives
- Describe different approaches to capacity planning
- Explain the content of theory of constraints

3.1 Control at Different Levels in a Company

Operations in a manufacturing company include management activities of many kinds. For a company to be efficient and competitive, all these activities must be co-ordinated as far as possible. With reference to the problems which are treated, the issues which are analysed, the decisions which are made and the measures which are taken to achieve such co-ordination, we usually distinguish between three different levels of control: strategic, tactical and operative. The area of manufacturing planning and control affects and is affected by all of these levels of control.

Strategic control is primarily aimed at positioning the company in its business environment. The types of issues and decisions involved are concerned with the field of business activity, goals and overall allocation of resources. Current strategic issues may include what products should be manufactured, which customer segments and markets products will be marketed on, and what production resources will be used internally and what production will be purchased externally from subcontractors. In the following list examples are given of issues which are included in, or have connections with, the area of manufacturing planning and control and which are of a strategic nature:

- Goals for production and materials management activities, for example delivery times to customers and service levels
- Decisions on product mix and breadth of variants in the product range
- Choice between having all production in one place or separately for each market
- Policy for capacity sizing
- Choice between making and delivering directly to order or from stock
- Decisions on supplier structure, i.e. whether to use one or several suppliers per purchased item

Tactical control is aimed at adapting and developing the structure of the company within the framework of the goals and general field of activity as prepared and established at the strategic control level. This may be an issue of the type of organisation, planning system or allocation of decisions. Tactical control also includes decisions and activities related to procurement of resources and usage of resources at the company. The following may be named as examples of issues and tasks from the manufacturing planning and control viewpoint which relate to tactical control:

- Preparation and establishment of sales plans and production plans for operations
- Planning of capacity and usage of capacity
- Choice of manufacturing layout
- Establishment of rules for determining order quantities and safety stocks
- Choice of centralised or decentralised planning organisation
- Selection of planning system and planning methods

The third and lowest control level is *operative control*, which is connected with ongoing activities and daily decisions and procedures. Operative control is aimed at putting into practice the selected strategic direction of operations within the framework of resources and structures determined by tactical control decisions. Several issues and tasks within operative control lie in the area of manufacturing planning and control. Some examples are given below:

- Setting delivery dates to customer orders
- Planning manufacturing orders and purchase orders/call-offs
- Short-term capacity and workload planning
- Assigning priorities to production in the factory/workshop
- Delivery monitoring
- Stock accounting

As shown in the above examples, there is also a difference between the control levels with regard to their time horizons. Strategic decisions related to manufacturing planning and control often have a time span of one year or more, whereas tactical decisions generally cover periods of six months to a year. At the operative control level it may be a question of days and weeks. The differences in time horizon are above all related to the time margin needed to carry out decisions. Strategic decisions, such as capacity expansion or change of product mix, take considerably more time to implement than operative decisions on calling off a delivery of materials. The working methods and planning methods treated in this book are primarily related to operative and tactical control levels.

3.2 A Concept and Structure for Planning Material Flows and Production

In simple terms, planning means making decisions about future activities and events. For material flows and production operations, such decisions may concern the next few hours or days, such as when a new manufacturing order is to be started on the shop floor. Decisions may also relate to activities and events that are six months or one year in the future, such as when to conclude a delivery agreement with a subcontractor or when there are large capacity investments in new plants.

The different decision situations which arise do not only vary in terms of time horizons, however. They also differ with regard to the precision and level of detail of information behind the decisions. To be able to start a new order in a few hours, exact information is required about the quantities to be manufactured, whereas a decision on capacity expansion in a year's time can be made even though information on quantities may be approximate in comparison. There are two main reasons for differences in precision and level of detail of information used as a basis for decisions. One reason is that it is less meaningful to have high levels of detail in bases of decisions related to the distant future since there are so many other factors that may influence the outcome. Another reason is that in most cases it is not possible to get more precise information. It is only possible to use the accuracy of data that is feasible and financially reasonable to obtain.

This problem with different time horizons and levels of precision is handled by planning material flows and production successively in a hierarchical structure of planning levels. Low precision and a low level of detail is used for planning levels with a long planning horizon, while planning levels with a short planning horizon use high precision and high detail information. The planning structure presented here contains four different planning levels: **sales and operations planning**, **master production scheduling**, **order planning**, and **execution and control**. The trade-off between different time horizons and level of detail is illustrated in Figure 3.1. A summary of the characteristics of each planning level is shown in Figure 3.2.

Sales and operations planning is the highest planning level, and has the longest planning horizon and lowest precision and degree of detail. This is a process at the top management level

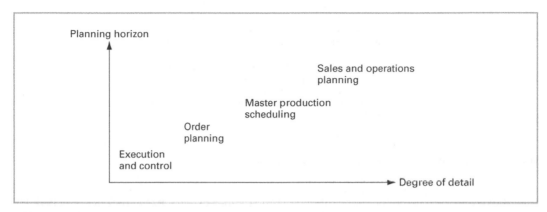

FIGURE 3.1 The relationship between degree of detail and planning horizon

Planning level	Planning object	Horizon	Period length	Rescheduling
Sales and operations planning	Product group Product	1–2 years	Quarter/month	Quarterly/ monthly
Master production scheduling	Product	0.5–1 year	Month/week	Monthly/weekly
Order planning	Item	1–6 months	Week/day	Weekly/daily
Execution and control	Operation	1–4 weeks	Day/hour	Daily

FIGURE 3.2 Characteristics of different planning processes

of the company and is aimed at preparing and establishing master plans for sales, deliveries and production.

The guiding lights for sales and operations planning are the company's business mission, its strategies and its overall business goals. Based on forecasts and other estimates of future demand for the company's products, plans are produced for expected sales and deliveries. After any adjustments for stock on hand, these plans represent a need to produce, and thus are the basis for establishing production plans. The planning objects for such **production plans** are often entire product groups, and the plans could be expressed in monetary or other units which are comparable between the product groups. Typical time horizons are one to two years. In many companies, sales and production plans produced correspond to sales and production budgets, respectively, used for financial control of the company.

The next planning level is called master production scheduling. At this level, delivery plans and production schedules for the company's products are drawn up on the basis of current **customer orders** and/or forecasts, with adjustments for any stock on hand. **Master production schedules** relate to quantities to be produced and delivered for each planning period for each product, i.e. products are planning objects. Master production schedules may also be expressed in the form of planned manufacturing orders with information on order quantities and delivery dates. Typical time horizons for master production scheduling are between six months and one year.

Order planning is the planning level related to materials supply, i.e. to ensure that all raw materials, purchased components, parts and other semi-finished items are purchased or

Strategic control	Sales and operations planning
Tactical control	Master production scheduling
Operative control	Order planning
Execution	Execution and control

FIGURE 3.3 The relationship between planning structures and planning processes

manufactured internally in such quantities and at such times that the master production schedules drawn up under the master production scheduling process can be carried out. The planning objects are those items used in the products, and the result is material plans in the form of planned manufacturing orders and purchase orders for each of these items in order to ensure the supply of materials. In the case of highly repetitive manufacturing, planning objectives may be stated in manufacturing rates per day and item rather than orders. The **planning horizon** at this level of planning often lies between two and six months.

For internally manufactured items, more detailed planning of manufacturing orders created at the order planning level takes place. This level is called execution and control. Planning objects are the operations, i.e. manufacturing steps, belonging to the manufacturing order planned to be carried out. The level of execution and control covers planning of new **order releases** to the shop floor, including material availability checks and sequencing in which order the various operations should be carried out in the available manufacturing resources. The planning horizon of execution and control is on the level of days or weeks.

Manufacturing in many companies is characterized by set-up times requiring order quantities bigger than the immediate needs, lead times being several weeks long, discrete and non-repetitive material flows, and varying capacity requirements. In such planning environments execution and control of manufacturing orders has to be carried out in an extensive way like this. In other, more lean, environments manufacturing may be rate based, lead times short, order quantities small and the material flow more or less pulled through the company by kanban cards or similar means. In such environments order planning and execution becomes an integrated function.

For items purchased, execution of orders planned at the order planning level includes placing the orders or call-offs with suppliers and monitoring deliveries. This is typically carried out by the **procurement** function at the company. The relationship between the planning structure and the levels of control discussed above is illustrated in Figure 3.3.

3.3 The Relationship between Different Planning Levels

For a structure of planning levels as described above to be able to function efficiently, two conditions must be fulfilled. The first is that decisions at one planning level must be made within the limits of decisions taken at the planning level above, i.e. the scope of action for planning decisions at one planning level is limited by planning decisions made at the level above. The second condition is that the planning methods used must be such that general decisions at one planning level can be transferred to subordinate planning levels, i.e. planning methods must contain functions so that the implementation of the first condition can be ensured.

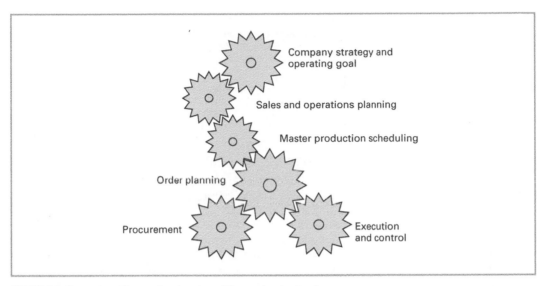

FIGURE 3.4 Illustration of interactions between different planning levels

If these conditions are not fulfilled, decisions at any specific planning level will be meaning-less, either because they are replaced by later decisions at a lower planning level or because they cannot be implemented due to the planning methods used. The relationships and interactions between different levels of planning are illustrated in general terms in Figure 3.4.

The fact that planning decisions made at one planning level must be taken within the framework of decisions at the next level up is illustrated in Figure 3.5 for the master production scheduling level. In the upper part of the figure, a production plan for decided quantities to be produced per product group and month, and made up at the sales and operations level, is shown. Within the limits of this overall production plan, production quantities per product and

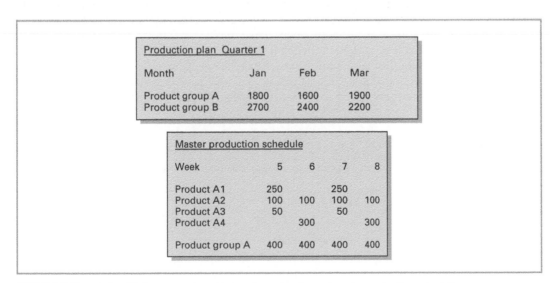

FIGURE 3.5 The relationship between sales and operations planning and master production scheduling

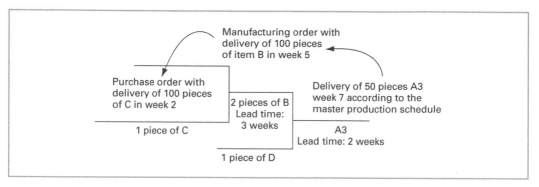

FIGURE 3.6 The relationship between master production scheduling and order planning

week at the master production scheduling level has been planned, as shown in the lower part of the figure. The sum of planned quantities to be produced of those products included in product group A during weeks 5–8 is 1600 according to the master production schedule. This quantity is equal to the quantity to be produced of product group A during February, according to the production plan at the sales and operations planning level. In practice, deviations are accepted between these quantities within certain tolerances.

The relationship between decisions at the master production scheduling level and order planning level is illustrated in Figure 3.6. The master production schedule according to Figure 3.5 shows, for example, the decision to produce and deliver 50 pieces of product A in week 7. If two pieces of item B are required for each product A, and it takes two weeks to assemble product A, then 100 pieces of item B will be required in week 5.

Thus, the order planning process must ensure that a minimum of 100 pieces of item B are available at the latest in week 5 for the master production schedule to be carried out. If the manufacturing lead time for B is three weeks, order planning must also ensure that manufacturing of B starts at the latest during week 2 by making sufficient quantities of item C available. The manufacturing order for B which is required to ensure the planned material flows, if there are no other requirements besides those derived from product A, should be scheduled by order planning to start in week 2, delivered to assembly in week 5, and should be for 100 pieces.

For execution and control, the consequences of decisions in order planning for item B is that the three operations involved must be scheduled for weeks 2, 3 and 4, i.e. within the limits of the start and due dates for the manufacturing order established at the order planning level – refer to Figure 3.7. If they are scheduled earlier there is a risk that sufficient quantities of item C will not be available; if they are scheduled later, the delay may jeopardise the manufacturing planned for product A.

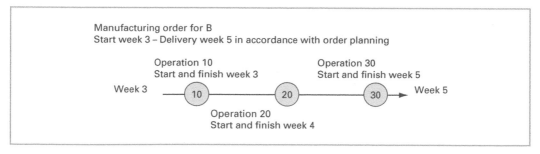

FIGURE 3.7 The relationship between order planning and execution and control

All these planning levels do not always exist in all companies. It is usual for sales and operations planning and master production scheduling to be included in one common planning process, often called **master planning**. Also, in practice the different planning levels often overlap each other. The planning activities included in the order planning and execution and control levels in certain companies may, for example, be organised within one common planning department. In other companies, a specific and separate execution and control level may not exist at all. This is, for example, the case in lean production environments where manufacturing may be initiated and material pulled through the company by kanban cards or similar means.

Irrespective of how planning activities are organised, planning is normally carried out at different levels in a company as shown in the structure described above. Even though the planning levels described are not always found in concrete forms within the organisation of a company, they can generally be identified as processes which are carried out in one way or another. It is important to understand the significance of these processes in order to understand problems involved in planning and the use of different planning methods.

3.4 Materials and Capacity Perspectives

At all of the above mentioned planning levels there are four common, general questions to be answered:

- How large are the quantities demanded, and for when?
- How much is there available to deliver?
- How large are the quantities that must be manufactured or purchased, and for when?
- What capacity is required to manufacture these quantities?

These four questions are related both to the planning of material flows within the company, and to the planning of capacity and capacity usage.

Planning from the materials perspective means establishing what products/items need to be delivered, what quantities of these products/items need to be delivered and when these quantities must be delivered. Planning from the capacity perspective means establishing what capacity is required to manufacture the quantities needed, and what capacity is available.

Planning from both the materials perspective and the capacity perspective is relevant at all planning levels. Planning of material flows and production in manufacturing companies is always a question of balance between the need to be able to deliver and the possibilities of being able to produce. A structure for planning from the materials perspective as described above has its equivalent in a structure of functions for planning from a capacity perspective. Common terms for the different levels for planning from a capacity perspective and their corresponding levels for planning from a materials perspective are shown in Figure 3.8. Planning in accordance with the concept illustrated in Figure 3.8 is sometimes called **manufacturing resource planning (MRP II)**.

The relative importance of planning from the materials perspective versus from the capacity perspective depends on the planning horizon. Questions from the materials perspective such as "What shall be produced when?" and "In what sequence shall manufacturing take place?" have a crucial significance in the short term. From a longer-term perspective it is often sufficient to determine quantities per rough time period without needing to specify any more detailed priority. With the exception of strategic sourcing, material planning is less significant in the long-term perspective. Instead, it is more important to ensure access to capacity. In a corresponding fashion, room to manoeuvre in order to adjust available capacity and current load is small in the short

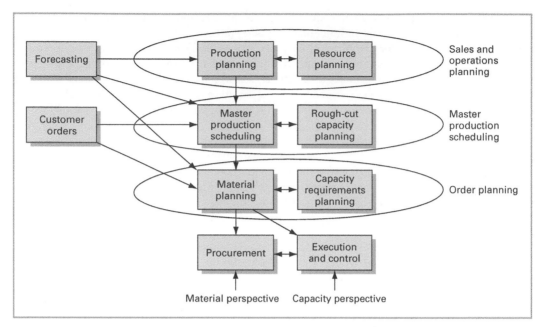

FIGURE 3.8 Planning from the materials and capacity perspectives

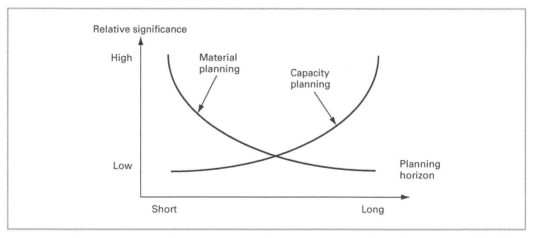

FIGURE 3.9 Planning as a balance between the need to deliver and the capacity to deliver

term and a more important question is how best to use the capacity which is available, i.e. setting priorities in material planning and shop floor scheduling. Capacity planning issues thus have comparatively less significance in the short-term perspective. The relationship between the planning perspective and the planning horizon is illustrated in Figure 3.9.

3.5 Approaches to Capacity Considerations

As described in the previous section, planning always involves a balance between what needs to be delivered and what can be produced or taken from stock. This balance means that consideration must be given to resource limitations of different types that exist within a production system. If consideration is not paid to existing resource limitations, plans will not be realistic and they will be impossible to put into practice. The primary limitation to be considered is access to manufacturing capacity but other types of limitations may be decisive for planning, such as access to supplier capacity, raw materials, tools, storage areas, transportation and handling equipment. These types of resource limitations also fall into the category of capacity and will not be treated separately for this reason.

In the long term, resource limitations do not generally have the same character related to planning. If the planning horizon is long enough there is always sufficient time to eliminate any limitations and to adapt available resources. The problem is more of an economic or financial nature. It is an economic problem in the sense that the utilisation of resources costs money and operations must be carried out in a profitable fashion. It is also a financial problem. To put into practice an expansion of resources, it must be possible to obtain capital in a financially acceptable way.

This is generally the situation at the planning level of sales and operations. At this planning level the problem of sufficient access to capacity to produce is primarily economic and financial. These issues lie outside the subject area at the core of this book, however, and will not be treated any further here. The capacity planning which is relevant at the sales and operations planning level is mostly intended to provide data for assessing what capacity requirements exist and to make decisions on future capacity adaptations based on planned sales and production volumes.

The situation is different at the levels of master production scheduling and order planning. Even if there are some margins for short- and medium-term capacity adjustments, consideration must still be given to limitations dictated by current access to capacity. It is thus necessary to balance desirable flows of material and access to capacity on the basis of current demand in order to achieve these flows of material. This balance should ideally be created in the form of simultaneous optimisation of material requirements and access to capacity. This optimisation is difficult to achieve in theory, and methods for solving it are only developed for very simplified models of the more complex underlying reality. The methods are typically based on advanced operations research theory. Planning systems capable of simultaneous optimisation of material flows and manufacturing capacity are typically called **advanced planning and scheduling (APS) systems**.

Yet another problem associated with finding a common optimal solution using analytical calculations is that of making cost estimates which can be used in optimisation models. Estimating costs for shortages and delays in deliveries to customers, which may arise from insufficient capacity, and weighing these against costs for increasing capacity through overtime work or subcontracting is extremely difficult. Besides, such costs are always specific to the situation and as such cannot be generally determined at a planning stage far in advance of the situation.

In practice the problem of optimisation is usually handled as a step-by-step solution. Two principally different approaches exist. In the first alternative, consideration is given to materials first and then to capacity, while in the second, consideration is first paid to capacity and then to materials.

Prioritize materials – resolve capacity requirements. The flows of materials are planned in accordance with existing demand or demand which can be forecast, i.e. without consideration to access to capacity. In the next stage, the consequences of the planned material flow on capacity are checked. If the capacity requirements exceed what it is possible to achieve during the time frame available, material flows are adjusted to an extent that available capacity is sufficient.

Prioritize capacity – resolve materials requirements. Manufacturing is planned for existing capacity to obtain the highest and most even workload possible. If available capacity is too small or too large in relation to what is required to satisfy forecast demand, an assessment is made of the extent to which capacity can be increased or decreased by extraordinary measures. The same is the case if available raw material is less than what is needed to maintain capacity-based planned production.

To put it simply, the first alternative means that material flows are given priority and any problems with access to capacity are considered as a secondary issue to be resolved if possible. This alternative may be characterised as material-driven manufacturing. It also represents a product and market orientation rather than a production orientation. In a similar fashion, the second alternative is characterised by the utilisation of capacity being given priority and demand and material flows being of subordinate importance to be solved as well as possible. This alternative represents capacity-driven manufacturing and is, in comparison with the previous alternative, more production oriented.

The first alternative is most common in Western industry and is also generally the most natural alternative in a market economy. Exceptions, in which the second alternative is used and is preferable, do occur, however. This may be the case in extremely capital-intensive operations such as the processing industry. Another example is in business activities with exceptionally high and prioritized set-up times in which sequences of material flows must be chosen from the capacity utilisation perspective rather than current materials and customer requirements. The following may be put forward as the main motivation for normally preferring the approach of the first alternative:

- Most companies nowadays must be customer focused and market oriented to be able to compete successfully on the market. Satisfying customers' desires and needs is crucial. How this is achieved is relegated to a mere consequence of the overall goal. The material-oriented approach to planning corresponds with this view.

- In most contexts, available capacity is not completely fixed and given in advance. By starting from the material requirements situation, an idea of the optimal capacity requirement is gained. From this stage it is possible to make correct decisions about possible and suitable capacity adjustments.

- Future access to capacity is in many cases not known exactly, especially in companies with labour-intensive manufacturing. With only an approximate idea of access to capacity, it is not possible to optimise material flows in an acceptable fashion.

- It is less complicated to adjust scheduled manufacturing than to try to eliminate expected disruptions in material supplies that arise because manufacturing orders have been planned to exceed available capacity. This is related to the fact that the material requirements situation is normally far more complex and difficult to analyse than the capacity requirement situation.

The planning methods presented and analysed in the following chapters are all primarily intended for and based on the application of the first alternative. Giving priority to material flows and subsequent assessments of consequences for capacity is carried out at every level in the planning structure as discussed above. This means that at any specific planning level, some consideration is always paid to capacity at the level above before planning at the level in question is carried out.

At the level of execution and control, the planning horizon is only a few days or weeks and the options for adjusting capacity are very limited. Planning must take place to a larger extent within existing capacity and material flows will be assigned priorities on the basis of these limits. At this level it is more reasonable to approach the planning problem by planning the utilisation of capacity in the first stage and subsequently give priority to manufacturing sequences and material flows based on this capacity planning. According to the planning structure shown above and illustrated in Figure 3.8, this will mean that at the level of execution and control the release of new manufacturing orders will be regulated after giving consideration to available capacity through some form of planned order release. The released orders will then be assigned priorities with the aid of **priority control** and priority rules.

3.6 The Theory of Constraints Approach

A well-known and recognized approach to considering capacity limitations when planning material flows is theory of constraints, developed by Dr. Eli Goldratt. Its origin is a finite capacity scheduling method and software called optimized production technology (OPT). The method is characterised by its focus on identifying and fully utilising bottlenecks along the material flow and subordinating the entire production system to these bottlenecks. The basic principle of the OPT was later developed into a general concept for constraint-based planning, scheduling and management.

A constraint is defined as everything that limits the performance of a system. Such a constraint can generally take one of three forms: physical, market or policy. A physical constraint exists, for example, if the capacity in a manufacturing resource is less than the demand on the product manufactured in the resource, or if available raw material is less than what is needed to manufacture demanded volumes. A market constraint exists if the demand on the marketplace is less than the volumes the system has capacity to manufacture. However, a constraint may also be policy based. This means that applied policies, formal or informal rules or just traditions limit the productive capacity of the system. Having identified a constraint, the objective simply stated is to synchronise production and the material flow with customer requirements. The core of theory of constraints is the following five-step process to accomplish such synchronisation and to continuously improve the throughput from the system:

1 *Identify the constraint.* Identify what part of the system constitutes the weakest link and represents the constraint for the output of the system as a whole. According to Goldratt, there are very few constraints in any production system.

2 *Exploit the constraint.* Once the constraint has been identified, determine the most effective means to exploit it. In this context exploit means to bring maximum efficiency from the constraint in its present configuration. For a manufacturing resource it might, for instance, imply eliminating all possible idle time at the resource or postponing preventive maintenance to after working hours. It may also imply working overtime or moving cross-trained personnel from under- to over-utilised work centres. A way to exploit a manufacturing resource is also to make sure that material necessary for the manufacturing process is always available, to avoid interruptions due to shortages, e.g. by keeping a buffer stock of material in front of the resource.

3 *Subordinate everything else.* The output of a system is limited by the rate of throughput at the constraint. Therefore the third step is to subordinate the rest of the system to the constraint. This means that the utilisation of other resources is adapted so that the material flow through these resources is synchronised with the flow through the constraint. Non-bottleneck resources are, in other words, consciously idle parts of the time. If non-bottleneck resources are not subordinated, the only outcome of them operating to full extent is increasing inventories and work in process.

4 *Elevate the constraint.* Should additional output be necessary, the fourth step involves elevating throughput by adding capacity to the system at the constraint. In the long term this may include adding another shift, buying an additional machine or increasing outsourcing.

5 *Go back to the first step.* If adding capacity in the previous step, the constraint has been broken and a new constraint will show up in another part of the system. The process then starts again from the beginning.

The theory of constraints approach is rather close to the approach advocated in the previous section, first considering the flow of material and then capacity. To first only consider the flow of material and what is demanded on the market irrespective of available capacity is the only way to find out if and where any throughput-limiting resource exists in the system. Having found the limiting resource, the next step is to focus on its capacity, try to use it as much as possible and adjust the flow of material to the extent capacity is available.

3.7 Summary

This chapter described control at the strategic, tactical and operative levels. The planning of material flows and production at these different levels is interrelated in a hierarchical structure of planning levels, each of which has different time horizons and levels of detail. The principles of the four most common planning levels were described in overall terms: sales and operations planning, master production scheduling, order planning and execution and control. It is within the framework of such a hierarchical structure of planning levels that the planning methodology later in the book is presented.

All planning involves balancing supply and demand, i.e. the need to balance the material flow to be able to deliver in line with demand and the capacity available to deliver. Two general methods of planning with consideration to material flows and capacity, considering materials first and then capacity and considering capacity first and then materials respectively, have been described. A well-known and recognized approach to address the issue of capacity constraints when planning the material flow, theory of constraints, has also been presented.

🔑 Key concepts

Advanced planning and scheduling (APS) system 39

Customer order 33

Execution and control 32

Manufacturing resource planning (MRP II) 37

Master production schedule 33

Master production scheduling 32

Master planning 37

Order planning 32

Order release 34

Planning horizon 34

Priority control 41

Procurement 34

Production plans 33

Sales and operations planning (SOP) 32

Theory of constraints (TOC) 30

Discussion tasks

1 One condition that must be fulfilled for the hierarchy of planning levels to work properly is that decisions at one planning level must be made within the limits of decisions taken at the planning level above. What may the result be if activities on the order planning level are outside the limits of decisions taken on the master production schedule level?

2 When planning manufacturing operations, materials as well as capacity issues have to be considered. Why is the capacity issue more significant in the long term? Which are the most significant long-term materials issues?

3 Within the theory of constraints concept three different forms of constraints can be identified. One is the market. Discuss possible actions when applying the five-step process in this case.

Further reading

Arnold, T. and Chapman, S. (2004) *Introduction to materials management.* Prentice-Hall, Upper Saddle River (NJ).

Fogarty, D., Blackstone, J. and Hoffman, T. (1991) *Production and inventory management.* South-Western Publishing Co., Cincinnati (OH).

Goldratt, E. and Cox, J. (1984) *The goal.* North River Press, Croton-on-Hudson (NY).

Nicholas, J. (1998) *Competitive manufacturing management.* Irwin/McGraw-Hill, New York (NY).

Oden, H., Langenwalter, G. and Lucier, R. (1993) *Handbook of material and capacity requirements planning.* McGraw-Hill, New York (NY).

Rahman, S. (2002) "The theory of constraints' thinking process approach to developing strategies in supply chains", *International Journal of Physical Distribution & Logistics Management,* Vol. 32, No. 9/10, pp. 809–828.

Tincher, M. and Sheldon, D. (1995) *The road to class A manufacturing resource planning (MRP II).* Buker Inc, Chicago (IL).

Vollmann, T., Berry, W., Whybark, C. and Jacobs, F. (2005) *Manufacturing planning and control for supply chain management.* McGraw-Hill, New York (NY).

Watson, K., Blackstone, J. and Gardiner, S. (2007) "The evolution of a management philosophy: the theory of constraints", *Journal of Operations Management,* Vol. 25, No. 2, pp. 387–402.

Wight, O. (1984) *Manufacturing resource planning: MRP II.* The Oliver Wight Companies, Essex Junction (VT).

PART 2
Preconditions for Manufacturing Planning and Control

The second part of the book contains Chapters 4–6 and presents the preconditions for conducting manufacturing planning and control. Chapter 4 explains how manufacturing planning and control affect a company's goals and performance. A number of performance variables of relevance for the area are outlined. Complete and correct information is of crucial importance for the performance of the manufacturing planning and control processes. One part of the information required is given the term "basic data". Basic data is fundamental information about the company's products, what item they consist of and how they are manufactured, and the company's production resources. Chapter 5 describes basic data. Different types of calculations are carried out when planning production and material flows. A number of parameters and variables are used in these calculations, for example ordering costs, inventory carrying costs, service level, annual value classification, issue frequency classification and lead times. In Chapter 6, parameters and variables used in manufacturing planning and control are defined and discussed.

Part Contents

04

Manufacturing Planning and Control Performance

Manufacturing planning and control is not a goal in itself. Its purpose is to contribute to improving performance in companies and thereby achieving a positive impact on profits. Performance can be expressed with the aid of different measurements, each representing performance in a certain respect for a certain variable. By measuring and following up performance it is possible to influence behaviour in a company and thus create the conditions for achieving a direction of operations that corresponds to the company's overall strategy and objectives, and which supports its competitiveness. Performance measurements are also tools for following up performance and for this reason are fundamental conditions for delegating responsibility and applying a decentralised form of organisation for logistics activities.

A number of different performance variables and measures related to logistics are discussed in this chapter. They can be roughly divided into three categories:

1 Variables which influence a company's revenues

2 Variables which influence a company's costs

3 Variables which influence a company's assets

Here, eight types of performance variables of relevance for manufacturing planning and control are covered. The first five (**stock service level**, **delivery precision**, **delivery reliability**, **delivery lead time** and **flexibility**) are customer-service related variables which influence a company's revenue. Logistics costs, the sixth variable, contain eight types of costs. **Tied-up capital** measures a company's assets and the variable capacity utilisation affects both costs and tied-up capital.

Performance variables are also intended to be used for sizing in the logistics system; for example, using service level requirements for sizing safety stocks. The sizing perspective of a number of variables is taken up in Chapter 6 regarding planning parameters and planning variables, but also in other chapters. Sizing of capacity is treated in Chapter 14.

Several of the performance variables presented here are of a contradictory nature and cannot be maximised simultaneously. It is often necessary to prioritise or in other ways ensure that performance variables used are aligned with the company's overall strategy. The attitude towards the importance of achieving improvements with respect to performance variables often varies between different functional areas in a company. The question of prioritisation of different performance variables and goal conflicts is briefly treated at the end of the chapter.

4.1 Stock Service Level

Stock service level is a part of the overall customer service and means the extent to which products can be delivered to the customer directly from stock. The performance measurements may refer either to an individual product, a product group or to the overall performance level for the entire product range in stock. The measurement is then based on the ability to deliver individual customer order lines from stock, since every **order line** represents one unique type of item. The measurement of service level can, however, also refer to the ability to deliver complete orders from stock. There are several different ways of expressing and measuring the stock service level in these respects. The description here will be limited to the following three commonly used alternatives:

1 Proportion of order lines delivered directly from stock
2 Proportion of completely delivered orders directly from stock
3 Value-related proportion of orders delivered directly from stock

Proportion of order lines delivered directly from stock – order line fill rate

The performance measurement of the proportion of order lines which can be delivered directly from stock, the order line fill rate, expresses the relationship between the number of order lines which can be delivered directly from stock according to customers' wishes and the total number of order lines delivered during a certain time period, for example one month. It is a relative measurement which can be expressed as a percentage. If, for example, 467 order lines out of a total of 531 during one month could be delivered directly from stock, the stock service level is equal to 467/531, or 88 per cent. The measurement concerns a product-oriented delivery capability, in that it can express delivery capability for each product separately. It could therefore also be used for expressing delivery capacity for a product group, or for an entire product range, by combining the separate product measurements.

In practical applications, different variants of the measurement are often used. The number of delivered order lines from stock may, for example, also include the customers' accepted split deliveries. Delivered order lines within a certain time frame may also be included, such as accepting all deliveries which can take place within a couple of days as direct deliveries from stock.

Proportion of completely delivered orders directly from stock – order fill rate

Compared with the previous performance measurement, order fill rate is more of a customer and customer-order oriented way of measuring delivery capability from stock. It states how many customer orders have been delivered completely from stock in relation to the total amount of delivered orders during a certain time period. The more order lines included on each customer order, the more stringent is this measurement compared with the order-line oriented service level. In a strict sense, split deliveries are not accepted at the order line level, nor are individual whole order lines back-ordered so that an order can be considered as delivered directly from stock.

In a similar way as in the previous performance measurement, different variants are conceivable. In this case, too, a few days' delivery lead time may be acceptable as delivery direct from stock. It is also possible to refine the measurement by not only including the proportion of orders that can be delivered directly, but also accepting the proportion of orders that demand a second or even third delivery to be complete.

Value-related proportion of orders delivered directly from stock

In certain contexts it may be desirable to include delivered value in delivery capability, especially when the company's products vary a great deal in value, i.e. a mix of low and high product prices. The performance measurement of value-related proportion refers to the value of goods delivered directly from stock in relation to the value of all deliveries during a time period. The measurement can be calculated by order line or by completely delivered order. The same variants as for the two previous service level measurements can be used.

Measuring stock service levels as above has a number of error sources. One of the more important sources of error is that the actual customer preference for delivery is not certain to be reflected. An inquiry about delivery which cannot be made directly from stock may result in the order being lost, and then there is typically no data to be used as a basis for calculating the service level for those particular orders. Measuring service levels purely on the basis of customer preferences is both difficult and time-consuming.

Another source of error of a similar nature is where a customer, not being able to obtain immediate delivery, accepts the delivery lead time proposed. It is not always clear whether these orders/order lines will be included in goods delivered directly from stock or if they will be considered as a shortage situation, especially not if the customer wants to have a delivery lead time without being persuaded.

4.2 Delivery Precision

Delivery precision refers to the extent to which deliveries take place at the delivery dates agreed with the customer. While the performance measurement of stock service level is intended to state the delivery capability from stock, the measurement of delivery precision is intended as an expression of delivery capacity for companies with assemble-to-order, make-to-order and engineer-to-order types of operation. The measurement may also be relevant to a certain extent for make-to-stock companies. It then refers to delivery precision for orders that have become back-ordered as a result of shortage situations, or deliberately planned to be delivered after a certain time.

Since an increasing proportion of companies in the manufacturing industry apply some form of customer-order initiated manufacturing, and many also work with philosophies such as lean production and just-in-time, this performance measurement has gained a growing significance. It is generally the case that customers prefer to have high delivery precision, even if it is at the expense of a longer delivery lead time, rather than promises of short delivery lead times which often cannot be kept.

A company's delivery precision is influenced by how delivery lead times are set and how the company manages to keep them. It is important to emphasise the significance of setting delivery lead times in this context. Many delays in deliveries are the result of wrongly set delivery lead times rather than, for example, disruptions in manufacturing or poor planning when executing customer orders. The quality of both delivery lead-time setting and delivery lead-time keeping can be influenced to a large extent with the aid of efficient manufacturing planning and control.

Commonly occurring measurements of delivery precision include the following:

Number of actual deliveries in relation to number of promised deliveries. The delivery precision measurement of the number of real deliveries compared with the number of promised deliveries during a time period is a relative measurement corresponding to the type shown for stock service level. If, for example, 73 orders have been promised for delivery during one month and only 67 of them have actually been delivered, the delivery precision is 92 per cent.

Number of not delayed deliveries in relation to the total number of deliveries. The previous measurement of delivery precision is rather coarse, especially for measurement periods longer than one week. A more correct measurement can be obtained by making comparisons per order rather than per time period. The number of orders delivered at the right time during a chosen measurement period is then divided by the total number of orders delivered during the same period. The same type of measurement as above is obtained, expressed as a percentage. Depending on the tolerances chosen, deliveries which take place one or two days early or late may also be accepted with correct deliveries.

If it is wished to further refine the measurement of delivery precision, it is possible to separately report the proportion of orders delivered too late and too early as a complement to the delivery precision measurement. A further refinement would be to follow up the distribution between too early and too late deliveries, for example by stating the proportion of orders with 1, 2, and 3 days' delay, and a corresponding measurement for those orders delivered too early.

In the same way as for the performance measurement of stock service level, there are difficulties in measuring delivery precision that are mainly related to back orders and changes of originally agreed delivery lead times. In the case of back orders, a new delivery date is agreed for the quantity not delivered. Under such conditions, the handling of deliveries carried out in accordance with agreed delivery lead times is not straightforward.

Another difficulty with measurements arises when delivery dates are changed in agreement with the customer company. It is especially problematic from the measurement viewpoint if delays or earlier deliveries are desired by the customer. It is difficult to draw up general rules for solving this type of problem, and each case must be dealt with on its own merits.

4.3 Delivery Reliability

The performance measurements as described above all relate to delivery quality in the sense of being able to deliver the desired quantity directly or at an agreed date. Another type of delivery service which also constitutes a measurement of performance for a logistics system is delivery reliability, meaning the extent to which the right products are delivered in the right quantities.

Delivery reliability as defined in this manner consists of two elements. The first refers to directly delivered quantities. The right quantity in this case is not a question of split deliveries, back orders or similar deviations from agreed quantities; the quantity intended for delivery and stated in the delivery document is actually delivered. The second element in delivery reliability is delivery of the right product. This reliability concept refers not only to delivery of the right product, but also to delivery of the right product in terms of quality. Delivery of the wrong product often occurs when different products have very similar physical appearance and are thus easy to confuse, for example small injection-moulded plastic items or electronic components.

If a company can achieve a high degree of delivery reliability in these respects, a large part of the work which otherwise must be carried out at the customer company's goods reception can be eliminated. Using terminology inspired from the Toyota production system, you can talk about avoiding waste. Efficient systems and procedures in manufacturing planning and control make large contributions to eliminating such waste. Other examples of conditions which influence delivery reliability are picking procedures, unreliable stock locations and stock addressing, as well as lack of automatic item identification such as bar codes. The following performance measurement can be used as an indication of delivery reliability:

> The proportion of customer orders without complaints from the customer in relation to the total number of delivered customer orders

This measurement gives the relationship between the number of customer orders delivered during a certain period which have not resulted in any comments on incorrect products, not-accepted quality levels or incorrect quantities, and the total number of delivered customer orders during the same time period. In the same way as for stock service level, this measurement may also be modified to relate to delivered order lines. In this case, the company has a follow up of its delivery reliability for each product in addition to a customer-oriented follow-up.

4.4 Delivery Lead Time

Delivery lead time is, from the supplier's perspective, the time between a customer order being received until delivery can take place. Delivery lead times for products are to a large extent dependent on product characteristics and manufacturing methods. They may also be influenced by the logistics system and planning methods.

Delivery lead time to a customer may be said to include two main elements, one related to operations and the other related to current workload from additional orders on hand. The first element consists of administration and order-processing time, waiting times of different types, time for engineering and manufacturing preparations, throughput time in manufacturing, and dispatch and transport time. Which of these categories are involved in a particular case will depend on the type of company in question. Of the time elements mentioned above, administration and order-processing time as well as dispatch and transport time can be highly affected by efficient logistics systems. The appropriate choice of stock points in the material flows can also contribute to reducing delivery lead times considerably.

The elements of delivery lead time related to current workload can also be influenced by the logistics system. Through access to reliable workload information from open orders and thereby controlling current available capacity, unnecessary safety margins for delivery lead times can be avoided. The logistics system can also contribute to a higher degree of manufacturing flexibility, so that the ability to temporarily increase available capacity can be improved in times of overload.

Delivery lead times are expressed in the form of calendar weeks, consecutive workdays and, in certain cases, hours, such as for sequential deliveries in the automobile industry. Follow-up measurements are generally average values per product or product group for all orders delivered during a certain time period, such as one month.

4.5 Flexibility

The concept of flexibility is complex and wide-ranging. In general terms it can be defined as the ability to quickly and efficiently react to changing conditions. It has become an increasingly important performance parameter as striving to control manufacturing activities through customer orders becomes more widespread. Here we limit the treatment of the concept of flexibility to those dimensions which are interesting from the perspective of manufacturing planning and control, i.e. **product mix flexibility**, **volume flexibility** and **delivery flexibility**. There are no simple units in which to measure flexibility. The flexibility dimensions that are discussed here are only considered as general terms for efficiency in the production and logistics system.

Flexibility costs money, but it can also be a source of revenues – above all, through increased competitiveness which can be achieved through an improved ability to adapt to customer preferences. The performance parameter of flexibility can therefore be said to influence both costs and revenues in a company.

Product mix flexibility

Product mix flexibility means the ability to rapidly adapt manufacturing and material supply to shifts in demand between existing products and product variants. In other words, it is the ability to rapidly change over to increased manufacturing volumes of certain products and fewer of other products than was originally planned, on the condition that the same capacity requirement exists in general terms. A company's product mix flexibility is influenced by a number of factors, including delivery lead times for purchased items, the size of manufacturing orders, throughput times in manufacturing and access to information.

Volume flexibility

From the viewpoint of resources, if a company has a reasonably uniform product programme the product mix flexibility is mainly an issue of flexibility on the materials side. In this case, the same manufacturing equipment can be used irrespective of which product variant is being manufactured. Volume flexibility may then be more critical and of greater interest. It is an expression of the ability to quickly increase or decrease manufacturing volumes in the company, whether or not mix changes take place simultaneously. Factors that influence a company's volume flexibility include delivery lead times for purchased items, size of direct materials inventory, throughput times and order quantities in manufacturing.

Volume flexibility has a clear relationship with the performance parameter of utilisation rate. If a low utilisation rate is accepted, there are greater margins for increasing manufacturing volumes without requiring further capacity investments. The parameter also has a close connection with costs of levelling stock. The greater the volume flexibility, the less will be the need to make to stock when demand is low.

Delivery flexibility

Due to deficiencies in customers' own material control systems and disruptions or other unpredictable events, the need to temporarily have shorter delivery lead times than normal may arise.

Similarly, there may be requirements for customers to change delivery lead times and order quantities for orders already placed. Delivery flexibility is an expression of a company's ability to achieve such delivery changes when necessary to adjust to customers' changing requirements. Factors that influence delivery flexibility include delivery lead times, throughput times in manufacturing, set-up times and manufacturing order quantities. The ERP system's ability to support rescheduling and to analyse the consequences of changes is also of critical significance.

4.6 Logistics Costs

In every company there are a number of cost items that are associated with, and that may be influenced by, the logistics system. From the perspective of manufacturing planning and control, these costs can be roughly divided into costs associated with material flow and costs associated with production. Material flow costs discussed here are transportation and handling costs, packaging costs, storage costs, shortage and delay costs, and administrative costs. Production costs discussed are capacity costs, costs for changes in rate of production and **set-up costs**. It is not uncommon that these logistics and logistics related costs amount to 20–30 per cent of products' invoice value. It is clearly of interest to be able to reduce them.

Transportation and handling costs

Costs for transportation and handling are relevant both externally and internally. External transportation and handling costs are primarily an issue for material supply and distribution in the logistics system. However, they are also influenced to a certain extent in the production system, for example through the sizing of manufacturing order quantities and thereby the number of transports and the varying degrees of shortages in stock and delays in deliveries from manufacturing which may require back-order deliveries. Internal transportation and handling costs are for internal transportation and movements of goods, including capital costs for transportation and handling equipment as well as for inputs and withdrawals from stock.

Packaging costs

Packaging costs consist of all costs associated with packaging materials, wrapping and marking of goods. These are mainly costs which lie outside the subject area of manufacturing planning and control and as such are not treated further in this book. However, certain packaging costs in the production system are of interest. An example of such costs is standard load carriers used for kanban-managed material flows in production. Load carriers and packages are sometimes shaped in special forms to facilitate physical inventory and item counting when issuing from stock. They may then constitute a factor to consider when manufacturing planning and control solutions are being worked out.

Inventory carrying costs

Inventory carrying costs in material flows are often quite large. They comprise three types: capital costs, storage costs and risk costs. Capital costs are normally calculated as an interest rate times the capital tied up in inventory. The interest rate is often expressed in the form of an internal interest requirement. Internal interest can then be set as equal to the interest paid when borrowing capital from a bank or other credit grantor, or as equivalent to the internal yield that would have been gained through another best alternative in which to invest capital. Storage costs include costs for warehouses, warehouse shelves, handling equipment, running costs, insurance and so on.

The third type of cost is related to risk and includes costs for insurance, wastage, scrapping and other decreases in value. The larger an inventory, the higher the insurance costs. Costs for wastage and scrapping normally increase with the size of stock levels, too. Every item that is put in stock runs a risk of being less saleable at a later stage. The larger the quantities put in stock and the longer they remain in stock, the greater the risk that they will become obsolete or depreciate in value. Depreciation may depend on limited shelf life. It may also be related to changes in fashion and improvements in design of different sorts, which may mean that there is no longer any demand for the product. Large values of obsolete items are scrapped every year in industry. These amounts may be seen as a measurement of the efficiency of the company's materials management from an obsolescence perspective.

Shortage costs and delay costs

Shortage costs and delay costs refer to costs which arise when a demanded product cannot be delivered due to shortages in stock or delayed delivery from manufacturing. Costs may be incurred by loss of earnings due to the absence of sales, costs for loss of goodwill, damages costs and costs for extra transportation or special transportation.

Shortage and delay costs also arise in manufacturing, such as costs for manufacturing disruptions due to shortages of raw materials in stock or delayed delivery from a supplier. There may also be extra costs for rapid transportation of raw materials and various purchased components to avoid a delayed start of a manufacturing order, or costs for forced manufacturing compared to plan with extra overtime pay to compensate for delays in inbound deliveries. With current increasingly capital-intensive manufacturing equipment in industrial companies, disruption costs of this type can be very significant. Shortage and delay costs are often very difficult to calculate and are thus difficult to use as measures of efficiency.

Administrative costs

Administrative costs are all those costs associated with control of material flows and production. First and foremost, they include costs for administrative personnel for order processing, planning, stock reporting and so on. There are also costs for running computer systems for logistics activities, for example ERP systems.

Capacity costs

The capacity costs in a production system consist of costs for capital invested in manufacturing equipment including write-down, and costs for maintenance and operation of equipment. From the perspective of production and materials management, capacity costs can be influenced by the degree to which equipment is utilised. Since capacity costs are mainly fixed costs, higher rates of utilisation will mean that the same cost can be divided over a larger number of units produced, thus decreasing unit product costs. As a performance measurement for a production system, the measurement of utilisation rate may be more useful than the absolute value of a company's capacity costs.

Costs for changing rate of production

In the review of different types of stock in Chapter 2, levelling stocks were mentioned, the purpose of which is to keep an even rate of production even when demand from the market varies, for example due to seasonal variation. Another approach is to vary the rate of production in the short term so that it is adapted as much as possible to the rate of demand, resulting in less need for levelling stock.

A company's rate of production can be adjusted in the short term, at least within certain limits, without making capacity investments. This can take place by changes in the workforce through more overtime or by subcontracting, i.e. temporarily using external suppliers to manage production peaks. In all cases there will be extra costs which must be weighed against the advantages and profits obtained by working with less inventory or by not losing sales during demand peaks. The costs associated with the adjustment of capacity are what are called changing the rate of production.

Set-up costs

When changing over the manufacturing of one item to another, a number of activities must be performed. Tools may need to be replaced, including put away and withdrawal from stock, adjustment of machines or other manufacturing equipment made, and trial runs carried out. All these activities incur costs. During the time that machines are being adjusted or set up and trial runs are taking place, they cannot be used for any other productive activities. There is accordingly a manufacturing downtime, which also represents a cost. All these costs together are usually called set-up costs.

Normally set-up costs are considered to be independent of the quantities manufactured, i.e. they are treated as specific fixed costs. It is therefore of interest to produce as much as possible in one batch to decrease set-up costs per item. This must be weighed against inventory costs which arise when manufacturing large quantities. This question of cost balancing is treated in more detail in Chapter 11.

4.7 Tied-up Capital

The efficiency measurements described above influence revenues and costs in the company. Manufacturing planning and control also affects a company's assets and the capital tied up in these assets.

A company's assets can be divided into fixed or capital assets and current or liquid assets. Fixed assets, in simplified terms, are assets intended to be used on a permanent basis. Examples are factory buildings, manufacturing equipment and other equipment such as fixtures and fittings. The efficient use of fixed assets is related to **capacity utilisation**.

Current assets are of a less permanent nature and are intended to be constantly renewed. In many manufacturing companies, current assets are in the same order of magnitude or even larger than fixed assets. There are generally two dominant items in current assets: inventories and accounts receivable. It is above all the capital tied up in inventory in different stages that can be influenced by the logistics system, and it is this type of tied-up capital which will be discussed here. It may be noted, too, that tied-up capital in accounts receivable may also be affected by the efficiency of logistics systems. For example, customers' settlement periods are influenced by the extent to which customer orders can be fully delivered, i.e. the delivery capacity – which the logistics system has a decisive influence on.

As described in Chapter 2, capital is tied up in the material flow itself as well as in different types of inventory along the flow. From the perspective of capital tied up, significant amounts normally occur only in inventories of three categories: raw material, purchased components and manufactured semi-finished items, i.e. direct material for manufacturing, work in process and finished goods inventories. For this reason, only performance measures related to these three categories of tied-up capital are treated here. Three different performance measurements are normally used as an expression of these types of tied-up capital: tied-up capital in monetary values, turnover rate and **run-out time**.

Tied-up capital in monetary values

The simplest way of reporting tied-up capital is to directly state the value of inventory in monetary value: one value for each of the above mentioned categories. If a more detailed division is required, inventory value per item group may be stated. It is normally possible to achieve this within existing ERP systems.

Inventory turnover rate

One disadvantage of using tied up capital in monetary values as a performance measure is that comparisons are very limited since capital tied up is not described in relation to anything, such as the scope of operations. This is true for comparisons with other departments, companies and industries as well as internal comparisons over time. To obtain comparable measurements of tied-up capital, the **inventory turnover rate (ITR)** can be used instead as a measurement of performance. The turnover rate can be defined in the following manner:

$$Inventory\ turnover\ rate = \frac{Cost\ of\ goods\ sold}{Average\ capital\ tied\ up\ in\ material\ flow}$$

Thus the inventory turnover rate expresses the relationship between the value of the total flow of material during a certain time period, often one year, and the average capital during the same period which is tied up in this flow, normally in the form of inventories.

To obtain a correct ratio, the same type of valuation should be used both in the numerator and the denominator. Capital tied up in material flows should be evaluated in the same way as above for tied-up capital in absolute figures. Outbound deliveries should be evaluated at the cost of goods sold, not at invoicing or sales prices. If the cost of goods sold during one year at a certain company was €178 million and the total average tied-up capital in material flows and inventories was €33 million, the inventory turnover rate will be 178/33, or approximately 5.4. This is usually expressed by saying "inventory is turned over 5.4 times". The higher the rate of turnover, the better is the ratio from the perspective of tied-up capital.

The measurement of inventory turnover can also be divided into inventory of direct material for manufacturing, work in process and finished goods inventory. The values of tied-up capital are determined in the same way as above. To obtain the best possible ratio, different delivery values should be used for each type of stock. The following three formulas may be used:

$$Turnover\ rate\ in\ inventory\ of\ direct\ material = \frac{Annual\ value\ of\ consumed\ direct\ materials}{Tied-up\ capital\ in\ inventory\ of\ direct\ material}$$

$$Turnover\ rate\ in\ WIP = \frac{Annual\ manufactured\ value}{Tied-up\ capital\ in\ WIP}$$

$$Turnover\ rate\ in\ finished\ goods\ inventory = \frac{Cost\ of\ goods\ sold}{Tied-up\ capital\ in\ finished\ goods\ inventory}$$

The calculation of turnover rates for individual groups of items or item types, for example purchased items, is of course also possible.

Run-out time

Another measurement of tied-up capital in stock which also expresses a relationship to the scope of operations is the inventory run-out time, sometimes called **cover time**. For total tied-up capital, the measurement can be expressed in the following way:

$$Run\text{-}out\ time = \frac{Average\ tied\text{-}up\ capital\ in\ material\ flow \cdot 52}{Cost\ of\ goods\ sold}$$

The use of the factor 52 means that the measurement shows how many weeks' deliveries the stock on hand corresponds to. Naturally, time units other than weeks and thus factors other than 52 may be used. The shorter the run-out time, the better from the perspective of tied-up capital. Using figures from the example above, a run-out time of 33 · 52/178 is obtained, i.e. 9.6 weeks. Put differently, stock on hand covers 9.6 weeks' delivery requirements in terms of value.

In this calculation, the same aspects of choice for the basis of valuation apply as with the calculation of inventory turnover rate. There is, in fact, a very close relationship between the two measurements. Run-out time in weeks may also be calculated with the aid of the following formula:

$$Run\text{-}out\ time = \frac{52}{Inventory\ turnover\ rate}$$

The measurements are thus very similar. The use of run-out time may, however, have certain advantages since the measurement is more easily grasped and gives a better indication of the size. In the same way as for turnover rates, run-out times can be calculated separately for inventory of direct material, work in process and finished goods inventory. The run-out time for each of the types of inventory is then an expression of how long the inventory of direct materials will be sufficient to cover requirements in manufacturing, how many days' manufacturing value work in process represents and for how long deliveries to customers can be covered by finished goods inventory.

The above three performance measurements are easily applicable as expressions of the amount of tied-up capital in material flows, which is of crucial significance for financial results in most manufacturing companies. The importance of tied-up capital for profitability in the form of return on capital employed (ROCE) can be illustrated with the aid of a profitability diagram.

A company's return on capital employed is the relationship between its profit and invested capital in a company. This relationship can be rewritten in the following way:

$$Return\ on\ capital\ employed = \frac{Profit}{Turnover} \cdot \frac{Turnover}{Invested\ capital}$$

The profit in relation to turnover is the profit margin of the company, and turnover in relation to invested capital is the rate of turnover. The relationship can accordingly also be expressed in the following way:

$$Return\ on\ capital\ employed = Profit\ margin \cdot Turnover\ rate$$

The relationship is illustrated in Figure 4.1 for three different levels of return on capital employed. The figure also illustrates clearly that an improvement in the company's return on capital employed can generally be achieved in two different ways: by increasing the profit margin or by improving the turnover rate for invested capital. Since tied-up capital in most manufacturing companies to a large extent is in material flows, logistics measures to reduce tied-up capital and increase the turnover rate are of great significance for achieving improvements in financial results. With less tied-up capital, the requirements for a high profit margin are reduced for unchanged return on capital employed.

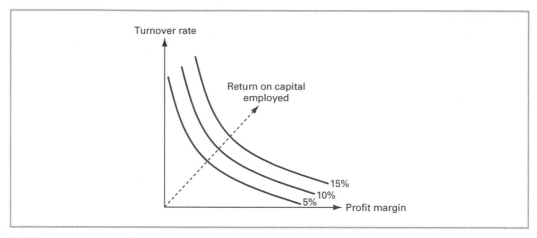

FIGURE 4.1 Example of profitability diagram showing the relationship between return on capital employed, profit margin and rate of turnover

4.8 Capacity Utilisation

A company's capacity utilisation has great significance for its capacity costs and thus production costs. In addition it is important for the size of fixed assets and consequently tied-up capital in the company, since a higher utilisation of capacity means lower capacity requirements.

Capacity in this context is an expression for how large production resources are from the perspective of volumes, i.e. what volumes can be produced. Even though the concept can be defined easily, its content is far from simple, especially with respect to performance levels referred to when discussing different degrees of utilisation. When discussing efficiency measurements for capacity utilisation it is generally most useful to start with nominal capacity, i.e. the capacity which a company is able to use under normal conditions.

The performance measurement of utilisation rate may be used to express capacity utilisation, defined in the following way:

$$Utilisation\ rate = \frac{Produced\ volume}{Nominal\ capacity}$$

For this performance measurement, too, it is important to use the same unit in the numerator and the denominator, i.e. produced volume and nominal capacity must be expressed in the same units, such as number of items. Produced volume and nominal capacity can also be stated in units of time, such as working hours or machine hours.

4.9 Optimisation of Performance Variables

Seeking optimal logistics solutions means trying to achieve the best possible results with respect to the above performance parameters. However, two general issues complicate such

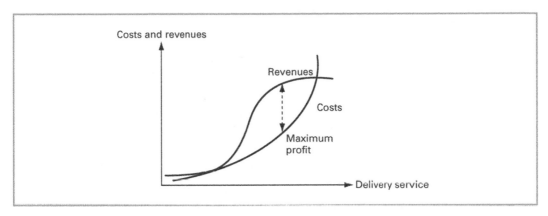

FIGURE 4.2 The relationship between costs, revenues and delivery service

an approach. The first issue is whether there is any limit on how much a company has reason to improve each parameter, and the second is how to deal with problems related to different variables that conflict with each other. These two matters of principle are treated briefly below.

To what extent a company should continue to perform better depends to a large degree on the performance of competitors. This is especially relevant to variables that influence revenues, since different forms of delivery service such as higher stock service levels and greater delivery precision will incur higher costs. The relationship between costs, revenues and delivery service is illustrated in general terms in Figure 4.2.

As shown by the revenue curve, the effect on demand is greatest within a certain delivery service performance. At lower and higher delivery service levels, the effects on demand, and thus on revenues, are less. This is a general phenomenon which is true of several performance variables in the logistics system and the conditions for delivery service can serve as a general example of the phenomenon. The phenomenon is usually called the law of diminishing marginal utility. In this case the law describes that when a certain level of performance has been achieved, further improvements will have a more or less marginal significance. Since improvements will incur further costs, they should be limited to the point at which they cause higher marginal costs than marginal revenues, and as such generate maximum effect on profits.

The gradient of the revenue curve in Figure 4.2, in reality the marginal effects of further improvements in performance, is determined by the competition on the market. The greatest increase in demand, i.e. the greatest effect on revenues, will be obtained close to the delivery service level of competitors.

The second general issue is related to managing performance variables that conflict with each other. For example, high service levels may be achieved at the expense of large inventories of finished goods. Figure 4.3 illustrates some typical examples of conflicting performance variables. Minus symbols mean that the variables are in conflict with each other, and plus symbols signify that they are complementary.

As can be seen in Figure 4.3, several of the performance parameters are conflicting. It is therefore important to make certain priorities when selecting working procedures and when drawing up logistics strategies and logistics systems. These priorities should be made on the basis of overall goals for the company's operations in order to strengthen the company's unique competitiveness. This idea is developed in the next section.

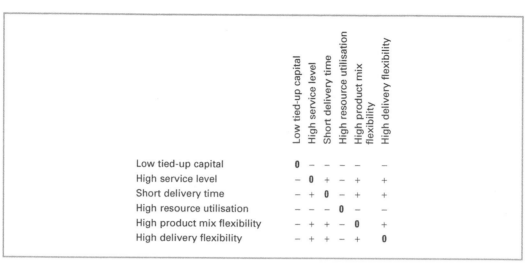

	Low tied-up capital	High service level	Short delivery time	High resource utilisation	High product mix flexibility	High delivery flexibility
Low tied-up capital	0	–	–	–	–	–
High service level	–	0	+	–	+	+
Short delivery time	–	+	0	–	+	+
High resource utilisation	–	–	–	0	–	–
High product mix flexibility	–	+	+	–	0	+
High delivery flexibility	–	+	+	–	+	0

FIGURE 4.3 Examples of opposing and complementary performance variables

4.10 Goal Conflicts and Priorities

The previous section explained that several performance variables clash with each other and that there are goal conflicts, also called trade-offs. The goal conflicts described were mainly of a financial character. There are also goal conflicts related to organisation, which are important to consider.

In most companies there are different opinions about which of the above performance variables are most important to maximise in order to achieve the best possible total influence on profits in the company. These opinions tend to be correlated with the different functional areas of an organisation, and are often rather similar from company to company. Managers in different departments usually have the following opinions about which variables are most important for their company:

- Marketing manager: High service level/delivery precision
  High flexibility in product mix and delivery
- Financial manager: Low tied-up capital
  Low total costs
- Production manager: High capacity utilisation
  Large order quantities

Despite the fact that they all belong to the same company, managers often have different opinions about what is most important overall for the company. These differences of opinion are often related to the fact that people always tend to perceive a situation from their own perspective, and in this case from their own area of management. Another explanation could be the way in which managers' performances are measured and followed up, as it is not unusual that their behaviour is influenced by these factors. Traditionally, a marketing manager's main task is to sell as high volumes as possible, and this is what is expected from him/her. Naturally, it is easier to achieve higher sales volumes if you are able to sell from a wide range of products with competitive and flexible delivery lead times, and are able to deliver with high precision.

With the knowledge that different performance variables may conflict with each other, and that their significance is often perceived differently by various decision makers in a company, it is important that the management of a company prioritise one or more of the variables and then endeavour to become as high performing as possible in these variables. Priorities should be made on the basis of the most important aspects that support the company's business concept and competitiveness. Secondary and other lower priority variables should then be maximised, but not at the expense of the primary variables. If, for example, a company's unique competitive strength is associated with always being able to deliver its products immediately, with great precision and from a wide product range, first priorities should not be reduced costs and slimmed stocks.

When discussing which competition variables should be given priority and in what areas measures should be taken to become higher performing, it may be an advantage to distinguish between order-qualifying and order-winning variables and performances. An order qualifier is a competition variable which requires a certain minimum performance level for the company to survive on the market and be able to compete at all. The competition variables which distinguish a company from its competitors, and which allow it to win orders and create further turnover, are called order winners. In this context, it is important to be aware that a performance or variable that has been an order winner for some time may well change into an order qualifier at a later phase of a product's life cycle. Setting priorities between efficiency variables is very much a dynamic process.

When issues are considered that affect a company's performance it is also useful to distinguish between its efficiency and effectiveness. Effectiveness refers to the company's ability to adapt to and take advantage of opportunities that exist on the markets in which it operates, while internal efficiency is the capacity to be efficient in operative activities within the framework of existing resources and products. If these concepts are distilled into a few words, efficiency is the ability to work right, whereas effectiveness is the ability to work with the right things. Both of these performance concepts are related to and dependent on each other. On certain markets, a low price may be crucial for successful competition, or for achieving high effectiveness. This may be achieved through efficient production, or higher efficiency in other internal processes.

Efficiency and effectiveness may also conflict with each other. By reducing inventory, tied-up capital can be decreased, thus leading to improved efficiency. But cuts in inventory may also give rise to poorer delivery capacity which may bring about a fall in competitiveness and a drop in effectiveness. Improvements in efficiency and effectiveness must, therefore, be balanced with each other. The variables that should be prioritized first are those which support effectiveness. This does not mean that efficiency can be neglected and other variables should be considered as more or less meaningless. In the long term, efficiency is important in maintaining effectiveness. It is more an issue of ranking goals when there are conflicts of interest.

4.11 Summary

The aim of manufacturing planning and control is to contribute to improving efficiency and effectiveness in a company and thereby achieving a positive effect on profits. This chapter has taken up a number of different logistics-related performance variables that influence the company's revenues and costs. Applications and definitions of measurements for stock service levels, delivery precision, delivery reliability and delivery lead time were presented. Tied-up capital was discussed in terms of absolute monetary values, inventory turnover rate and run-out time. Capacity utilisation and utilisation rate were treated. A number of types of logistics costs were described, and three types of logistics-related flexibility.

Performance variables were characterised from two perspectives: for follow-up purposes, and for sizing in the logistics system, such as the use of service level requirements for sizing of safety stock.

Several of the performance variables presented here conflict with each other and are not easily maximised at the same time. It is often necessary to prioritise and ensure in other ways that the performance variables comply with the company's overall strategy and direction. The perception of the importance of achieving improvements in individual performance variables often varies between different management areas in a company. Issues affecting priorities between different performance variables and goal conflicts were treated briefly at the end of the chapter.

🔑 Key concepts

Capacity utilisation 55	Order line 48
Cover time 57	Product mix flexibility 52
Delivery flexibility 52	Run-out time 56
Delivery lead time 47	Set-up costs 53
Delivery precision 47	Shortage costs 54
Delivery reliability 47	Stock service level 47
Flexibility 47	Tied-up capital 47
Inventory carrying costs 53	Volume flexibility 52
Inventory turnover rate (ITR) 56	

 ## Discussion tasks

1 Delivery precision means the extent to which deliveries take place at the agreed, and by the suppliers confirmed, delivery dates. In what basic ways can the supplier influence this measure of customer service?

2 Discuss what it might mean that one company has an inventory turnover rate of 7 and another company an inventory turnover rate of 14. Is the second company twice as good as the first one?

3 In many companies, managers from different departments have different opinions of what is important for the company to be profitable even though they all work in the same company. Discuss why it might be like this and what the consequences might be.

Problems[1]

Problem 4.1

The annual turnover measured as cost of goods sold (COGS) at Cycle Pro Inc. is €520 million. Tied-up capital in the finished goods inventory is estimated to €80 million.

a) What is the inventory turnover rate for the finished goods inventory?

b) There are 250 delivery days to customers per year. How many inventory days (run-out time) does this inventory turnover rate correspond to?

c) How much should the tied-up capital in the inventory be reduced to increase the inventory turnover rate to 10 times?

Problem 4.2

If the run-out time is calculated per item and the items are sorted in ascending order, i.e. starting with the item with shortest run-out time and ending with the item with longest, a graph with the shape as in the figure can normally be drawn.

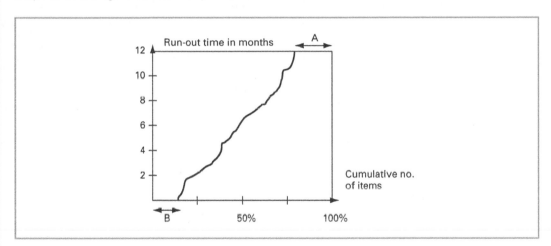

a) What characterises the items in part A of the graph?

b) What characterises items in part B of the graph?

c) How should the graph be shaped in an optimally controlled inventory?

Problem 4.3

Ventilation Products Inc. makes indoor ventilation products to stock. The turnover, measured as cost of goods sold, is €800 million per year. The average tied-up capital in finished goods stock and work-in-process in the final assembly manufacturing is €80 million, in the semi-finished goods stock of in-house manufactured components €70 million, the stock of purchased components to be used in the final assembly is €60 million and the stock of purchased raw materials to be used in the in-house component manufacturing €40 million. The company decides to establish a new plant for component manufacturing in a low-cost country. The turnover in this

1   Solutions to problems are in Appendix A.

plant, valued as cost of goods sold is estimated to be €450 million. Deliveries of components from the new plant to Ventilation Products' main plant are triggered by direct assembly requirements. Therefore, there is no need for keeping a stock of these components at the main plant. The size of the different types of stock are not expected to be changed due to the outsourcing.

a) What was the inventory turnover rate before the outsourcing?

b) What will the inventory turnover rates be at the main plant and new plant, respectively, after the outsourcing decision?

Problem 4.4

During week number 38 Selma's Relax Products has delivered the following customer orders.

Order 240	Promised delivery date: Tuesday		
	Row 1 Item 35983	Quantity 100	Delivered 100 Tuesday
	Row 2 Item 34628	Quantity 30	Delivered 20 Tuesday
Order 246	Promised delivery date: Wednesday		
	Row 1 Item 75394	Quantity 70	Delivered 70 Friday
	Row 2 Item 48333	Quantity 30	Delivered 20 Wednesday
	Row 3 Item 23349	Quantity 200	Delivered 200 Wednesday
Order 249	Promised delivery date: Thursday		
	Row 1 Item 35983	Quantity 50	Delivered 50 Thursday
Order 250	Promised delivery date: Thursday		
	Row 1 Item 44883	Quantity 10	Delivered 10 Thursday
	Row 2 Item 23349	Quantity 45	Delivered 35 Thursday
	Row 3 Item 43219	Quantity 90	Delivered 50 Friday
Order 254	Promised delivery date: Friday		
	Row 1 Item 56732	Quantity 35	Delivered 35 Friday
	Row 2 Item 66687	Quantity 150	Delivered 100 Friday

a) What is the delivery precision, measured as order line fill rate in percentage of that week?

b) What is the delivery precision, measured as order fill rate in percentage during that week?

Further reading

Ballou, R. (2004) *Business logistics/supply chain management.* Prentice-Hall, Upper Saddle River (NJ).

Christopher, M. (2005) *Logistics and supply chain management – creating value added networks.* Prentice-Hall, Upper Saddle River (NJ).

Cox, T. (1989) "Toward the measurement of manufacturing flexibility", *Production and Inventory Control Journal*, Vol. 30, No. 1, pp. 68–72.

Hill, T. (2000) *Manufacturing strategy: text and cases.* Palgrave, New York.

Jonsson, P. (2008) *Logistics and supply chain management.* McGraw-Hill, New York.

Maskell, B. (1991) *Performance measurement for world class manufacturing.* Productivity Press, Cambridge (MA).

Lambert, D., Stock, J. and Ellram, L. (1998) *Fundamentals of logistics management.* Irwin/McGraw-Hill, New York (NY).

CHAPTER

05

Basic Data for Manufacturing Operations

A ll rational decision-making is characterized by access to information. Since decision-making is an essential part of control processes, access to information is therefore vital to the control of manufacturing operations. The control of material flows and production processes is no exception in this respect. Complete and correct information is of crucial significance for the quality of decisions made and for a company to fulfil objectives for its operations.

One part of the information required for controlling manufacturing operations is usually given the generic term of "**basic data**". In simplified terms, basic data may be defined as fundamental information about the company's products, what items they consist of and how they are manufactured, and the company's manufacturing resources, i.e. those resources at the disposal of the company for manufacturing the products. This chapter describes the types of information that constitute basic data in this respect, and to which access is necessary for efficiently controlling manufacturing operations. In addition, different issues are treated which affect the adding, maintaining and use of such information.

LEARNING OBJECTIVES

After studying this chapter, you should be able to:

■ Describe what types of basic data are necessary for conducting manufacturing planning and control

■ Explain how item data is used in manufacturing planning and control

■ Explain how bill-of-material data is used in manufacturing planning and control

■ Explain how **routing** data is used in manufacturing planning and control

■ Explain how work centre data is used in manufacturing planning and control

■ Explain the purpose and strategies of analysing basic data

5.1 Types of Basic Data

The concept of basic data was defined above as fundamental information on the conditions and circumstances of a company's products and its manufacture of those products. Starting from this definition, the following main types of basic data are usually identified:

- Item data, with information that identifies and characterises items, including item number, description, unit of measure, weight, etc.
- Bill-of-material data, with information describing how products and other assembled items are structured, i.e. which raw materials, purchased components and manufactured semi-finished items they consist of
- Routing data, with information describing how products and other internally manufactured items are manufactured and what use of resources is required for this manufacturing
- Work centre data, with information describing what manufacturing resources are available and what capacity and performance these resources have

In order for this information to be practically maintained and used for different types of applications, it is stored in a database in ERP systems which virtually all manufacturing companies normally have access to. Four basic data files are usually referred to, corresponding to the above basic data types: item files, bill-of-material files, routing files and work centre files.

The relationships between the different files are illustrated in Figure 5.1. The connection between item files and bill-of-material files enables the structural composition of products and other assembled items to be specified and described. The connection between item files and routing files ties together manufacturing specifications for each product and other internally manufactured items. The connection between routing files and bill-of-material files enables incorporated materials to be related to the manufacturing steps where they are required for products and other assembled items. Finally, the connection between routing files and work centre files brings together different stages in the manufacture with the manufacturing resources in which the manufacturing stages are carried out.

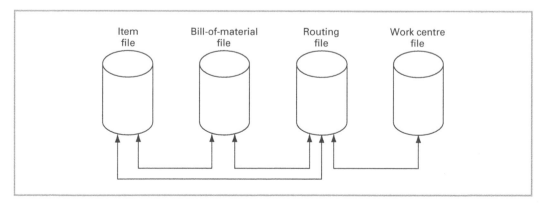

FIGURE 5.1 The relationships between different basic data files

5.2 Areas of Use

A complete and appropriately designed and maintained basic database is a prerequisite for a number of functions in the efficient control of material flows and production in a manufacturing company. Basic data information has also great significance for a number of other units in companies, including design, production engineering, order entry and purchasing. Some important areas of use for each unit are shown below.

Design and product development

In the area of design and product development, with the help of a basic database it is possible to specify, store and report information about the composition of products. It also enables analyses of different types, for example where-used analyses to analyse in which products and assemblies a certain item is included. Searches can be carried out to produce lists of standard dimensions used, or similar previously engineered products. In this way it is easier to handle the issues of variant items and avoid problems with item assortments that are too wide. With the aid of a rationally structured basic database it is also possible to facilitate and organise procedures for **engineering changes**. Examples of such procedures are scheduling of phasing in and phasing out new materials and components, and monitoring ongoing engineering changes.

Production engineering

The production engineering unit can specify, store and report routings and run-times in a basic database in the same fashion. There are also many possibilities for different types of analysis, such as the analysis of set-up times, throughput times and critical paths in manufacturing networks. The critical path refers to the longest total lead time for procurement of materials and manufacturing in a product structure. Analyses and information about which items and operations use specified work centres can also be obtained. Using a basic database, it is possible to create basic information for the grouping of manufactured items through, for example, flow analysis or various group technology methods when creating manufacturing cells.

Material control

With the aid of stored basic data, it becomes possible to automatically allocate materials to manufacturing orders. Access to information on quantities available in stock can then be gained, and material control accordingly improved. Efficient and commonly used material control methods, such as material requirements planning (MRP), are completely reliant on the bill-of-material file in a basic database. Without this information such methods would be impossible to use at all. In certain types of companies it is rational to apply backflushing (see Chapter 18) when reporting inbound deliveries of orders on finished products. For companies with repetitive manufacturing this is virtually a necessity. Access to the **bills of material (BOM)** from a basic database is indispensable for this.

Production control

Access to a basic database is crucial in a number of different ways for production control. Capacity requirements, as a basis for control of manufacturing, require access to basic data about operations and operation times. The same is true of priority in the workshop. For ordering purposes, the so-called shop packet (see Chapter 16) often plays a large role. Most of the information in a shop packet stems from basic data about the products.

Purchasing

A company's purchasing activities are also dependent on information which is available in a basic database. This is true, for example, of information about possible suppliers of purchase items and delivery lead times, purchasing units of measure and weight information. Different price information is also part of the basis for purchasing which may be obtained from a basic database.

Customer-order processing

Apart from item information such as item numbers, descriptions and units of measure, the connections between customer-order processing and basic data are not especially strong. However, there are some examples of cases where information from a basic database can provide considerable potential for rationalisation. When selling products made up of unassembled components, it is not uncommon that customer-order processing and invoicing is for whole kits, while picking and outbound transportation is for components of the kit. This problem can be solved with the aid of pick list bills in a basic database. There are corresponding types of issues for spare parts kits. Such kits can be joined together with the aid of special spare part bills in a basic database.

For companies with a large number of variants and modularised products, specifying variants when processing customer orders can be done with the help of specially structured information in a basic database. The selection of alternatives is carried out with the help of so-called product configurators, which provide a basis for automatically generating the desired product variants from bill-of-material files in the basic database.

5.3 Item Data

An item can be generally defined as an identifiable physical unit. In this context, items refer to products, raw materials, purchased components, manufactured parts, semi-finished items and indirect materials. Fundamental information on such items can be divided into a number of different types as shown below. The different types are illustrated in Figure 5.2 with the help of examples of typically occurring item information for each type.

- Base data, covering information which describes, characterises and specifies items as physical phenomena. This information is obtained internally, such as descriptions, and externally, such as supplier information on a purchased item. The item number provides an identity for the item.

- Price and costing data, covering information on prices, overhead rates and costing. This information is obtained externally, such as prices of purchased items, and internally, such as overhead rates for costing. This type may also include different cost drivers that are used for **activity-based costing**.

- Forecasting data, covering forecast data and information for forecasting of demand for items.

- Sales data, covering information on sales volumes.

Basic data

Item number	Unit of measure	Item type
Description	Supplier	Stock location

Price and cost data

Latest price	Average price	Standard price
Actual cost price	Price conversion factor	Overhead percentage

Forecasting data

Forecast value	Forecast error	Standard deviation
Control limit	Smoothing factor	

Sales data

Delivery statistics	Order statistics	Sales data

Planning data

Lot sizing method	Order quantity	Safety factor
Ordering cost	Lead time	Re-order point

Inventory data

Stock on hand	Allocated quantity	Open order quantity
Date physical inventory	Physical inventory discrepancy	Number of issues per year

FIGURE 5.2 Example of item data for a manufacturing company

- Planning data, covering information for planning of procurement, manufacturing and dispatch of items.
- Inventory data, covering information on inventory status of items. This information is mainly generated internally.

Certain of the above data are mainly fixed in character, i.e. they do not change during the life cycle of an item. Such data includes descriptions and units of measure. Other data changes during the item's life cycle, but relatively seldom, and can thus be characterised as semi-fixed or low-variation data. This could include prices, stock positions, forecasts and planning parameters of different kinds. A third category of data is characterised as highly variable and changes very frequently, in many cases several times a day. An example of this type is stock on hand.

With respect to direct materials, i.e. items excluding consumables or indirect materials, we can distinguish between four item types: end products, semi-finished items, purchased components and raw materials. End products refer to items that are manufactured and ready to be delivered to customers, whereas semi-finished manufactured items are items that are manufactured but are included as parts of another semi-finished item or an end product. Semi-finished items may also be products in the sense that they are items that are sold and delivered to customers. This is true of manufactured items sold as spare parts. Final assembly and sub-assembly are common alternative designations in the mechanical industries. End products and manufactured semi-finished items all involve bills of material, i.e. they in turn consist of other items. Raw materials and purchased components are items that are included in the manufacture of end products and manufactured semi-finished items. They normally have no bills of material, and are treated as indivisible units, even though they consist of incorporated parts at the company they are purchased from. Figure 5.3 shows an example of item data in an ERP system.

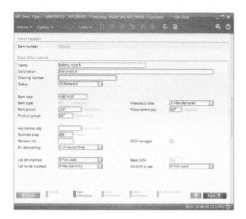

FIGURE 5.3 Example of item data in ERP systems

The picture represents a function from the ERP system Lawson M3

5.4 Bills-of-Material Data

Characteristic for all manufactured items is that they are produced through some form of refinement of direct materials. They always consist of some other items, which may be purchased externally or manufactured, and, if manufactured, in turn consists of still other items. The relationship between the items is called bills of material. A bill of material for a product describes how it is composed of other items, and information about this is a precondition for being able to calculate its manufacturing cost and for estimating the requirements of raw materials and components needed in its manufacture. The term "bill of material" will be used consistently in this book. In different industries there are a number of other designations for the same thing, such as formulas or recipes in certain pharmaceutical or chemical process industries.

Information in bills of material plays a crucial role for many applications in the area of manufacturing planning and control. It is a specification of how different items and products are interrelated, and it is the basis of product costing. By using bills of material, material allocations can take place automatically, and bills of material are necessary to be able to use material planning methods of the MRP type at all.

Bill-of-material definition and content

To be able to define a bill of material, it is necessary that structural interrelationships are specified, i.e. how the items incorporated are connected, from purchased raw materials and components to end products. These interrelationships are called bill-of-material relations, or parent–component relations, in which the component item is part of the parent item. For example, a simple wooden chair may consist of legs, a seat and screws. The legs, in turn, may be composed of wooden parts joined together from purchased wooden parts. In the same way, the seat may be manufactured from purchased wooden parts. To describe these stages the terms one-level bills of material and **indented bills** of material are used. A one-level bill of material contains information about direct items incorporated in a composite item – in the case of the chair, information that it consists of legs, a seat and screws. An indented bill of material also shows which items are included at lower bill-of-material levels, i.e. it shows the complete structure of a composite item as far as the raw materials and/or purchased components. An indented bill of material for a snow shovel is illustrated in Figure 5.4.

Permatron Inc.		Indented bill of material	21.9.07
Item no 234 764		Snow shovel	
Level	Item no	Description	Quantity per
1	259 355	Scoop assembly	1
2	444 732	Handle coupling	1
2	908 552	Scoop	1
2	468 332	Blade	2
2	578 449	Rivet	16
1	675 003	Nail	4
1	757 889	Handle assembly	
2	476 099	Welded handle	

FIGURE 5.4 An indented bill of material for a snow shovel

It is clear from the bill of material in Figure 5.4 which bill-of-material level each incorporated item belongs to. This is expressed in the form of a level code, which states the structural level relative to the upper level of the product which the item is part of. Common practice is to call this upper structural level "level code 0", the level under that "level code 1" and so on. For simple products, the number of bills of material is often in the order of two or three. More complex products may have many more levels, and there are cases with up to 15 levels. In most ERP systems the level code is automatically updated when maintaining bills of material.

The same incorporated item may occur in different bill-of-material levels in the different products of which it is a part. For certain applications it is then necessary to know the lowest level at which an item can occur, such as in the case of obtaining the correct level-by-level explosion in material requirements planning. Basic databases in ERP systems contain information for each item on its lowest-level code. This states at what lowest level an item is used in a product, this giving it the highest number. An item that is used in three bills of material at levels 3, 4 and 2 thus has the lowest level code of 4.

For a bill of materials to be completely defined, in addition to the structural connections and the structure level, it must also specify quantities required per parent item. In the example of the chair, there must be information on how many legs the chair should have and how many screws are required to fasten the legs to the seat. In most cases there are few problems encountered in establishing this quantity information. This is the case with discrete components for assembly, such as the legs and screws of a chair. In other contexts a great deal of calculation and production engineering work must be carried out to determine quantity information. It is often difficult to establish exact quantities. Examples of cases in which difficulties arise are stamping out sheet metal parts, die casting, injection moulding of plastic parts and cutting lengths of bars.

In the mechanical industry quantities per unit of the parent item are stated, while in the chemicals industry, the pharmaceutical industry and the foodstuffs industry, it is more common to specify required quantity per batch size. One batch size is the smallest order quantity which is used in manufacturing. The order quantity may also be a multiple of the batch size. It is often defined on the basis of capacity restrictions in production; for example, in the mixing stage in a manufacturing process a tank of 3000 litres is used. The quantities of input materials are then stated per batches of 3000 litres of the parent item.

A number of other data types are also common in bill-of-material files. Effectivity date from and effectivity date to are stated for every parent–component relationship in bills of materials for products. Using such effectivity dates, it is possible to efficiently plan the introduction of engineering changes so that an item is replaced by another item on a desired date. By stating the corresponding manufacturing operation in bills of material, items can be connected to the manufacturing step where they are used. This means, among other things, that the time point when a raw material or component is required in manufacturing can be determined individually for each operation and not generally for all material at the start of the manufacturing order. It means, too, that material can be picked and moved directly to the work centre where the operation will be performed.

Design bills and manufacturing bills

There are many areas of use for bills of material for a product. Each one of these may put special demands on the bill-of-material structures, which may be difficult to combine. The compromises which must then be made may reduce the quality of use and efficiency of control. This is a frequent problem, especially in the mechanical industry where design and manufacturing interests must be weighed against each other in the structures of bills of material. For this reason, the terms "**design bill**" and "**manufacturing bill**" are used.

A design bill means a bill of material for a product which is formed to describe how a product is composed from a design and functional point of view. In a design bill, the product's structure is generally considered from a functional perspective. We speak of function groups, specification groups and design groups of items, which often represent functional units such as the cooling system in an engine.

A manufacturing bill refers to a bill of material for a product which reflects how it is intended to manufacture the product, the structure of material flows, which assembly and sub-assembly steps are required and so on. Here, instead, we speak of assembly groups or material groups of items. Somewhat simplified, we can say that one level of a manufacturing bill represents one step in the manufacturing process, while one level in a design bill represents one step in the design process.

The grouping of items into design groups in a design bill may deviate considerably from the grouping of items in assembly groups in the manufacturing bill for the same product. Figure 5.5 illustrates such differences in one part of a bill for a mechanical product. Seen from above, the bill refers to a design-oriented grouping into functional groups, while seen from below it illustrates the corresponding manufacturing bill grouped for assembly stations. For the designer to be able to use the bills in a bill of a material file, they must be structured as they are seen from above in the figure. This is to ensure functional qualities in the design, but also because design work is often organised so that an individual designer is responsible for an entire function. On the other hand, the bills cannot be used for controlling manufacturing and flows of materials in an efficient manner if they do not reflect how, when and where material requirements arise.

Another frequent difference between a design bill and a manufacturing bill relates to the quantities per parent item. Consideration must be taken in manufacturing bills of the anticipated total material consumption if they are to be used for material requirements planning and product costing. This means also that material consumption through inevitable wastage must be considered; for example, when wooden staves are sawed to be processed into legs for the wooden chair in the example given above. Attention must also be paid to any scrapping involved. Such extra materials requirements are not of primary interest to designers, but are a big issue for material requirements planning.

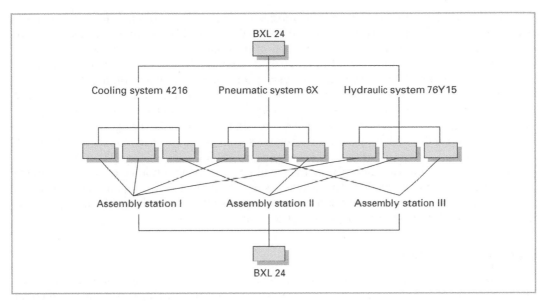

FIGURE 5.5 Illustration of the difference between a design bill and a manufacturing bill

In many cases the differences between the two bill types are not too large to prevent solutions being found which will work for both parties, especially in the case of structurally simple products. In other cases, such compromises are not possible and it may be necessary to work with separate bill-of-material files, one oriented towards design which will cover the needs of designers and product developers, and the other oriented towards manufacturing which will satisfy needs from manufacturing and material flow aspects.

Bill-of-material principles

For use in manufacturing planning and control, bills of material for products must be oriented towards manufacturing and material flow. To achieve efficient control with the aid of an ERP system, there is virtually no room for compromise when adding and maintaining bills of material so that specific design interests can be provided for. The following summary describes some important principles for adding and maintaining bills of material to obtain suitable manufacturing bills:

1 The manufacturing bill must be a model of how manufacturing is laid out and material flows structured.

2 The manufacturing bill must be structured so that forecasting and master production scheduling can be carried out with as few items as possible.

3 The manufacturing bill must be structured so that it supports customer-order processing when manufacturing and delivering customer-order specific product variants.

4 The manufacturing bill must, at a minimum, cover all the items in products that are manufactured or purchased.

5 The manufacturing bill must be structured so that adding and maintaining bills of material are facilitated.

The first requirement means that the bills must reflect the way in which manufacturing takes place and the structures of material flow, otherwise the bill will be of limited use as a basis for time phasing material requirements, allocation of material or printing out pick lists. Requirements 2 and 3 are primarily related to cases where the number of variants is large, meaning that the bill-of-material structure must be given special attention, otherwise the database may become unmanageably large. The fact that the bills, in accordance with requirement 4, must include all items in products is necessary so that product costing can take into account all material costs. It is also a requirement if all material needed for a manufacturing order is requested to be included on the pick list used for withdrawals from stock. All items to be controlled with the aid of material requirements planning must also be included in the bills of material for products. Different bill-of-material principles will cause varying amounts of work for adding and maintaining bills of material. This is why, as in requirement 5, a company should strive to choose principles by which the work required will be as little as possible without jeopardising the utility of the bills from the perspective of planning and costing.

Engineering change

There are a number of reasons why bills of material must be changed. Apart from changes that are corrections of faulty registration, they may be the result of normal product development or manufacturing engineering development, breakdowns or near-accidents with products, quality problems in manufacturing and so on. In many companies there are well-developed procedures for engineering change, from initiating an engineering change request, via a decision on the change, the release of an engineering change order and through to the execution of a change in the database and in manufacturing. From the perspective of basic data, it is possible to differentiate between three categories of changes: those which must be executed immediately, those which can be executed when it is financially appropriate, and temporary changes.

The first of these change categories refers to changes which must be made due to breakdowns, risk of personal injury, legislation and so on. Since these changes must be carried out immediately, they represent no specific basic data problem. When changing items it is a question of executing ordinary bill maintenance. Temporary changes are caused by disruptions in manufacturing and material supply, for example when an item normally used in a product cannot be procured from the supplier during a certain time period and must be temporarily replaced by another item. Changes of this type are, by their very nature, impossible to plan for and do not therefore constitute any true problem with respect to basic data processing.

For the second category, however, processing of basic data is of great significance. If engineering change procedures are not executed in a planned and controlled fashion, there is a great risk that unnecessary costs of various types will arise. An example of this type of cost is scrapping materials due to engineering changes being introduced when there are still large quantities of the item remaining in stock. With the aid of engineering change procedures, which are based on the use of effectivity dates in bills of material, it is possible to plan flows of material and manufacturing so that much of this type of cost can be eliminated.

Phantom items in bills of material

The previous section described some different item types. These item types represent all physical phenomena. In ERP systems, item types are also used to represent administrative phenomena. These item types are called **phantom items**. The use of phantom items varies somewhat, but there are two general areas of use: when adding and maintaining bills of materials, and for creating more rational planning procedures and material flows.

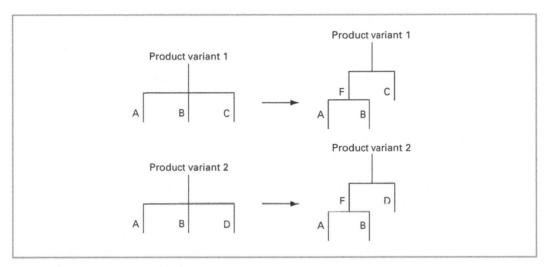

FIGURE 5.6 Example of the use of phantom items

The first area of use is related to adding and maintaining bills of material. Phantom items represent a group of items which are processed together but which do not have any corresponding items in manufacture. Items are linked to phantom items with the help of bills of material, and the aim is to rationalise the maintenance of files. If, for example, a number of items are common to a number of product variants, it is possible to link them together with a phantom item which is part of each product variant, as illustrated in Figure 5.6. In this figure, F is the phantom item used for grouping items A and B. When a new product variant comes up, it is not necessary to make bills of material for all of these common items since only one bill for the phantom item is required. From the point of view of planning, these phantom items are treated as if they do not exist. They are not shown in the inventory balance and there are no manufacturing orders or purchasing orders put on them. When a manufacturing order is released for an item at the bill-of-material level above the phantom item, only the items under the phantom item are allocated and not the phantom item itself.

The other area of use for phantom items is in units for creating more rational planning procedures and material flows. Phantom items may be used as an aid to handling engineering changes when a new item is introduced in the bill of materials when the old item is out of stock. Phantom items can also be used in just-in-time types of manufacturing, for example to manage items which are ordered on the shop floor without any manufacturing order in the ERP system. Another example of their use is to enable forecasting and planning of groups of items instead of individual items. By using bills of material, it is possible to connect a number of products belonging to a certain product group to a phantom item which then represents this product group, and forecasting can be carried out for the product group and individual forecasts created through explosion from the phantom item. This is described in more detail in Chapter 7.

5.5 Routing Data

All manufacturing of products and other items can be described as the refinement of materials in a manufacturing process. Information about material contents is specified in the form of a bill-of-material file. In much the same way, information about refinement in the manufacturing process is specified in a routing file, partly in the form of a description of how and in what steps the manufacturing process takes place, and partly in the form of what resources are required.

Routing data has a number of very important areas of use within different units of material and production control. To be able to carry out product costing, it is not only information on material consumption and material prices that is required. Information is also needed about resource requirements in the form of man-hours, machine-hours and costs for this time. Time information for this purpose is obtained from routing data in a routing file.

The same type of time information is also used for capacity requirements planning based on information about capacity requirements for manufacturing from the routing file. This is a key condition for capacity requirements planning in the same way as bills-of-materials data is a key condition for material requirements planning. Another area of use for routing data is in supplying information in the form of work descriptions and manufacturing instructions.

A routing file contains a number of operation records, one for every operation required to produce each manufactured item. Every operation, meaning every manufacturing step in the routing which must be carried out to produce the item, is numbered. A number series with intervals between each successive operation number is often used for numbering operations, such as the series 10, 20, 30, etc. By using this convention, the addition of extra operations in manufacturing is made easier and new operations in re-engineering can be added without needing to change subsequent operation numbers. An operation added after operation 10, for example, can be given number 15 and will therefore take place before operation 20.

The work centre where each operation will be carried out is specified. This is done by stating the work centre number in question. For reasons of information and instructions, the operation is also given a description. Standardised abbreviations are used in many companies for these operation descriptions. There are two reasons for this: rationalising work with adding and maintaining routing data, and limited space on screens, routing lists, labour tickets and so on.

Alongside the routing number, description and work centre, the most important data in a routing file is the operation time, i.e. the time taken to achieve refinement in the operation in the manufacturing process. This operation time can be divided into two parts: a quantity-independent set-up time, and a run time, which is equal to the manufacturing quantity multiplied by the run time per piece. The relationship between these time concepts is illustrated in Figure 5.7.

Set-up time refers to the time required to prepare for and conclude an operation. It may include time for preparing and setting up tools, setting machines, removing tools and cleaning machines before and/or after manufacture. This time is normally fixed per manufacturing event, irrespective of the number of manufactured units.

For certain types of operations and work centres, it may be inappropriate or of limited value to determine times for operations in a conventional manner. This is, for example, the case of inspections, heat-treating operations and surface treatment operations. To be able to plan

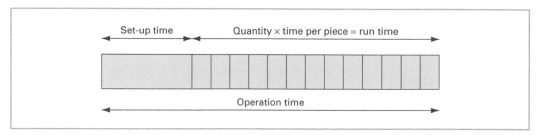

FIGURE 5.7 Operation time and its constituents

preceding and subsequent operations despite this, and to be able to calculate total lead times and throughput times, operation lead times are determined for such operations. These lead times include set-up times, run times and, in some cases, queue times and transportation times in production.

Other information which is usually part of routing files includes data on tools required to carry out operations and manufacturing instructions. There may also be information on alternative work centres for carrying out an operation, in case the one normally used is not available.

5.6 Work Centre Data

Manufacturing resources in a manufacturing company consist mainly of machines and machine equipment of various types, and operators. An operator's role may be to manufacture something without any mechanical tools. It may also be to perform manufacturing in combination with machines and mechanical equipment, or it may only involve monitoring machines which automatically execute manufacturing.

Operators and machines are usually grouped into manufacturing units called work centres. A work centre can be generally defined as a manufacturing unit consisting of one or more operators and one or more machines, which from the planning perspective are considered as an indivisible unit which can independently perform an operation. For example, a work centre may consist of one machine and one operator, one operator and three machines if the machines are to some extent automatic and the operator can serve three of them, one machine and several operators if several operators serve one machine, or one operator at a workbench. The term machine should be interpreted in a wide sense here. To be able to identify work centres, each is given a unique identification number which corresponds to the item number for items. It is this work centre number that is stated for every operation in the routing file in order to link operations with the manufacturing resources which are intended to perform them.

The most central information for production control in a work centre file is that related to available capacity. The capacity of a work centre is a measurement of how much it can produce. The most common units for capacity are man-hours per period or machine-hours per period. Other units such as pieces, kilograms and volumes per period are also used in some cases. The important thing is that the unit is representative for the activity in the work centre and that run times for the operations performed by the work centre are expressed in the same unit.

There are generally four types of data in a work centre file for calculating its capacity: the number of machines or other production units in the work centre, the number of shifts per day, the number of hours per shift and the degree of utilization which can be expected. Information on the number of working days per planning period is also required for capacity calculations. This information is typically obtained from a shop calendar.

In general, work centre files also store information on times for internal transportation between work centres as well as queue times, i.e. the lead times for orders awaiting processing at each work centre. Information about these queue times and transportation times is necessary for capacity planning and for the calculation of finish dates for operations and orders. Storing transportation times in work centre files can provide acceptable accuracy in activities with simple and uniform flows. It is then taken as the average transportation time from all work centres which are immediately before the work centre in question in terms of sequence of operations. If material flows are more complex and irregular, however, this approximation may not be good enough. Transportation times should then be stored in transportation timetables which are linked to the work centre file. These transportation timetables will store transportation times from all work centres to all other work centres. Work centre data also includes information about which department each work centre belongs to and information about alternative work centres. Machine-hour costs and man-hour costs are usually stored as a basis for cost calculations.

5.7 Analysing Basic Data

The primary aim of adding and maintaining basic data in the form of item files, bill-of-material files, routing files and work centre files is to obtain access to information for planning manufacturing and material flows. Information that is available in a basic database is also useful for analyses of various types. Some commonly used basic data analyses are described below. Most often, these are available in ERP systems.

Bill-of-material analysis

A bill-of-material analysis is a list of raw materials, components and semi-finished items which a certain product or semi-finished item consists of. It is often presented in the form of a printed list or an image on a screen. We usually distinguish between a single-level bill-of-material analysis and an indented bill-of-material analysis. A single-level bill-of-material analysis contains information about items that are direct parts of other items and their quantities.

An indented bill-of-material analysis shows, level by level, all items incorporated in a product. The structural level and quantity per parent item is stated for every item in the structure. An indented analysis of a shovel is illustrated in Figure 5.4.

Where-used analysis

A where-used analysis can, in simple terms, be described as an inverted bill-of-material analysis. This means that it is possible for every item to analyse in which other items it is used. Thus, the product structure tree is observed from below and up, instead of from above and down as in the bill-of-material analysis. Such an analysis makes it possible to study where a certain item is used. This is of interest in design contexts when an engineering change is planned, or during replacement of items, and it is necessary to study what products will be affected by the change. This type of analysis may also be of interest when long-term shortage situations arise. With the help of a where-used analysis it is possible to gain information about which products will be affected so that appropriate preventive measures can be taken.

The where-used analysis exists in different variants. Single-level where-used analysis only shows items immediately above the current level in the structure. An indented where-used analysis, on the other hand, shows items at all structural levels above the item in question. Using this variant of the analysis, it is clear which items are directly affected by the replacement of an item, and which are indirectly affected. If an item is in several higher bills of material, it will be repeated every time it occurs in the same way as in an indented bill-of-material analysis.

Work centre where-used

The **work centre where-used** type of analysis corresponds to operations as where-used analysis corresponds to items. It includes all operations, and through that indirectly all items, which are processed in a certain work centre. Analyses are made with reference to work centres.

Permatron Inc.		Lead-time analysis		21.9.07
Item no	234 764	Shovel		Ack lead time 45 days
Level	Item no	Description	Lead time	Ack lead time
1	259 355	Scoop assembly	12	32
2	444 732	Handle coupling	15	15
2	908 552	Scoop	15	15
2	468 332	Blade	20	20
2	578 449	Rivet	20	20
1	675 003	Nail	9	9
1	757 889	Handle assembly	10	
2	476 099	Welded handle		

FIGURE 5.8 Illustration of lead-time analysis

The work centre where-used analysis is primarily of interest from the production engineering perspective when making changes in production. It can answer questions of the type "Which operations and items will be affected if a work centre is replaced?" and "For which operations must we find alternative resources in the case of a machine breakdown?"

Lead-time analysis

Lead times play a central role in the context of logistics. This is true of both item-individual lead times as well as accumulated lead times for manufacturing complete products. With the aid of lead-time analyses, a tool is obtained for identifying items with especially long lead times and for identifying critical paths for accumulated lead times in products. This information can be used to prioritise production-planning measures and as a basis for changes in the structure of manufacturing with the objective to reducing lead times.

In the lead-time analysis in Figure 5.8, the accumulated lead times are expressed as the shortest time it takes to produce a certain product.

5.8 Summary

This chapter has described aspects of basic data, i.e. information required in the area of planning and controlling manufacturing operations and materials flows. It involves basic information about the company's products, including the items they consist of and how they are manufactured, as well as the company's manufacturing resources, i.e. the resources which are available to manufacture the products.

Four types of basic data were distinguished in the chapter: item data, bill-of-material data, routing data and work centre data. The structure and common areas of use of basic data were described, as well as a number of issues related to adding, maintaining and using of such data.

Key concepts

Activity-based costing 68	Indented bill 70
Basic data 65	Manufacturing bill 72
Bill of material (BOM) 67	Phantom items 74
Design bill 72	Routing 65
Engineering change 67	Work centre where-used 78

Discussion tasks

1 An item master file contains lots of data originating from different departments in a company. How do you suggest that the work regarding collecting all these data should be organised and carried out when a new item is designed and is to be added to the item master file?

2 It is common to use effectivity dates to control the replacement of one item with another in a bill of material in cases when engineering changes can be planned in advance. How can you determine these effectivity dates to minimise the risk for shortages? What are the prerequisites to do so?

3 The item file, bill-of-materials file, routing file and work centre file are related to each other in the database in an ERP system. What is the reason for relating a parent–component relation in the bill of material to an operation in the routing file? What is the advantage of doing so?

 ## Problems[1]

Problem 5.1

The Table Plant Corp. makes wooden tables and chairs. One of its tables, model EUF, contains a circular table top and four legs. The table top is circular sawn and ground from rectangular 120 × 220 cm sized chipboards. The legs are sawn, planed and ground from 200 cm long rectangular wood poles. The legs are affixed to the table top using four consoles and four thumbscrews per console. Table tops and legs are made to stock because both items are used also for making products other than the table EUF. Chipboards, wood poles, consoles and thumbscrews are procured in batches from suppliers and put in a raw material stock. The tables are delivered to customers assembled but not painted.

a) Develop an indented bill of material with quantities per item for table EUF.

b) The tables are assembled in batches of 100. How would the bill of material look if making the corresponding quantities of tops and legs from chipboards and wood poles in a common component manufacturing and assembling work centre?

---

1 Solutions to problems are in Appendix A.

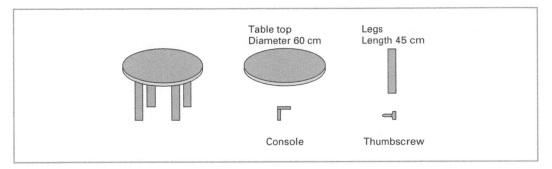

Table top
Diameter 60 cm

Legs
Length 45 cm

Console

Thumbscrew

Problem 5.2

The company Products Etc. makes and sells a large number of various products, for example mouse traps. The assembly drawing for one of these mouse traps is shown below. Measures in the drawing are in millimetres (mm). The bait holder is made from 30 mm long sheet metal bands cut from rolls of sheet metal with the same width as the holder. Each roll of sheet metal contains 30 metres of sheet band. The spring is made from 1.1 mm steel thread and the release mechanism from 1.3 mm thread. The thread is cut in lengths of 250 mm and 70 mm, respectively, from the reels. For both dimensions the reels contain 50 metres of thread. The plate is made from 2 metre long and 50 mm wide red deals. Rolls of sheet, clamps, reels of steel thread and wood are procured from external suppliers.

a) Draw a bill of material for the mouse trap that shows how the different parts are connected.

b) Make a bill of material for the mouse trap, including quantities per parent item.

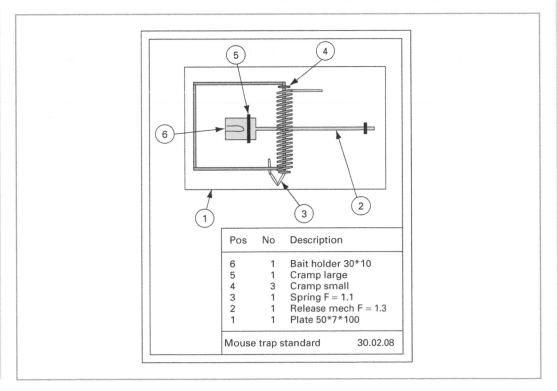

Pos	No	Description
6	1	Bait holder 30*10
5	1	Cramp large
4	3	Cramp small
3	1	Spring F = 1.1
2	1	Release mech F = 1.3
1	1	Plate 50*7*100

Mouse trap standard	30.02.08

Problem 5.3

The bills of material for the two different products, P1 and P2, are described in the figure. The numbers in brackets are the number of pieces included in the respective parent item.

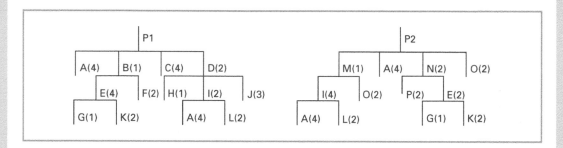

a) Develop a single-level where-used analysis for item A.

b) Develop an indented where-used analysis for item E.

c) What quantity of item K is consumed if making 100 pieces of P1 and 50 pieces of P2?

Problem 5.4

Sakpro Inc. is a company making safety products for the automotive industry. One of its products, PFB, contains component A1. In order to improve the product functionality and reliability of PFB, the company plans to exchange A1 for the new component A2. The new A2 will be phased in on 15 April when the first delivery is estimated. A1 and A2 are not fully exchangeable because the functionality of PFB is affected by the engineering change. The components are also sold as spare parts. Because it is a safety function, the traceability of what components are included in what products is very important at spare part exchange.

At 1 April, there are 650 pieces of A1 in stock and the company plans to start to manufacture a batch of 500 pieces of PFB on 6 April and another batch of equal size on 20 April.

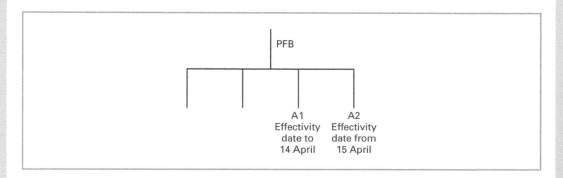

a) How many of item A1 will remain in stock on 25 April given manufacturing has been carried out according to plan?

b) What should be done with pieces remaining in stock on 25 April?

c) How many items would remain in stock if A1 and A2 are fully exchangeable and do not affect the function and reliability of product PFB?

Problem 5.5

At Danish Factories Ltd a specific product is made from components according to the figure. Numbers within brackets show required lead times in weeks to make or buy the respective item.

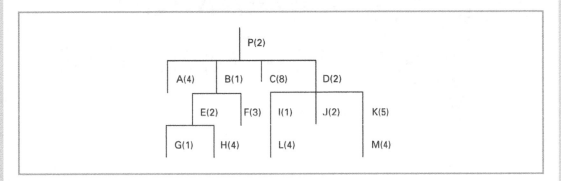

a) Conduct a lead-time analysis for the product.

b) How long is the cumulative lead time for the product?

c) For what items is it most important to decrease lead times in order to decrease the cumulative lead time for the product?

Further reading

Garwood, D. (1995) *Bills of materials: structured for excellence*. Dogwood Publishing Company, Marietta (GA).

Hamilton, S. (2002) *Maximizing your ERP system*. McGraw-Hill, New York (NY).

Huang, G. and Mak, K. (1999) "Current practices of engineering change control in UK manufacturing industries", *International Journal of Operations & Production Management*, Vol. 19, No. 1, pp. 21–37.

Mather, H. (1987) *Bills of materials*. Dow Jones-Irwin, Homewood (IL).

Plossl, K. (1987) *Engineering for the control of manufacturing*. Prentice-Hall, Englewood Cliffs (NJ).

Scott, R. (1997) "Bills of materials and routings". In Greene, J. (ed.), *Production and inventory control handbook*. McGraw-Hill, New York (NY), pp. 5.1–5.30.

Tacvar, J. and Duhovnik, J. (2005) "Engineering change management in individual and mass production", *Robotics and Computer-Integrated Manufacturing*, Vol. 21, No. 3, pp. 205–215.

Vollmann, T., Berry, W., Whybark, C. and Jacobs, R. (2005) *Manufacturing planning and control for supply chain management*. Irwin/McGraw-Hill, New York (NY).

Wacker, J. (2000) "Configure-to-order planning bills of material: simplifying a complex product structure for manufacturing planning and control", *Production and Inventory Management Journal*, Vol. 41, No. 2, pp. 21–26.

Planning Parameters and Variables

I n order to plan material flows and production, different types of calculations are performed in the ERP system which companies use in their business operations. A number of parameters and variables of different types are used in these calculations. To simplify matters we can say that these parameters and variables direct the calculations and thus the results of planning in the form of plans and information for decisions. The quality of the parameters and variables used is therefore of great significance, and to a large extent the values used are decisive for how efficiently planning can influence the company's profitability and competitiveness.

In this chapter, the following common parameters and variables that are generally used for planning purposes are described:

- **Ordering costs**
- **Inventory carrying costs**
- **Service level**
- Procurement lead time
- **Manufacturing lead time**

Methods for annual value classification and picking frequency classification are also presented. They are used to differentiate methods for forecasting and material planning. Other parameters and variables that are specific for forecasting and material planning are treated in later chapters.

LEARNING OBJECTIVES

After studying this chapter, you should be able to:

- Explain how to determine ordering and inventory carrying costs
- Define two measures for stock service level
- Explain the main lead-time elements
- Explain how to conduct annual value classification and picking frequency classification

6.1 Ordering Costs

Ordering costs refer to all the incremental costs that are associated with performing the ordering process for procurement from an external supplier or internal manufacturer. These costs are called incremental since they only arise to the extent ordering processes take place. In general terms, ordering costs are not usually dependent on procured or manufactured quantities per order event. The parameter is used in lot sizing, for example, in formulas for the calculation of economic order quantity.

Ordering costs for purchased items

Those activities normally associated with performing a procurement process are shown in the list below. The costs which the activities incur in each case must therefore be included in the calculation or estimation of an item's ordering costs when being procured from an external supplier. The costs concern primarily personnel costs, data processing costs and transportation and handling costs.

1 Request for quotation
2 Supplier negotiations
3 Selection of supplier
4 Purchase order/order proposal
5 Purchase order processing
6 Delivery monitoring
7 Other supplier contacts
8 External transportation

9 Goods reception
10 Inspection
11 Put away in stock
12 Delivery reporting
13 Internal transportation
14 Invoice check
15 Payment

Of these 15 types of cost, not all are ordering costs in all purchase situations. For procurement which is in the form of a call-off based on some form of agreement with a supplier, the costs for activities 1–3 will disappear. Activities 1 and 2 also disappear in the case of procurement of low value items and indirect materials through either the application of simplified selection of supplier or the use of pre-selected suppliers.

Inspection costs in the form of quality control are only included in ordering costs in cases of acceptance sampling of delivered orders. Costs for 100 per cent inspection of a delivered batch of items shall not be included in ordering costs since these costs are not incremental costs. Costs for external transportation are often included in the item price, i.e. items are bought on the payment conditions CIF (cost, insurance, freight). They are not considered as ordering costs in this case, either. In certain cases transportation costs are so dependent on the quantity shipped that they cannot be considered as an incremental cost.

In a simplified manner, the ordering costs for an item can be calculated as an average incremental cost. Calculating average incremental costs means that all incremental costs for procurement and goods receipt activities are included in the calculation of ordering costs and these total ordering costs over a period of one year, for example, are divided by the number of order lines during that time period.

Example 6.1

The total incremental costs for procurement and goods receipt activities at a company have been estimated at €983 000 per year. The number of order lines per year is estimated at 22 600. The ordering cost per order will then be 983 000 / 22 260, or €44 per order.

Ordering costs for manufactured items

Those activities over and above the direct value-adding process that are required to perform a manufacturing order can be divided into administrative activities, material flow activities and production activities. The sum of the ordering costs incurred by these activities then constitutes the total ordering costs.

The following administrative activities may be relevant when performing a manufacturing order, and will then incur administrative incremental costs. The costs are mainly personnel costs and data processing costs.

1 Processing of order proposal
2 Planning of manufacturing order
3 Preparing manufacturing instructions
4 Preparing the shop packet
5 Ordering
6 Reporting of material withdrawals
7 Reporting of work hours
8 Order monitoring
9 Reporting of deliveries
10 Actual costing

Ordering costs incurred by these activities can generally be considered as equally large for all manufacturing orders, irrespective of which items are to be manufactured. It is thus possible to allow the use of an administrative ordering cost common to all items. The administrative ordering cost may then be regarded as an average value incremental cost in the same way as for purchased items.

Those material flow activities which incur ordering costs are as follows:

1 Withdrawal of materials from stock
2 Transportation to manufacturing
3 Quality and quantity controls
4 Transportation of manufactured item to stock

Quality and quantity controls are not always ordering costs – only in cases of acceptance sampling of delivered orders. Costs for 100 per cent inspection of a delivered batch of items are not included in ordering costs since these costs are not incremental costs.

The scope of these activities often varies considerably from item to item, especially in activity 1. Finished products often consist of a larger number of various components than semi-finished items, and thus require more picking work for every order. If there are high requirements for precision in the estimation of ordering costs, there may be reason to differentiate ordering costs for material flow activities per item type, for example, by using different ordering costs for finished products and semi-finished items.

The third category of ordering costs for manufactured items stems from manufacturing activities. The following activities may be included:

1 Issuing necessary tools.
2 Set-up/adjustment of machines
3 Initial manufacturing and check
4 Start and stop scrap

5 Put away tools etc.

6 Cleaning of machines/equipment

7 Internal transportation and handling

The time these activities take is often called **set-up time** and represents the time taken for preparing and completing a manufacturing order in every step of the operations. The set-up time is virtually always different for each item, and must be estimated separately for each item manufactured. In simple terms, the ordering cost for set-up is an estimated sum of the set-up time multiplied by the man-hour costs and/or machine-hour costs in the work centre where the set-up is carried out. It should be noted that the machine-hour cost is dependent on the current workload situation, but in practice it is difficult to take this into consideration.

6.2 Inventory Carrying Costs

Inventory carrying costs refer to all those costs that are associated with and that arise from items being carried in stock. They are typically incremental costs, i.e. they vary as stock levels change and they decrease if less inventory is carried. In general, inventory carrying costs have a variable component and a fixed component. The parameter of inventory carrying costs is used in lot sizing for determining the economic order quantity for items carried in stock.

The following types of cost may arise as a result of items being carried in stock. To the extent that they are relevant in specific cases, they should be considered in the calculation of inventory carrying costs.

1 Capital costs

2 Costs for premises

3 Costs for shelves, racks, etc.

4 Handling equipment costs

5 Handling costs

6 Insurance costs

7 Costs for depreciation

8 Scrapping costs

9 Wastage costs

10 Costs for physical inventory

11 Administrative costs

12 Data processing costs

Capital costs for inventory consist of the costing interest for the capital tied up in stock. It is thus a variable cost that increases linearly with the amount of tied-up capital. It is most often treated as a policy variable established by the company management. Cost type 7 is also variable, while other costs in the medium term are more or less fixed or at least half-fixed.

The sum of the above costs is usually expressed in the form of an **inventory carrying factor** and as a percentage in relation to the capital tied up in stock. This means that the inventory carrying cost is assumed to have a linear relationship with the size of stock and the inventory carrying factor may be regarded as a type of interest. The inventory carrying cost for an individual item is calculated as the inventory carrying factor multiplied by the inventory value for the item at the time of calculation.

Example 6.2

The capital costs for goods in stock at a company have been established as 20 per cent. The total incremental inventory costs, excluding capital costs, have been calculated at €1 570 000. During the coming years the inventory value is expected to be €25 million on average. This means that the inventory carrying factor will be 20 + 6.3% (20 + 1570 / 25 000 · 100), or in round figures, 26 per cent. For an individual item with a manufacturing cost of €70, the inventory carrying cost per year will be equal to 70 · 0.26, or €18.20.

6.3 Service Level

The service level is an expression of the capability to deliver from stock. It may be generally defined as the probability that an order can be delivered from stock in accordance with a customer's wishes, or that materials can be picked from stock for a manufacturing order in accordance with plans made.

A number of different service level definitions are used in industry, with somewhat different significance. The two most common definitions are as follows:

Probability of being able to deliver directly from stock during one inventory cycle – **cycle service**. An inventory cycle means the time from one stock replenishment to the next.

Proportion of demand that can be delivered directly from stock – **fill rate** service.

The concept of service level is used in companies in two different respects. It is used as an efficiency variable to measure the performance level in stores, i.e. it has a follow-up purpose. In this respect, the concept was treated in Chapter 3. However, the service level is also used as a parameter for sizing safety stocks, i.e. it has a planning purpose. The latter aspect of service level is treated here.

For obvious reasons, the service level concept should be defined in the same way whether it is used for follow-up or planning and sizing of safety stocks. However, in most cases this is not completely possible. A definition which is appropriate and useful from the follow-up point of view may be impossible to use from the sizing perspective, due to limitations in the theoretical models available for sizing. On the other hand, a definition which is easy and practical to use, for which there are supporting theoretical calculation models and which is thus appropriate for sizing purposes, may have great deficiencies with respect to use for follow-up purposes.

Cycle service

Cycle service is the probability of being able to deliver directly from stock during an inventory cycle. In terms of calculations, the service level measurement probability of being able to deliver directly from stock during an inventory cycle is defined as follows:

$$Service\ level\ in\ \% = \left(1 - \frac{Number\ of\ inventory\ cycles\ with\ shortage}{Total\ number\ of\ inventory\ cycles}\right) \cdot 100$$

Using this service level definition, high-turnover items with many inbound deliveries per year will face more shortages than low-turnover items if the same service level is used. A service level of 90 per cent means, for example, that an item for which stocks are replenished once a year will only be in shortage every tenth year, whereas an item for which stocks are replenished 20 times per year will be in shortage twice a year. When using this service level measurement, high-turnover items with short inventory cycles should accordingly be given higher service levels than low-turnover items.

Example 6.3

For an item with an annual demand of 500 pieces, a replenishment lead time of two months and demand variation corresponding to a standard deviation during the replenishment lead time of 25 pieces has a desired service level established at 95 per cent. The order size for each replenishment is 100 pieces. Based on the cycle service level definition, the safety stock will be equal to 41 pieces. How to carry out this calculation is described in Chapter 13.

Fill rate service

The service level defined as the proportion of demand satisfied directly from stock, i.e. how much of the total demand can be delivered directly from stock, is defined as follows:

$$Service\ level\ in\ \% = \frac{Part\ of\ demand\ that\ can\ be\ delivered\ directly\ from\ stock}{Total\ demand} \cdot 100$$

While the previous service level definition expressed the delivery capability in one inventory cycle, this definition expresses delivery capacity over time.

Example 6.4

For the same item as in Example 6.3 but based on fill rate service, the safety stock will be equal to 12 pieces. The service level of 95 per cent using this service level definition thus results in a considerably smaller safety stock than when using the previous service definition, i.e. the cycle service definition.

6.4 Procurement Lead Time

The concept of lead time generally refers to the calendar time required to carry out an administrative process. For the procurement process, the lead time is equal to the time from when the material need occurs, such as the need to replenish stocks, until the materials delivered are available for use.

The parameter of procurement lead time is used from the logistics viewpoint in a number of contexts. It is used when determining re-order points in the re-order point system and when comparing current run-out time in run-out time planning. It is also used for calculating when new purchase orders should be released in material requirements planning.

The total procurement lead time from need to satisfied need consists of a number of main elements, as shown in Figure 6.1. The time from point 1 to point 2 represents the time elapsed between two consecutive material planning occasions, and may be called the planning interval. If, for example, a re-order point system is run once a week, then the planning interval will be one week. The time from point 2 to 3 will be the order processing time, and refers to the time from when a planned order is generated by the material planning system, or from a purchase requisition, until a purchase order has been sent or a call-off made to a selected supplier. The time from point 3 to 4 is the supplier's delivery time, and includes time for transportation from the supplier to the company. Finally, the time from point 4 to 5 is the time taken for goods

FIGURE 6.1 Main elements of the procurement lead time

receipt, quantity and quality inspection, and for putting away into stock or internal transportation directly to manufacturing.

If material requirements planning is used instead of some usage-oriented systems, such as the re-order point system, planned orders create a time margin in advance of the needed order release date. If this time margin is longer than the time from point 1 to 3, this elapsed time does not need to be included in the procurement lead time.

When determining procurement lead time, it may be advantageous to divide it into two parts: an internal part which represents administrative time and is equivalent to the material planning time, order-entry lead time plus goods reception time, and an external part which is equivalent to the real **delivery lead time**. The administrative part of the lead time is in many cases fairly constant and equal for all items, while the external lead time, the delivery lead time, is often more changeable over time and also specific for each item. For every item, then, the procurement lead time is equal to the common administrative lead time plus the item-specific delivery lead time. If there are large variations in the administrative lead times, a differentiation of the procurement lead time may also need to be made in this respect. Such differences may occur between items called off based on some form of delivery agreement, compared with items for which a more traditional procurement process is carried out. Variations may also be caused by the fact that different items require differing times for quality inspection in conjunction with goods receipt.

6.5 Manufacturing Lead Time

Manufacturing lead time is the calendar time required to carry out a manufacturing process. For standard items which are already prepared for manufacture, it is equal to the time from a material need arising, such as the requirement to replenish stock, until delivered materials are available for use. For items which are not designed and prepared for manufacturing in advance, calendar time for engineering and manufacturing preparations will have to be added. The description below refers primarily to manufacturing lead times for standard, repetitively manufactured items that are ready for manufacture and which have prepared routings in the ERP system.

The parameter manufacturing lead time is used from the logistics perspective in a number of contexts. It is used for determining re-order points in the re-order point system, for calculating the number of kanban cards in the kanban system, and when making comparisons of current run-out time in run-out time planning. It is also used to calculate when new manufacturing orders should be released when using material requirements planning (MRP). In addition, manufacturing lead times are used to determine dates for allocating materials to manufacturing orders, and time phasing material requirements.

The elements of manufacturing lead time

In a manner corresponding to procurement lead times, manufacturing lead times consist of a number of the main elements, as shown in Figure 6.2. The time from point 1 to 2 represents the time elapsed between two consecutive material planning occasions and is called the planning interval. If, for example, a re-order point system is run once a week, this material planning time will be one week. The time from point 2 to 3 is the order processing time, or the time taken to create a manufacturing order in the ERP system and check that materials and required tools are available. The time from point 3 to 4 is the start-up time. This is the calendar time which, under normal workload conditions, elapses before there is sufficient capacity to start the manufacturing order. It may also include time delays that arise due to normally occurring disruptions in material supply, access to tools and so on.

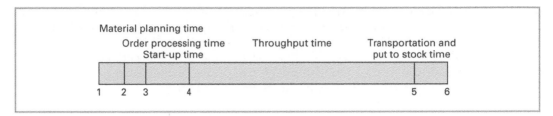

FIGURE 6.2 Main elements of the manufacturing lead time

The time from point 4 to 5 is the real **throughput time** in manufacturing and is treated in more detail in the next section. Finally, the time from point 5 to 6 is the time taken for internal transportation, and possibly quality inspection, and for put away to stock or internal transportation directly to manufacturing.

If material requirements planning is used instead of some usage-oriented systems, such as the re-order point system, planned orders can be created in advance with parameter-controlled time margins. If used time margin is longer than the time from point 1 to 4, this elapsed time does not need to be included in the manufacturing lead time.

In the same way as when determining procurement lead times, manufacturing lead times can be divided into two parts: an internal part which represents material planning time, order processing time, start-up time plus goods reception time and a manufacturing part which is equivalent to the throughput time. The administrative part of the lead time is in many cases fairly constant and equal for all items, while the manufacturing part is often more changeable and depends on the workload situation. It is also specific for each item. For every item, the manufacturing lead time is equivalent to the shared administrative lead time plus the throughput time specific to each article. If there are large differences in the administrative part of the manufacturing lead time, it may be necessary to differentiate this aspect. Such differences may be caused by different items requiring different amounts of quality inspection in conjunction with goods receipt.

The elements of throughput time

A manufacturing order consists of a number of operations. Each operation consists in turn of four elements: **queue time**, set-up time, run time and transportation time, as illustrated in Figure 6.3. These elements of throughput time can be calculated with the aid of information from the basic data stored in the ERP system. The total throughput time for a manufacturing order can be obtained by adding the times for all its operations.

The queue time is the waiting time for an operation before it is started in a work centre. It arises as a result of interference between different operations, since different operations may require the same manufacturing resources at the same time. In the calculation of throughput

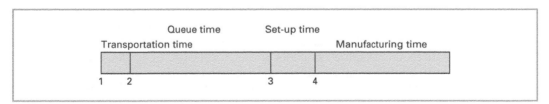

FIGURE 6.3 The elements of throughput time

times, various normally occurring disruptions in activities in a work centre are included for practical reasons, such as maintenance of machines. The length of the queue time will also depend on the current workload situation.

The set-up time is the time it takes for all activities required to initiate and close a manufacturing order, and is in most cases independent of the order quantity. Information about set-up time is normally available in the routing file in the ERP system.

Example 6.5

There are four operations in one manufacturing order. The order quantity is 100 pieces. Current operation data in hours is shown in Table 6.1. With an eight-hour working day, the throughput time for the manufacturing order will be $64 / 8 + 26 / 8 + 100 \cdot 0.5 / 8 + 16 / 8 = 19.5$ days.

Operation	Queue time	Set-up time	Unit time	Transport time
10	16	4	0.10	2
20	8	4	0.15	4
30	24	12	0.05	8
40	16	6	0.20	2
Total	64	26	0.50	16

TABLE 6.1 Times for four different operations in a manufacturing order

The manufacturing time is the time taken for the value-adding activities. It is calculated as the order quantity multiplied by the run time per piece. These data are normally stored in the routing file of the ERP system. For certain types of operation it is neither practical nor possible to estimate time per piece. In such cases a fixed standard lead time is normally used instead, which includes set-up time and the total manufacturing time. Such standardized lead times are typically also stored in the routing file.

The transportation time refers to the time at which an operation is completed until the material has been transported to the next operation's work centre, or to stock, or to quality inspection if it is a final operation. Transportation times may be stored in the work centre file in the ERP system, and they represent then average transportation times to the next operation's work centre, wherever that is. Transportation times may also be stored in special transportation timetables.

6.6 Annual Value Classification

An item's annual value is the product of one year's consumption and the item price or item cost. Annual value is also called volume value. Annual value classification means dividing items in a range into different annual value categories, which correspond to different brackets of annual value. These annual value categories are often called A, B, C, etc., and the analysis correspondingly **ABC analysis**. Annual value classification is built on Pareto's so-called minority principle, which means that in every group of items there is a small proportion which accounts for a large part of the effect. This principle is often called the eighty/twenty rule since it is often the case that 20 per cent of a number of items account for 80 per cent of the effect; for example, 20 per cent of a company's products account for 80 per cent of total sales.

Annual value classification is used as a tool for differentiating material control activities, in order to allocate most time and resources to those items which have the greatest significance for the business operation. Examples of different areas of use for annual value classification are the sizing of safety stocks by setting different service levels for different annual value categories, sizing of safety times, sizing of control limits in control and follow-up of forecast errors, and in selecting time intervals for physical inventory, i.e. the determination of inventory quantity by actual count.

The following working procedure may be applied when calculating annual value and determining annual value classes:

1 Calculate the annual consumption for each of the items included in the annual value classification.

2 Calculate the annual value for each of the items by multiplying its annual consumption by its sales price or standard cost/average cost.

3 Calculate the total annual value for all items.

4 For each item, calculate the annual value in per cent of the total aggregate annual value, i.e. each item's proportion in per cent of the total annual value.

5 Rank the items by percentage of the total annual value, from highest to lowest.

6 Analyse the distribution of items and divide them into a suitable number of different groups by specifying a number of different annual value brackets. Every bracket should correspond to one annual value classification, A, B, C, etc.

7 Assign to each item the annual value class that corresponds to its annual value.

Example 6.6

The annual value of 10 items is expressed in euros, and their annual value shares in percentages have been calculated as in Table 6.2. The items are ranked by their proportion of total annual value.

Item	Annual value	Percentage	Accumulative percentage
7	34 000	46.7	46.7
1	25 000	34.3	81.0
9	4 500	6.2	87.2
5	3 360	4.6	91.8
2	2 200	3.0	94.8
6	1 950	2.7	97.5
3	640	0.9	98.4
4	560	0.8	99.2
10	480	0.6	99.8
8	120	0.3	100.0
Total	72 810	100.0	

TABLE 6.2 The annual values and annual value shares of 10 items

Class	Items	Share of items in %	Share of annual value in %
A	7, 1	20	81.0
B	9, 5, 2, 6	40	17.5
C	3, 4, 10, 8	40	2.5

TABLE 6.3 Annual value classification of items

Based on this annual value calculation, three different annual value brackets can be defined: A equivalent to annual values larger than €10 000, B equivalent to annual values between €1000 and €10 000, and C equivalent to annual values less than €1000. An annual value classification for the 10 items is then obtained as shown in Table 6.3.

As can be seen in Table 6.3, the 40 per cent of items classified in C account for only 2.5 per cent of the total annual value, while the 20 per cent of items classified as A account for 81 per cent.

6.7 Picking Frequency Classification

The picking frequency of an item refers to how many withdrawals per time unit it is subjected to. Classifying items by picking frequency means that they are assigned to different categories according to the number of withdrawals per time unit.

Picking frequency classification is used as a tool for differentiating methods of forecasting and controlling inventories. Items with high picking frequency, and accordingly many usage events, are easier to forecast and control than items with few usage events. By dividing items into picking frequency classes, more optimised methods for forecasting and controlling inventories may be used for each class, and those parameters that influence the functions of these methods can be selected and sized more optimally. Picking frequency classification is also used when designing warehouses and choosing stock locations in a rational manner for picking, so that items with many withdrawals are located closer and are more easily accessible than items with few withdrawals. Picking frequency classification is also used when developing physical inventory procedures, so that items with many withdrawals are counted more frequently than items with few withdrawals.

The following working procedure may be applied when determining each item's picking frequency class:

1 Retrieve information on number of withdrawals per year from the ERP system.

2 Draw up a frequency diagram for the number of withdrawals. The aim is to gain an overview of the situation to facilitate division into brackets.

3 Establish suitable limit values with the aid of the frequency diagram for a number of picking frequency brackets.

4 Assign each item to a picking frequency class using the brackets determined.

Example 6.7

For a group of 20 items, the number of withdrawals during the last year is illustrated in Table 6.4.

Item	No. of withdrawals per year	Item	No. of withdrawals per year
1	6	11	25
2	19	12	17
3	37	13	9
4	4	14	11
5	7	15	13
6	19	16	28
7	10	17	22
8	8	18	5
9	5	19	11
10	17	20	29

TABLE 6.4 Number of withdrawals for 20 items

Based on these statistics of stock withdrawals, three picking frequency classes can be created with the brackets 0–12, 13–24 and more than 24 withdrawals per year. The number of items will be 10, 6 and 4 respectively, from the lowest to the highest frequency class.

6.8 Summary

This chapter has defined and described parameters and variables used in calculations for the planning of material flows and production in companies. These parameters and variables govern calculations and thus the results of planning in the form of plans and information on which to base decisions. The quality of parameters and variables used has great significance, and the values used are to a large extent decisive for how effective planning can influence a company's profitability and competitiveness.

Ordering costs for purchased items and manufactured items, as well as inventory carrying cost, are used in lot sizing of items in stock. The different activities included in the ordering process and an approach to estimate the corresponding costs was described. The service level as a parameter for sizing safety stocks, and accordingly used for planning purposes, has also been covered and the two most common definitions called cycle service and fill rate service were specified.

The principles of annual value classification and picking frequency classification were described. Annual value classification is used as an aid to differentiate material management activities and control, for example when sizing safety stocks. Picking frequency classification is used as a tool for differentiating methods of forecasting and controlling inventories.

A section of the chapter treated procurement and manufacturing lead times. The different elements of these lead times respectively and their areas of use in logistics and planning were discussed.

Key concepts

ABC analysis 92	Manufacturing lead time 84
Cycle service 88	Ordering costs 84
Delivery lead time 90	Queue time 91
Fill rate 88	Service level 84
Inventory carrying costs 84	Set-up time 87
Inventory carrying factor 87	Throughput time 91

Discussion tasks

1 Quality control costs when items are received from manufacturing or from an external supplier should not always be included in the ordering costs. Discuss under what circumstances this is the case and what the reason is for not including them.

2 Two different service level definitions are typically used when determining safety stocks: cycle service and fill rate service. Discuss advantages and disadvantages with each of these definitions.

3 Five main elements of the manufacturing lead time can be identified. In addition to this, the throughput time may be divided into four basic elements. Discuss which of all these elements represent value-adding activities. Discuss also in what way and to what extent the activities behind the rest of the elements represent waste.

 ## Problems[1]

Problem 6.1

Electronic Products Inc. is a supplier of electronic components to the mechanical engineering industry. When implementing a new ERP system the company has decided to review its ordering costs used to determine economic order quantities. For purchase items, especially electronic components, which were purchased based on annual price and delivery contracts with the suppliers, last year's purchase related costs were the following:

Costs for negotiation and supplier selection	€55 000
ERP costs for purchase order proposal generation	€4 000
Costs for purchase order generation from proposal	€26 000
Costs for supplier monitoring	€5 000
Costs for freight transport	€73 000
Costs for goods reception and put away in stocks	€22 000
Costs for quality control	€21 500
Costs for delivery notice	€9 000
Costs for invoice control and payment	€12 000

---

1 Solutions to problems are in Appendix A.

There were in total 3000 purchase orders last year.

The electronic components are very small and have low weight so the transportation costs are not dependent on the ordering quantities. What ordering cost should be used for the purchased items?

Problem 6.2

Electro Pipe Inc. makes a specific electronic system to the mechanical engineering industry. In order to determine economic order quantities for purchase items and manufacturing items they have to estimate the size of the inventory carrying costs. Much effort has therefore been spent on identifying the following inventory related costs (figures are annual costs):

Storage costs including costs for electricity and warming	€35 000
Depreciation costs for shelves, stacks, etc.	€90 000
Depreciation costs for materials handling equipment	€120 000
Costs for materials handling for stock put away and retrieval	€350 000
Insurance costs	€22 000
Obsolescence costs	€52 500
Costs for wastage	€13 000
Physical inventory counting costs	€45 000
General administrative and data costs	€23 000
Personal management costs	€50 000

Obsolescence mainly occurs for purchase items due to the fast technical development and consequently short life cycle for electronic components. The warehouse used is utilised to about 80 per cent. The tied-up capital in inventory during last year was on average €1.5 million, of which about half was due to purchase items and half was in-house made semi-finished components. The capital cost used for investments in the company is 15 per cent. How large should the inventory carrying factor be for purchased items and manufactured items, respectively?

Problem 6.3

MoveAlong Inc. has decided to use differentiated inventory control principles for low volume and high volume items. Therefore, an annual value analysis will be carried out on the stock assortment. The information in the table, about the stocked items, is collected:

Item no.	Standard price (€ per piece)	Annual demand (pieces)	Annual volume value (€)
101	47	410	19 270
102	9	1000	9 000
103	8.50	2000	17 000
104	850	80	68 000
105	10	290	2 900
106	1.50	1500	2 250
107	56	1000	56 000
108	19.50	820	15 990
109	0.20	3300	660
110	100	88	8 800
111	11	5330	58 630

The aim is that roughly 20 per cent of the items should be A items. Based on the identified data, what items should be A items?

Problem 6.4

Conduct an ABC analysis for the items in the following table.

Item	Turnover (pieces per year)	Value (€ per piece)	Contribution margin (%)	Picking events per year
A	100	2	70	100
B	30	59	65	15
C	50	65	85	50
D	2000	0.50	70	1000
E	100	5	75	25
F	3000	0.50	50	115
G	5000	1	65	1000
H	1500	6	70	450
I	2000	1	80	400
J	500	19	75	190
K	2000	2	45	550
L	300	43	65	75
M	2000	1	70	650
N	1500	1.50	75	420
O	2000	2	80	1100

a) Assume that the items are purchased items kept in factory stores and used as start-up materials for internal production. The purpose of the analysis is to determine safety stock levels for the items and should accordingly be sorted based on turnover.

b) Assume that the items are products that are stocked in finished goods stores. The purpose of the analysis is to divide the warehouse into different picking zones and should accordingly be sorted based on number of picking events.

c) Assume that the items are products that are stocked for sales in a finished goods stock. The purpose of the analysis is to determine safety stock levels for the items and should accordingly be sorted based on contribution margin.

Further reading

Fogarthy, D., Blackstone, J. and Hoffmann, T. (1991) *Production and inventory management.* South-Western Publishing Co., Cincinatti, (OH).

Selen, W. and Wood, W. (1987) "Inventory cost definition in an EOQ model application", *Production and Inventory Management Journal*, Vol. 28, No. 4, pp. 44–47.

Silver, E., Pyke, D. and Peterson, R. (1998) *Inventory management and production planning and scheduling.* John Wiley & Sons, New York (NY).

Vollmann, T., Berry, W., Whybark, C. and Jacobs, R. (2005) *Manufacturing planning and control for supply chain management.* McGraw-Hill, New York (NY).

PART 3
Forecasting and Master Planning

The aim of Part 3 (Chapters 7–10) is to outline the theoretical background of process and planning methods related to the customer demand and long-term production and resource planning. It is also to describe the design of related planning processes and choice of planning methods in various situations. Chapter 7 covers theories and methods of forecasting as a basis for a company's manufacturing planning and control. Chapter 8 treats the customer order process from the perspective of demand and logistics. Chapters 9 and 10 contain processes and planning methods for sales and operations planning and master production scheduling. The chapters also describe the specific characteristics and design of systems, procedures and processes for order and planning processes in companies that make to stock, make to delivery schedule, assemble to order, make to order and engineer to order, respectively. The relationships and interactions between demand management, sales and operations planning and master production scheduling are also emphasised.

Part Contents

Forecasting

Planning is making decisions about future activities. The area of manufacturing planning and control is largely involved with such planning and the activities to be planned affect manufacturing and material flows. To be able to make these decisions about the future in a qualified manner, information and assessments are required about how the future operation can expect to be influenced by external factors. Information and assessments of the future situation are also needed to make appropriate decisions about present activities in those cases where decisions and subsequent activities will have repercussions in the future. A decision made today to start manufacturing a certain product may have far-reaching consequences for sales, delivery service and tied-up capital in the future. Much of the decision making which affects logistics issues is closely tied to future assessments of this type.

A future assessment of external factors that can be expected to influence the company and which the company itself cannot fully influence is defined here as a forecast. A company may have some influence on some of these external factors. Market demand for a certain product may be partially influenced by its pricing and marketing. Under these conditions, too, we speak of forecasts. This chapter treats theories and methods of forecasting as a basis for a company's planning and decision making on issues that affect production and materials management. Three methods based on assessments and three based on calculations are described. In addition, we discuss approaches to how consideration may be given to trends in demand and seasonal variations, and approaches and methods for checking and following up forecasts in relation to real outcome – **forecast errors**.

The overall relationship between forecasting and the other planning processes is illustrated in Figure 7.1. The relationship with sales and operations planning refers typically to forecasts for product groups, while the relationship with master production scheduling refers to individual products, product models or product types. As in the planning hierarchy illustrated in Figure 7.1 and described in Chapter 3, the material planning level covers items such as raw materials, purchased components and internally manufactured semi-finished items. The relationship with forecasts at this planning level concerns these items and is relevant in companies where such items, as well as being used in the manufacturing of finished products, are also sold directly to customers – which is the case when items are also spare parts. The relationship is also relevant in companies where usage-oriented methods such as re-order point systems are used for material planning instead of requirements-oriented such as material requirements planning, MRP.

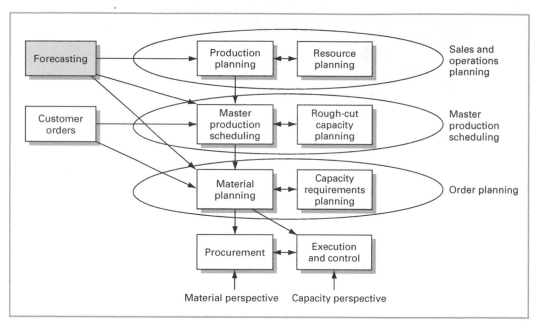

FIGURE 7.1 The relationship between forecasting and other planning processes in the planning hierarchy

LEARNING OBJECTIVES

After studying this chapter, you should be able to:

- Explain when and when not forecasts are needed
- Explain common reasons for defective forecasts
- Explain how the design of a forecasting system affects the efficacy
- Describe and compare common qualitative forecasting methods
- Describe and compare common quantitative forecasting methods
- Explain how to measure forecast errors and conduct **forecast monitoring**

7.1 Forecasting in General

Forecasts in a company may concern strategic, tactical or operative decision making. This chapter is only concerned with forecasts related to tactical and operative issues. These types of forecasts are related to decisions regarding procurement and usage of resources, and not least to decisions regarding operative activities. In this context it is mainly a question of assessments of future demand for the company's products, or demand forecasts. The size of demand is of crucial significance both for controlling material flows and for control of manufacturing. The following questions illustrate situations in which demand forecasts influence tactical and operative decisions:

- What capacity in terms of personnel and machines is required to manage next year's production plan?
- What quantities of item XX need to be purchased next year as a basis for making call-off agreements with suppliers?
- What quantity of item YY can be expected to be used next year, as a basis for determining a suitable manufacturing order quantity?
- How long will the current stock on hand last and when must replenishments be ordered?

Forecast – plan – budget

The concepts of forecast, plan and budget are all related to expected future conditions. It is therefore important that their meanings are kept separate. As described above, the term forecast refers to the assessed future demand for a company's products. It was also stated that the company to some extent can influence this demand through pricing and marketing campaigns. When forecasting, consideration should be given to the possible effects on demand of marketing activities already carried out or planned. Forecast assessments should not, however, be influenced by what a company can or wants to sell or manufacture. Expectations which are closer to hopes than realities must also be excluded. As much as possible, forecasts should only relate to volumes demanded on the market.

The term plan in this context refers to volumes decided. A sales plan describes volumes planned to be sold, a delivery plan describes volumes planned to be delivered and a production plan describes volumes planned to be manufactured. Plans may be identical with the forecast. They may also deviate from the forecast, perhaps due to an inability to obtain sufficient capacity to manufacture at the rate corresponding to demand. A plan may therefore be considered as a processed and decided forecast in which consideration has been given to limitations or further ambitions within the company.

The budget can be said to be a plan expressed in monetary terms. It therefore has the same relationship to the forecast as the plan does. An important difference between a plan and a budget is that the budget should be fixed during the year, while a plan is allowed to follow current changes in demand to the extent that sales and manufacturing can be adapted. At the time of budgeting, the budget and plan may be identical.

What needs to be forecast?

All demand for products and other items which must be delivered in a shorter lead time than they can be manufactured or acquired must, in principle, be forecast. However, this is not the same as stating that all items in a company must be forecasted with the help of what is normally called forecasting. A distinction is normally made between independent and dependent demand. Independent demand refers to demand for an item which is not directly related to the demand for other items. This is often the case for end products, spare parts and other items sold on a market. Dependent demand is characterised by the demand for an item being directly dependent on the demand for another item, and occurs primarily for raw materials and semi-finished items that are used in end products. For items with dependent demand, future needs may be estimated with the aid of calculations from planned manufacturing of the end products in which the items are incorporated through their bill of materials. If, for example, it is planned to manufacture 100 four-legged chairs, there will be a need for 400 chair legs. Assessments of future demand for this type of item can thus be replaced by calculations from production plans and master production schedules determined for end products.

Traditional demand forecasting is mainly intended for items with independent demand. In companies with low degrees of end-product standardisation and where engineering and preparation is to order, bills of material are not fully known before customer orders are received. Accordingly, it may not be possible to derive items to defined end products, and thus requirements for these items cannot be calculated from end-product manufacturing plans. In these cases, too, forecasting may be relevant in a similar way as for items with independent demand. The same may be the case in companies which manufacture standardised products with such a large number of variants that forecasting at the product level is not feasible. The conditions are not conducive to assessing future requirements of items in the product structures through dependent demand.

The extent to which items with derived demand need to be forecasted is also a question of what material planning methods are used and what type of planning system is available. If a company wants to use the re-order point system for material planning and is not able to explode requirements from items higher in the bill of material, traditional forecasting must be used even though demand for the items is dependent.

Reasons for inaccurate forecasts

Since a forecast is an assessment of future demand, in principle it will never coincide exactly with actual demand. A forecast, by definition, will always contain some degree of error. In other words, defective forecasts must be accepted, but the defects must be minimised.

There may be a number of different reasons for a forecast being defective and forecast errors being unacceptably large. The following are some common reasons for low forecast quality:

- Ineffective forecasting methods
- Misleading forecast data
- Not sufficiently combining automatic forecasting and manual assessments
- Unrealistic expectations
- Low acceptance level
- Conflicting interests
- Lack of forecast responsibility and forecast follow-up

One reason may be that the forecasting methods used are not effective enough. This may be true for computerised and automatic forecasting methods as well as procedures and approaches used with manual forecasting that are based more on experience and estimation. Another reason for defective forecasts is that the factual bases for forecast calculations are misleading, such as when deliveries of a large one-off order are allowed to influence usage statistics and thus influence forecast calculations and forecast estimates.

Automatically calculated demand forecasts of the type described here are always built on projecting the future through events in the past, or retrospective thinking. In an environment with volatile demand it is difficult, and in some cases almost impossible, to obtain good forecasts only through automatic calculations based on historic demand. By combining manual future assessments, for example at the product group level, with automatic calculations based on historic facts it is possible to achieve higher forecast quality.

The fact that forecasts are defective does not generally only depend on errors. There are also a number of psychological reasons involved in this context, such as unrealistic expectations. As mentioned above, forecasts are by definition erroneous. It is not meaningful to expect anything else and to make plans as if forecasts were exact.

Another reason why forecast quality is not, or is not perceived as, satisfactory may be that there is a vicious circle. If, for whatever reason, personnel start to doubt the quality of forecasts and start to make their own forecasts instead, there is a risk that the entire forecasting process will degenerate and the acceptance level will be low. The personnel whose job it is to make forecasts, in the marketing department for example, will have less motivation to produce good forecasts. The result is that forecasts become even worse and that even fewer personnel use them. It is not unusual in industry that planning and production personnel make forecasts parallel to those made at the marketing department for just this reason.

One further reason for defective forecasts is the conflict in interests which may exist between different departments in a company. The company's sales department may have an interest in submitting over-optimistic forecasts as a means of ensuring that manufacturing can deliver the quantities which they thought they could sell and deliver. The production department, on the other hand, may have an interest in keeping forecasts low to avoid risking low workloads on manufacturing resources sized on the basis of high forecasts. The existence of overall forecasting and master planning processes in the company are important to avoid this type of problem.

7.2 Demand Uncertainty

The higher the demand uncertainty, the more difficult it is to generate accurate forecasts. Some of the uncertainty is demand amplifications generated as a result of companies' behaviours as part of supply chains. Such "unnecessary" demand uncertainty should be identified and as much as possible reduced. Demand uncertainty is also a direct result of the increasing rate of change in the market and ever-shorter product life cycles. The requirement to react faster to market changes has therefore become more and more tangible. Managing demand uncertainty calls for increased collaboration and integration of different functions' and organisations' resources and processes. **Collaborative planning forecasting and replenishment (CPFR)** is a concept focusing on collaboration in this sense.

Demand amplifications

In most areas of business, the short-term variations in demand which occur in companies delivering to end-customer markets are rather moderate, with the exception of seasonal fluctuations. Short-term variations in demand tend to be considerably larger for companies upstream in the supply chain, i.e. raw material producers at an earlier stage in the supply chain. These variations in demand are reinforced for different reasons in the direction of the raw materials producers. Such amplifications are to a large extent generated by companies' behaviour in relation to each other. The amplification effect on demand increases at each step in the supply chain, and consequently the problem is greater the longer the supply chains. Rough estimates indicate that variations in demand double with each step in the supply chain. Cascade effects or **bull-whip effects** are terms used to describe this. The bull-whip effects are illustrated in general terms in Figure 7.2. The figure shows how small demand variations at the local store tend to amplify into larger demand variations further upstream in the supply chain. The main reasons for the bull-whip effect are:

- Large order quantities
- Few customers
- Non-aligned planning and control

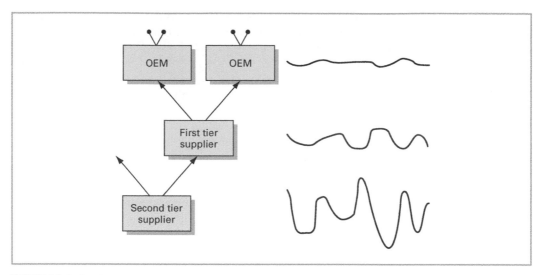

FIGURE 7.2 Bull-whip effects in supply chains

- Not sharing point-of-sales (POS) data
- Price fluctuations and promotions
- Rationing and shortage gaming

Amplification of variations in demand from company to company upstream in supply chains depends to a large extent on *large order quantities*. A material planning method generates a purchase order proposal of a certain quantity when the stock balance is below a certain level. A more or less even end-consumer demand at the retailer thus generates lumpy demand for the supplier. Order quantities tend to increase further upstream in the supply chain when purchase items become more standardised which further adds to the bull-whip effect.

The current development towards fewer and fewer production units due to economies of scale has led to suppliers having *fewer customers*, which in turn may further increase variations in demand. Many small customers will always give a more even demand than a few large customers. Larger customers in general have larger order quantities, which for the same reason will increase variations in demand.

Non-aligned planning and control activities between customers and suppliers also contribute to increasing variations in demand in supply chains. If a method is used to calculate economic order quantities, and a new calculation is made on each ordering occasion, self-propagating oscillations will occur in the supply chain. Assume that an upward trend in demand has been identified. The materials planning system does not only try to deal with the increase in itself, but also tries to increase the size of stock so that it will measure up to the new demand level. If the EOQ formula (see Chapter 12) is used to calculate the economic order quantity, a 20 per cent increase in demand will lead to an approximately 10 per cent increase in order quantity, i.e. the total increase in demand before the system has adjusted to it will be around 30 per cent. Similar effects occur if safety stock quantities are calculated dynamically at the same rate as demand changes. The materials planning system does not only try to replenish stocks on the basis of a new level of demand; it also strives towards increasing the size of the safety stock for the new demand, which is momentarily perceived as an increase in demand and, by extrapolation, results in forecasts which are higher than actual demand. Replanning a company's activities also contributes to creating instability and changes in demand upstream in the supply chain.

Even relatively small changes in current production plans can cause large disturbances in the supply chain, especially if the exchange of information between supplying and consuming parties is limited and changes in plans are communicated less frequently. In order to synchronise the material flow in the supply chain, the materials planning should be aligned between several actors in the supply chain. This could, for example, be by using vendor-managed inventories (VMI) or customer-managed ordering (CMO) as described in Chapter 17.

Another reason for the bull-whip effect is when a supply chain actor only acts on the orders placed by its immediate customers. The received customer order information is used to update forecasts but the company has little knowledge about what has actually triggered the order at the customer and what the true end-consumer demand actually is. If information about the reasons for demand changes is communicated upstream in a supply chain, there is a lesser risk that this will give rise to incorrect forecasts, temporary overstocking and resulting bull-whip effects. Communicating daily electronic *point-of-sales (POS) data* from retailers and upstream in the supply chain is an example of making more accurate demand information available in the supply chain.

Temporary sales price changes or sales promotions can increase sales volumes in the short-term. If price discounts take place frequently, buyers will stop buying when prices are high and only buy when discount prices are offered. This will, for example, result in capacity costs for overtime and undertime for employees and machines, require large warehouse volumes and inventory carrying costs because of large purchasing quantities when buying, and further contribute to the bull-whip effect. Eliminating sales price fluctuations may decrease several of the above-mentioned effects. Some retailers have therefore adopted a constant so-called everyday low price.

Another source of demand variations being generated in supply chains is *rationing and shortage gaming*, which is of a more psychological nature and caused by poor communications between customers and suppliers. If a temporary increase in demand occurs and a company is forced to extend delivery times to adapt to this, customers will perceive the supplier as having a delivery problem. To safeguard against this, companies then order larger quantities than usual and/or place their orders earlier. Without information about the real cause of demand, the supplier sees events as an increase in demand. When the supplier can resume more normal delivery times after increasing production capacity, many customers try to cut down on their orders on hand, or simply cancel them. Other customers, in accepting what they have ordered, will have covered their requirements far into the future since the increased order quantities were not matched by a real increase in demand from the end-customer market. The whole sequence of events was principally a result of lack of relevant information and a perceived risk of not being able to secure future supplies required. This customer behaviour leads to demand being perceived by suppliers as declining, which in this case does not correspond to conditions on the end-customer market.

Collaborative planning forecasting and replenishment

Collaborative planning forecasting and replenishment (CPFR) is a concept built on information exchange and collaboration between players in a supply chain. It was developed in the USA by players in the fast-moving consumer goods trade. Among the pioneers were large multinational companies such as Proctor & Gamble and Wal-Mart.

Collaborative planning forecasting and replenishment is aimed at creating collaborative relationships between suppliers and customers through common processes and a structured exchange of information to achieve increased sales, more cost-effective material flows and less tied-up capital. The concept is built on the following five principles:

1 Collaboration within the framework of partnership relations and mutual trust. For such collaboration to take place, it must be based on overall common goals and activity plans.

2 Using common and agreed forecasts which both parties in the customer and supplier relationship use as a basis for their planning and activities. This is an important prerequisite for synchronising material flows between the companies.

3 Exploiting core competencies in the supply chain irrespective of which company they are located in. The party most suited to performing an activity in a chain of refinement should do that, no matter which company owns the resource. Use of vendor-managed inventory (as described in Chapter 17) is an example of this procedure.

4 Using a common performance measurement system in the entire supply chain based on customer demands. This creates a uniform and shared focus for all companies to serve the consuming customer.

5 Sharing risks and utilities which arise in the supply chain. This is not only a consequence of far-reaching collaboration and commitments to other companies – it also brings about positive mutual influence on behaviour in the companies involved. If the total tied-up capital in a supply chain is reduced, retailers are motivated to improve their forecasting accuracy. Correspondingly, manufacturers are motivated to improve their delivery precision.

Collaborative planning forecasting and replenishment can be said to represent a management philosophy that differs from the traditional. Different companies have traditionally created their own individual forecasts and plans, and there have not been any mechanisms to ensure that they correspond to those of other companies. The opportunities to synchronise processes and material flows have not been great. If implemented well and used efficiently, CPFR can change this.

The process for achieving shared handling of plans and forecasts may be practically applied in the following way: the retailer shares his/her historic sales statistics which are obtained by reading off bar codes. The information is transferred via electronic data interchange (EDI) or equivalent technology upstream in the supply chain. The retailer also shares information on planned campaigns and other activities which may be expected to influence future demand. The forecasts which the retailer produces are also transferred to the supplier, and these forecasts are compared with the supplier's own order forecast. Any discrepancies are discussed by the two parties and adjustments made so that both forecasts correspond with each other. This procedure ensures that companies involved can use one single forecast throughout the entire supply chain.

7.3 Time Series and Demand Patterns

A basic concept in demand forecasting is that of **time series**. A time series is a collection of historic demand data which shows historic demand volumes period for period. More strictly defined, a time series is described as a sequence of chronologically ordered data with constant periodicity. An example of a time series is shown in Figure 7.3. The time series refers to deliveries per month of a certain product. The total sum of deliveries during the year was 3000 pieces and the quantity delivered per month on average was 250 pieces.

January	271	July	108
February	262	August	189
March	253	September	287
April	245	October	339
May	114	November	325
June	165	December	442

FIGURE 7.3 Example of time series of demand data

When working with such time series as a basis for forecasting, it is important to be able to distinguish between various demand patterns which are hidden in the sequence of demand data. Commonly occurring demand patterns of this type may be random variations on basically unchanging demand, **trend** changes when demand increases or decreases period for period, and **seasonal variations** in which demand varies with the time of year. Figure 7.4 illustrates these three demand patterns. When forecasting for several years ahead, it may also be necessary to consider cyclical demand variations that depend upon the general business cycle.

When forecasts are made from time series data, assumptions made about underlying demand patterns have a great effect on forecast calculations. To illustrate this fact, the following three forecast calculations from the time series in Figure 7.3 may be studied.

If it is assumed that deliveries during the previous year varied in a random fashion, it is reasonable to expect that the coming year's deliveries will be 250 pieces per month, i.e. 750 pieces in the first quarter. If it is assumed that there is a systematic trend in the time series, a completely different forecast will be obtained. If the time series is totalled every six months, a delivery volume for the first half year will be 1310 pieces and for the second half, 1690 pieces – a difference of 380 pieces, or an increase of 190 pieces per quarter. One conceivable forecast for the first quarter of the following year would then be 1690 / 2 + 190 = 1035 pieces.

A third possible assumption is that the variations during the previous year were seasonal. If the time series is converted to demand per quarter, the result in Table 7.1 will be obtained.

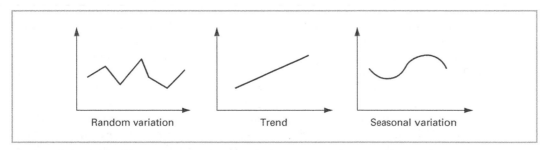

Random variation   Trend   Seasonal variation

FIGURE 7.4 Illustration of three common demand patterns

	Total	Per month
Deliveries in quarter 1	786	262
Deliveries in quarter 2	524	175
Deliveries in quarter 3	584	195
Deliveries in quarter 4	1106	369

TABLE 7.1 Demand per quarter for time series in figure 7.3

A reasonable forecast would then be 786 pieces for the first quarter next year. The same time series in all three calculations gives three completely different forecasts, depending on the assumptions made about demand patterns. These alternatives illustrate in a simple way some of the problems involved in all forecasting. The different forecast outcomes also illustrate the significance of supplementing forecasting methods based on quantitative calculations with manual assessments based on experience.

7.4 Designing Systems and Procedures for Forecasting

Selecting a suitable forecasting method is essential to obtain a forecasting system that works well. However, there are also a number of other significant factors that influence the efficacy of the forecasting system. These factors are described in this section.

Forecast data

One of the most important factors in obtaining a satisfactory forecasting system is the underlying data used for making forecasts. Apart from the fact that it should be as updated as possible, the main requirement for underlying data is really very simple to formulate. Data used as a basis for forecast calculations must represent the same variable as that intended to be forecasted. This may seem rather obvious, but in practice it is not always easy to comply with.

Since the aim is to forecast demand, historic demand data should be the basis of forecasts. However, obtaining actual demand is problematic from the aspects of quantity and time. Differences between measured demand and actual demand often arise, both in the form of quantity loss and time lag.

Historic data normally accessible in an ERP system consists of order entry statistics, delivery statistics and invoicing statistics. None of these fully represent actual demand. Differences may arise when stock shortages lead to loss of sales or delays in sales. General knowledge among customers that there are delivery problems may also mean they do not contact the company to place orders. The company may not even perceive existing demand in this situation. There may also be differences between desired, promised and real delivery dates, while delivery statistics in general only reflect real delivery dates. Which factor out of order entry, delivery or invoicing statistics should be chosen as an approximation to demand must be determined from case to case, partly depending on what statistics are available in the system. In general it can be said that the more a company has sales with delivery lead time, the more appropriate it is to choose delivery statistics or invoicing statistics. If a company has rapid invoicing procedures, invoiced volumes will correspond to delivered volumes without delays, or only a few days' time lag. With slower invoicing procedures, and thus a longer time between delivery and invoicing, delivery statistics are preferable. Most companies use either delivery statistics or invoicing statistics as a basis for forecasting.

The length of the forecast period

Demand forecasts are expressed as quantities or values per period. A forecast period could be one week, one month, one quarter or one year. What is suitable as a time period is mainly determined by the intended application of the forecast. For example, if it is intended to be used for calculating economic lot sizes or as a basis for annual delivery agreements with suppliers, a suitable forecast period might be one year. If, on the other hand, forecasts will be used for production schedules, it may be more appropriate to work with forecast periods of one month or one week.

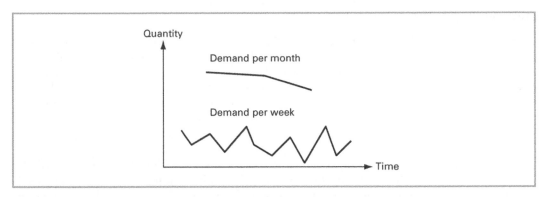

FIGURE 7.5 The significance of the length of forecast period for smoothing demand variations

A suitable length of forecast period will also depend on the delivery frequency of the items involved. The lower the delivery frequency of the items, the longer will be forecast period lengths. If delivery only takes place twice a year it is almost impossible, and at the same time rather meaningless, to make forecasts each week. Since items in a complete range often have great variations in delivery frequency, it may be appropriate to use different period lengths for different items.

The potential precision of forecasts depends very much on the period length chosen. It is easy to imagine that it is simpler to forecast sales over an entire year than sales per week during the year. The selection of a forecast period is influenced by forecast difficulty and demands on forecast precision. The significance of the length of forecast periods for difficulties in forecasting is illustrated in Figure 7.5. As can be seen, demand variations are considerably smaller if one month is used instead of one week. The length of the forecast period has a smoothing effect on demand variations.

One further aspect of the selection of periods in a forecasting system is the fact that months are not wholly divisible into weeks. If it is necessary to forecast and plan in weeks and a connection to the monthly financial accounting and follow-up is desired at the same time, there may be problems with accounting periods. In many companies this dilemma is solved by replacing the periodic term of one month by four weeks. Certain companies work with 13 four-week periods per year, while others join together the two summer and holiday periods, in general four-week periods 6 and 7, into one four-week period and can then continue to use one year divided into 12 periods.

Forecast horizon

The length of the **forecast horizon**, too, will depend on the applications intended for forecasts. If forecasts are used for production plans, forecasts of one or several years will be required; whereas forecasts for operative material supply and call-offs may be limited to three to six months.

The length of the forecast horizon is also significant for potential forecast precision. The longer the horizon, meaning the farther in the future that must be forecast, the more difficult it will be to avoid forecast errors. By reducing throughput times to allow a shorter forecast horizon, measures to cut throughput times in material flows are effective methods of improving potential forecast precision.

Forecast frequency

How often forecasts should be made largely depends on the selected planning frequency and the length of forecast periods. Forecast frequency should in principle be as high or higher than planning frequency. If there are no new forecasts on which to base planning, it is not always meaningful to prepare new plans. Forecasting more frequently than planning is a dubious policy, since the company has no use for the new forecasts if they are not followed by new plans. A forecast frequency of one per quarter may be appropriate for budgeting and budget reviews, while a forecast frequency of once a month may be more suitable for monthly recurrent master production scheduling.

The optimal forecast frequency is also related to the selected length of forecast periods. Generally, the time between consecutive forecasts should be as long as the periods selected, i.e. the periods' length determines to a large extent the most suitable periodicity. There are no obvious reasons why forecasts should be made per week if the period length is one month.

Aggregation level

Forecasting single products is considerably more difficult than forecasting groups of products. A smoothing effect is obtained through grouping products, comparable with that obtained when using longer forecast periods. The effects are illustrated in Figure 7.6. In this figure, curves 1, 2 and 3 represent demand variations for three different products separately, while the fourth curve reflects demand variations for the three products together. The lower degree of variation in demand at the group level enables a higher forecast precision than for each of the incorporated products separately.

Only forecasting at the product group level may be sufficient if the forecasts are intended for tactical and overall planning, for example in sales and operations planning. For operative materials control it is necessary to forecast at the single product level. There are two main options in this case. Forecasts may be made directly either at the product level or at the product group level. Forecasts at the group level are then divided into forecasts for the products incorporated in each product group. The division of forecasts at the group level may be made on the basis of sales statistics for the individual products so that each product is allocated a share of the group forecast corresponding to its sales in relation to total group sales. These types of forecasts are sometimes called explosion forecasts, and the explosion from product group forecasts to incorporated product forecasts is carried out with the help of forecasting bills of material. The quantities per parent item in these forecasting bills of material are derived from the shares of each product that are estimated to make up the aggregate demand for the product group.

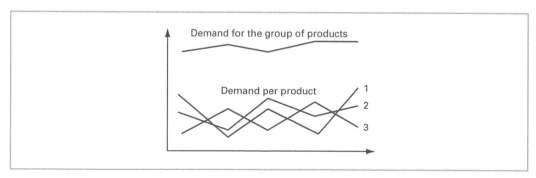

FIGURE 7.6 The significance of product grouping for smoothing of demand variations

Example 7.1

A furniture company manufactures and markets office furniture, including desks. These desks are available with tops in oak, beech or birch, while the legs and frames for all models are made in chrome-plated steel. During the current year, the number of oak desks sold was 2350, beech 6890 and birch 3630, the total for the product group of desks being 12 870 sold. This means that the oak top accounted for 18 per cent of sales, the beech top for 54 per cent and the birch top for 28 per cent.

For the next year, the total demand for the product group of desks is forecast at 14 000 pieces. The proportional demand for the different types of wood tops is not expected to change compared with the previous year. Forecast sales of the three products for next year will then be 18 per cent of 14 000, i.e. 2520, 54 per cent of 14 000, i.e. 7560, and 28 per cent of 14 000, i.e. 3920 for oak, beech and birch desks.

In addition to forecasting at the product group level and the individual product level, companies also forecast at the bill-of-material level under the product level. This is sometimes carried out when the number of variants of a special end product is so large that they cannot be forecast individually. A large number of variants may arise through the possibility of combining different standardised modules in many different ways. Forecasting at the underlying bill-of-material level then means that these modules are forecasted. Forecasting of products with many variants may also be carried out for each product model, meaning here a group of variants which represents all the different possible combinations of modules. From the forecast per model, the forecast values are then distributed in percentages between different modules and variants in a similar fashion to that used for product groups. If in Example 7.1 above, legs and frames were also available in different variants such as chrome-plated and painted, desk models would consist of the modules wood type and leg/frame type. The total forecast for models would then be distributed between the different variants of each module.

The three different grouping levels of product family, product and product model are illustrated in Figure 7.7. These groupings are made with respect to product characteristics. Product groupings for forecasting purposes can also be made on the basis of markets and geographical coverage, for example by a customer segment, region or country.

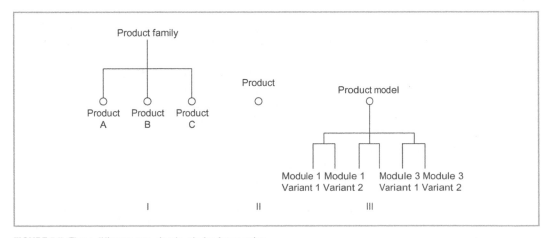

FIGURE 7.7 Three different grouping levels for forecasting

In companies of the type make to order and engineer to order, products are not fully defined before orders are received and there are in principle an infinite number of variants possible. These variants are often based on a number of different basic designs, called product types. Every unique order-specific product is created by adding, removing or modifying this basic design. If the amount of additions, removals or modifications is reasonable, forecasts of these product types may represent all available unique variants. When planning materials, items that are accessories to these product types are forecast separately in order to obtain a complete picture of overall materials requirements. From the forecasting point of view, product types are treated as a product consisting of items which are always or very often included in order-specific products, while those items which only sometimes occur are not included at all.

Forecast units

The unit for demand forecasts is usually units carried in stock and/or planned, such as pieces, metres, litres or kilograms. The reason for this is that forecast volumes are often translated into production plans and, before manufactured volumes are decided in these plans, quantities already available in stock are deducted. For budgeting purposes, forecasted quantities are translated into monetary values, valued at sales price, standard price or similar, depending on the application.

If forecasting takes place at the product group level, it may be necessary to use monetary values instead of quantity units. This is the case when products within a group are too different to be added using the units used for inventory purposes. Instead, the standard price is used as a conversion factor to make it possible to add the volumes within these groups.

In a few cases, hours are used as a forecast unit, such as in the case of companies that do contract manufacturing. These companies have neither their own products nor products defined by customers, and often they do not procure material supplies themselves. For them it is manufacturing capacity which is in demand instead of products per se. Hours may be used as a forecasting unit for planning workloads in these cases.

7.5 Forecasting Methods

Classification of forecasting methods

The forecasting methods that are used for demand forecasting in the area of manufacturing planning and control can be roughly divided into qualitative methods and quantitative methods. Qualitative methods refer to those that are built on subjective assessments by personnel with good knowledge of the market and market development. This category covers everything from "crystal ball" methods and rough guesses to systematically performed qualified assessments. The method is characterised by its insignificant use of quantitative calculations.

In contrast to qualitative methods, quantitative methods are based to a much larger degree on calculations that are more or less advanced. These methods can in turn be divided into extrinsic and intrinsic forecasting methods. Extrinsic forecasting methods are characterised by making a model of the relationship between the variables to be forecast and a number of explanatory variables which the forecast variable is dependent on, and which are easier to forecast or obtain information about in advance than the forecast variable itself. The relationship is determined analytically, often with the help of regression analysis. Extrinsic forecasting methods require more underlying data and more extensive forecast work than intrinsic forecasting methods. An example of the use of extrinsic forecasting for operative demand forecasting is the purchase of weather reports from meteorological services by breweries.

In methods belonging to the category of intrinsic forecasting methods, data is only analysed for the variable to be forecast. If the aim is to forecast future sales of certain products, calculations from previous periods' sales are made. Future sales are then calculated by extrapolation from earlier sales. All the methods described below fall into this category.

Even though the methods can be divided into qualitative and quantitative types, in practice it is often necessary to work with a combination of the two. This is primarily due to intrinsic forecasting methods being based on historic data. To obtain a future assessment it may therefore be desirable to supplement extrapolation with manual assessments of the future.

Requirements for forecasting methods

As pointed out earlier, forecasts by definition are erroneous to some degree. The aim is to minimise forecasting errors relative to actual demand. However, forecast accuracy must be weighed against the man-hours required to achieve it and what costs this work will incur. In general it can be stated that the costs for achieving improvements in forecasting accuracy must not exceed the value which the improved accuracy can provide.

The calendar time it takes to produce forecasts is also important for the choice of forecast accuracy. A high level of accuracy will require more work and therefore more time for it to be achieved. This extra time required means that forecasts will always start from older historic data than would be the case if forecast work could be carried out in a shorter time. The older the historic demand which the forecast is based on, the greater is the risk that forecast accuracy will deteriorate. Thus, there is a compromise between forecast accuracy and the time it takes to make new forecasts.

A good forecasting method, in addition to being easy to work with, must also fulfil the following main conditions:

- It must be stable with respect to random variations in demand.
- It must react quickly to systematic demand changes.
- It must be able to predict systematic changes.

There is a conflict between the first two requirements, which is something of a dilemma for forecasting methods. If the method reacts quickly to systematic changes in demand, it is difficult to prevent the same method from overreacting to random demand changes.

Qualitative forecasting methods

Qualitative forecasting methods are those which are mainly built on individuals' experience and more or less well-qualified subjective judgements of future demand. They are also characterised by being based on very few formal calculations based on sales statistics or other assessment data. Qualitative methods include everything from subjective estimates by a sales manager to detailed, planned and formal procedures and approaches with many individuals involved.

In comparison to quantitative methods for forecasting, qualitative methods are preferable in situations where the number of products to be forecast is small. They are generally also better when the number of periods to be forecast is small, such as forecasting annual demand rather than weekly demand. They are often preferred when the forecast horizon is long and there are demands for forecasts for long-term planning, and if demand is to a large extent influenced by the company's own marketing activities of different types. Qualitative forecasts also have advantages if they concern products recently being introduced to the market or going to be phased out, since in these cases there is a limited amount of information in the form of sales statistics on which to base forecast calculations.

FIGURE 7.8 Example on screen of manual forecast display
The picture represents a function from the ERP system Lawson M3

Even though assessment methods for forecasting do not generally rely on any automatic calculation procedures, different forms of computerised support may be used to efficiently supply demand information as a basis for assessment and to facilitate necessary calculations of various types. Figures 7.8 and 7.9 show two examples illustrating this type of support. Figure 7.8 illustrates an example with data in the form of earlier forecasts, actual delivered quantities and deviations between forecasts and actual sales from earlier time periods. Graphic illustrations of this kind of data are also used to facilitate forecasting work.

Figure 7.9 provides another example of how support for forecasting may be obtained in an ERP system. The example illustrates how a manually produced annual forecast can be divided into different time periods using a distribution curve. The distribution curve shows the percentage of the total demand on a yearly basis that is obtained in each of the year's 12 months. Tools of this type make it possible to rely more on manual assessment methods for forecasting since it is then sufficient to make forecasts on an annual basis for each product. The distribution curves can often be shared for all products in a product group and may be automatically produced from stored sales statistics. A special case of such periodic division of annual forecasts is made when there are seasonal variations in demand.

Two judgemental approaches to forecasting are described below, both of a more general character: the **sales management approach** and the **grassroots approach**. Another approach is also shown which combines both these approaches and at the same time includes a method of reconciling group-based forecasts with individual-based forecasts.

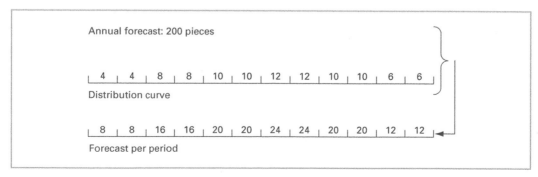

FIGURE 7.9 Division of an annual forecast using a distribution curve

The sales management approach

One approach for producing judgemental forecasts is that management personnel are gathered in one or more meetings to discuss and determine forecasts of future sales or related issues. Sales statistics and other types of assessment data are compiled, processed and distributed in preparation for these meetings. There are two possible variants. The first involves the company's management group meeting to produce forecasts. The composition of the management group will naturally vary depending on the size of the company. For a smaller company it may consist of the managing director, marketing manager, production manager, logistics manager, purchasing manager and financial manager. This variant is above all useful when forecasting is for more long-term and overall sales and business activity development at the strategic and tactical levels. The second variant involves only leading personnel within the marketing and sales organisations. These meetings may then include sales group managers, product managers, regional managers and so on. The principle is the same, but the members participating in the forecasting meetings are different. This variant is most useful for short-term and medium-term forecasting of expected future sales and delivery volumes.

The advantage of these management group methods is primarily their speed in producing new forecasts and that they allow a common assessment and consensus on the forecast. In this way it is easier to avoid different individuals and departments planning and acting on their own. One weakness with the method is the risk that the management personnel who have the greatest influence and most authority may have an unreasonably large influence on judgements in relation to those individuals who have most knowledge and information about what is taking place on the market. For the same reasons as for all judgemental approaches, there is also a risk that forecasts will tend to reflect wishful thinking more than realistic assessments.

The grassroots approach

The grassroots approach means that all salespeople and other personnel in direct contact with the market make their own assessments separately and work out their own proposals for forecasts. These forecasts may relate to total sales or only sales in the person's region or product area, depending on how sales activities are organised. These assessments are then collected and processed centrally. After adjustments and processing of the collected material it is compiled into a common forecast for the whole company.

The main advantage of the grassroots method is that forecasts are produced by those people in the company that have most contact with events on the market and thus should be best qualified to make correct future judgements. The method also has the benefit that the responsibility for forecasting falls on those who also have responsibility for realising the forecasts. In addition, the forecasts will be automatically divided into different regions, sales groups, product groups, etc. There is no post-processing required to break down total forecasts into different areas of responsibility, which is the case when using the management group approach. One drawback with the grassroots approach is that it takes more time and requires more data processing than the management group approach. Salespeople are not always known to be good forecasters. They are often either over-optimistic or over-pessimistic, depending on the results of the last few weeks' sales.

Pyramid forecasting

The above approaches can be combined by using so-called **pyramid forecasting**. This means that the total forecasts produced by the company management or sales management are considered as the most reliable. If the total forecasts compiled from the grassroots approach deviate from the management forecast, all individual and group forecasts are proportionally adjusted to this deviation so that the total forecasts are the same in both cases. Assume, for example, that the total forecast for a product produced by the company's marketing manager and central marketing department is 2000 pieces per year, while the sum of forecasts from the five different markets where the product is sold is 2200 – according to the forecasts produced by sales personnel on each market. The forecasts for each different market will then be reduced by approximately 9 per cent so that their totals tally with the overall forecast.

The benefit compared with only using a grassroots approach is that the managers of a company are often better at judging the overall financial development and business cycle, and thus are better equipped to make accurate overall forecasts. The downside of this combined approach is that the different salespeople or product managers behind the individual forecasts are not tied to their forecasts in the same way.

Quantitative forecasting methods

Quantitative forecasting methods are based completely or almost completely on calculations, in turn based on time series of sales, usage or other types of historic data. As described above, we distinguish between extrinsic and intrinsic quantitative forecasting methods. Only two commonly occurring intrinsic forecasting methods for operative forecasting will be taken up here: **moving average** and **exponential smoothing forecasting**. In addition to this, a simulation-based intrinsic forecasting method is included.

Moving average

The simplest form of intrinsic forecasting methods is to assume that demand for the next period will be equal to the demand for the latest period. However, such a method runs the risk of being too sensitive to random variations, especially if the forecast period is short. If the length of time periods involved is one year or more, it may be more useful.

If the average value of demand during a number of periods is taken as a measurement of future demand instead, a method is obtained which will fulfil greater demands for stability. The number of periods which should be included in moving average forecasts must be determined from case to case, and relates to the length of time periods used and the size of variation in demand, among other factors. In general it may be stated that if demand values for many periods are used, good stability is obtained with respect to random fluctuations, but one consequence is less responsiveness to trends and other systematic changes. The longer the time periods used, the fewer periods need to be included.

Example 7.2

Monthly sales of a product during the previous year are shown in Figure 7.3. Future sales at the company are forecast using 12 months' moving average. Forecast for January the following year is then 250 pieces, or equal to the average of the previous 12 months' demand. If the actual demand for January the following year is 422 pieces, the forecast for February will be one-twelfth part of the sum of the demand values from February of the previous year until January of the current year. One way of calculating this forecast value is to subtract the old January figure from the new one, divide by 12 and add this amount as a forecast adjustment to the old forecast. The following calculation has then been carried out:

$$Forecast\ adjustment = \frac{422 - 271}{12} = 13$$

$$New\ forecast = 250 + 13 = 263$$

This method of forecasting means that the oldest period's demand value is replaced by the latest period's demand value. For this reason the method is called moving average. The calculation is made with the aid of the following formula:

$$F(t + 1) = \frac{(D(t) + D(t - 1) + \ldots + D(t - n + 1))}{n}$$

where $F(t + 1)$ = forecast demand for period $t + 1$
$D(t)$ = actual demand during period t
n = number of periods in moving average forecast

The choice of the number of periods is decisive for how the method will operate. If a large value of n is chosen, meaning that many historic periods are included in the average calculation, forecasts will be stable with respect to random variations in demand and at the same time they will react slowly to systematic fluctuations in demand. A small n value will lead to the opposite forecast characteristics. The selection of n is accordingly associated with a balance between the capability to react quickly to systematic changes in demand and the avoidance of overreactions to random variations.

Using the moving average method assumes that systematic trends in demand are negligible. As an illustration of the consequences if this condition is not fulfilled, refer to Figure 7.10. To show the effects of trends clearly, all random variations have been excluded.

As can be seen from the time series, demand is constant for the first five periods. Subsequently there is a trend of 10 pieces increase per period. The table also indicates that the forecast with five periods' demand values has a time lag of two periods, while the forecast with three periods' demand values has a time lag of one period compared with demand. These time lags remain constant as long as the trend is unchanging. With systematic trends, the forecast value calculated with moving average will always lag behind. The following general relationship applies between time lag and the number of periods included in the calculation of averages:

Time lag in periods = 0.5 · (Number of periods included − 1)

Thus, a five-period forecast will result in 0.5 · (5 − 1) = 2 periods' time lag, while a twelve-period forecast will give 0.5 · (12 − 1) = 5.5 periods' time lag.

Period	Demand	Forecast on 5 periods	Forecast on 3 periods
1	100		
2	100		
3	100		
4	100		100
5	100		100
6	110	100	100
7	120	102	103
8	130	106	110
9	140	112	120
10	150	120	130
11	160	130	140
12	170	140	150

FIGURE 7.10 Illustration of time lag when using moving average

Exponential smoothing

The moving average method means that all demand values are given the same weight in calculations, irrespective of their age. We speak of a rectangular distributed weighting of demand values in time series. It is not difficult to imagine other weighting distributions, such as the most recent demand values being given a higher weighting than older demand values.

This could be achieved by using some form of triangularly distributed weighting. One typical triangularly weighted distribution over four periods is obtained by using weightings of 0.4, 0.3, 0.2 and 0.1, where 0.4 is used for the most recent demand value and 0.1 for the oldest. In principle, the forecast calculation can be carried out in the same way as in the previous section except that individual demand values are multiplied by a weight before they are totalled. However, this calculation procedure is rather cumbersome.

The typical triangularly distributed weighting means that the difference between two consecutive weights is constant. If instead of having a constant difference there is a constant relationship between two consecutive weights, we talk about exponential distribution of weights. With the relationship of 0.5 the distribution of weights will be as follows:

$$0.5 - 0.25 - 0.125 - 0.062 - 0.031 \qquad \text{and so on.}$$

This is an infinite geometric series whose sum is equal to 1 if the sum of the highest weight and the constant weight relationship is equal to 1. As a designation of this constant relationship, the Greek letter α is normally used, called the smoothing factor. Note that the value of α must always be less than or equal to 1.

Expressed in simple terms, using exponential smoothing and with an α value of 0.2, the most recent demand value will be given a 20 per cent weight while the old forecast will be given an 80 per cent weight.

It may seem that this method of weighting demand values is more complicated and requires the storage of an almost infinite number of historic demand values. In fact, this is not the case. It is possible to prove mathematically that exponential weighting is obtained by applying the following simple formula for calculating forecasts:

$$F(t + 1) = \alpha \cdot D(t) + (1 - \alpha) \cdot F(t)$$

where $F(t + 1)$ = forecast value for period $t + 1$
   $D(t)$ = actual demand during period t
   α = smoothing factor

As shown by the formula, only the latest actual demand values and the previous forecast need to be stored. The method of exponential smoothing is accordingly more attractive than the moving average method in this respect.

The selection of the smoothing constant has the same significance for exponential smoothing as the selection of the number of periods for moving average calculations with respect to its stability in the case of random variations and responsiveness to systematic changes in demand. A high α value will give better responsiveness to systematic changes in demand, but will also result in greater instability to random variations. A commonly chosen α value in industry is 0.1.

With the same average age for included demand values the relationship between the number of periods in moving average and the α value in exponential smoothing will be as follows:

$$\alpha = \frac{2}{n + 1}$$

An α value of 0.1 will then correspond to including 19 periods' demand values when using moving average for forecasting.

Example 7.3

According to Example 7.2, the forecast for January in year 2 is 250 pieces, and the actual demand for the same month is 422 pieces. If exponential smoothing is used with $\alpha = 0.3$, the forecast for February will be as follows:

$$\text{New forecast} = 0.3 \cdot 422 + 0.7 \cdot 250 = 302$$

Focus forecasting

Exponential smoothing forecasting was developed during the early days of computer systems when access to memory and processing capacity was considerably less than at present. Under these conditions it was important to find methods which put small demands on calculations and which did not require too much memory for access to large quantities of historic demand data. **Focus forecasting** is a method which can be said to represent the opposite approach. It assumes that there is large access to computer capacity and that this capacity can be used as a tool to achieve good forecasts.

Focus forecasting is based on two very simple basic hypotheses:

1 Simple forecasting methods are often just as good as more mathematically advanced methods, not least because they are understood by personnel involved and they are similar to manual forecasting approaches of which people already have experience.

2 If a forecasting method has worked well in the previous period, the probability is great that it will also work well in the current period. By testing how different methods would have worked if they had been used in the previous period's forecasting for which the outcome is known, the best of these can be selected for the next period's forecasting.

In the original version of focus forecasting the following seven forecasting methods were used, based on a forecasting period of one month:

1 The forecast for the next month is equal to the actual demand for the same month in the previous year.

2 The forecast for the next month is 110 per cent higher than the actual demand for the same month in the previous year.

3 The forecast for the next month is one-sixth of the total of the actual demand for the previous six months.

4 The forecast for the next month is one-third of the total of the actual demand during the previous three months.

5 The forecast for next month is equal to one-third of the actual demand during the previous three months of the previous year multiplied by a factor equal to the relationship between the latest three months' demand and the demand during the same three months of the previous year.

6 If the demand during the latest six months is less than 40 per cent of the demand during the previous six months, the forecast for next month is set at one-third of 110 per cent of the demand for the same three-month period the previous year.

7 If demand during the latest six months is more than 2.5 times larger than the demand during the previous six months, the forecast for the next month is set at one-third of the demand for the same three-month period the previous year.

The percentages used may be seen as examples. It may be noted that in methods 3 and 4, the moving average used is for six and three periods, respectively, that method 5 is a method of taking into consideration any trends and that methods 6 and 7 include consideration to any seasonal variations from a low season and a high season, respectively.

The focus forecasting method means that for every forecast occasion, when information about actual demand has been obtained, tests are made by running simulation programs to find which of the methods included would have worked best during the previous period or periods. This best working method is selected for use when a new forecast for the next period is to be calculated. Every item is treated individually in this way. The approach is illustrated in Figure 7.11. In this example, four different forecasting methods are tested. The forecasts calculated for periods 5 and 6 using those four methods are compared with the actual demand for the same periods. Using this comparison, it is clear that forecasting method 2 worked best during the two most recent periods. This method is then selected for forecasting in period 7.

There is nothing in the method of focus forecasting that requires that just the above-mentioned particular forecasting methods are used. Any forecasting method within reason can be used, including exponential smoothing. The only provision is that they should be relatively simple methods, partly because this is one of the basic concepts behind focus forecasting and partly because complex methods would create processing problems in simulation evaluations.

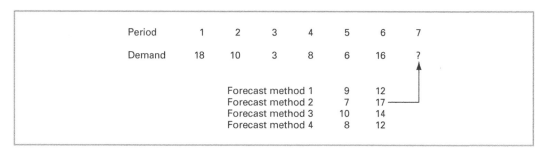

FIGURE 7.11 Illustration of the focus forecasting method

Consideration to trends

If there are systematic trends in demand, it is not difficult to take these into consideration within the framework of forecasting methods based on judgemental assessment. If instead one of the forecasting methods described in the previous section is used, it may be necessary to use supplementary calculations if the trend cannot be considered as negligible. These correcting calculations may mean that there is a faster correction of forecasts when actual demand shows a trend-related change, i.e. they contribute to decreasing the forecast time lag described above. One recommended formula for correcting forecasts for trends is as follows:

$$F(t+1) = BF(t+1) + \frac{1-\alpha}{\alpha} \cdot T$$

where $F(t + 1)$ = trend-corrected forecast demand for period $t + 1$
           $BF(t + 1)$ = basic forecast for demand during period $t + 1$ without consideration to any trends
           α = exponential smoothing factor
           $(1 - \alpha)/\alpha$ = correction factor
           T = the estimated trend through comparisons of forecasts period for period, calculated as in the formula below for example

It is possible to prove mathematically that this trend correction is optimal for linear changes in demand. Trend corrections of this type should be used with great caution, however. Demand in reality is never linear and there is a large risk that random variations will give rise to large and uncontrollable shifts in forecasts. It is difficult to work with trend corrections in general, and especially in cases with low usage frequency items where random variations may be as large as or larger than the average forecast.

Taking trends into consideration is especially important if forecast values are extrapolated far in the future, i.e. when forecasting is not only for the following period, such as one month ahead. If a forecast is desired for the next 12 months, some consideration to trends will often be necessary to gain reasonable accuracy in the forecast. This type of extrapolated forecast will be useful as a basis for drawing up production plans, among other things.

Consideration may be given to trend changes in the calculation of periodised forecasts with a long forecast horizon by adding to each period an estimated trend change, called an additive trend (see Figure 7.12). If the forecast obtained with the aid of a moving average or exponential smoothing is called a basic forecast and is 100 pieces, and the trend is an increase of 10 pieces per period, the extrapolated forecast for two periods ahead will be 110. The forecast for the nth period may be expressed using the following formula:

$$F(t + 1 + n) = F(t + 1) + (n - 1) \cdot T$$

where $F(t + 1)$ = forecast for the first period
           T = the additive trend, or the increase in demand per period
           n = the number of periods ahead for the forecast

Figure 7.13 illustrates an extrapolated forecast with a basic forecast of 290 pieces and a trend of plus 5 pieces per period.

It is common to assume that there is an additive trend in this type of short-term forecasting, but there are also multiplicative trends where demand increases by a certain percentage per period, instead of by an amount.

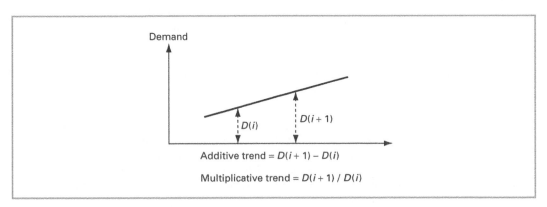

FIGURE 7.12 Additive trends and multiplicative trends

Period	1	2	3	4	5	6	7
Demand	290	295	300	305	310	315	320

FIGURE 7.13 Illustration of extrapolated forecast with trend

Additive trend changes can be calculated using moving averages. In the simplest type of case, the trend can be calculated as the difference between two consecutive demand values. The risk associated with this approach is an over-sensitivity to random fluctuations. As a rule it is more suitable to make trend estimates on the basis of changes over more than one period. From the example in Figure 7.3, the trend can be calculated as the change between the first half of the year and the second half of the year. Average demand during the first half of the year is 218 and during the second half 281. This would mean a positive trend of 63, and the trend-corrected forecast for January next year would be 250 + 63, or 313.

Trends can also be calculated with the aid of exponential smoothing. This avoids the problem of the excessive influence of random fluctuations on the trend calculated from consecutive forecast values. The following calculation formulas can be used:

$$BF(t+1) = \alpha \cdot D(t) + (1 - \alpha) \cdot (BF(t) + T(t))$$
$$T(t+1) = \beta \cdot (BF(t+1) - BF(t)) + (1 - \beta) \cdot T(t)$$
$$F(t+n) = BF(t+1) + T(t+1) \cdot n$$

where $BF(t+1)$ = basic forecast for period 1 without considering trends
 $BF(t)$ = basic forecast for current period t without considering trends
 $T(t)$ = trend for current period t
 $T(t+1)$ = trend from period 1
 $F(t)$ = forecast demand for current period t
 α = exponential smoothing factor for basic forecasting
 β = exponential smoothing factor for the trend
 n = number of future periods covered by the forecast

Both smoothing factors are selected between 0 and 1. The motives for choosing higher or lower values are the same for the β factor as for the α factor, i.e. that values near 0 place less emphasis on the latest periods' trend and instead even out the trends of several periods when forecasting future trend influence. Two initial values are required to carry out forecasting with the above method. One is $BF(1)$ which could be set equal to $D(0)$. The other is $T(1)$ which is often set equal to the trend of the last period, $(D(1) - D(0))$, or alternatively as an average trend change during the latest periods.

Example 7.4

Sales per month of a product during the previous year are described by the figures in Figure 7.3. From July onwards, the company uses the above method to forecast future demand. Calculations take place period by period and extend six periods ahead in time. The α value is set at 0.3, the β value at 0.3, BF(August) at 108, which is the same as D(July). T(August) is calculated as D(August) − D(July), i.e. will be 81, and F(August) is 108 + 81 = 189, i.e. equal to D(August). The table below shows the values calculated for $BF(t+1)$, $T(t+1)$ and $F(t+n)$. Examples of calculations:

$$F(\text{November}) = BF(\text{November}) + T(\text{November})$$
$$BF(\text{November}) = 0.3 \cdot 339 + (1 - 0.3) \cdot (275 + 83) = 352$$
$$T(\text{November}) = 0.3 \cdot (352 - 275) + (1 - 0.3) \cdot 83 = 81$$
$$F(\text{November}) = 352 + 81 = 433$$

$$BF(\text{December}) = 0.3 \cdot 325 + (1 - 0.3) \cdot (352 + 81) = 401$$
$$T(\text{December}) = 0.3 \cdot (401 - 352) + (1 - 0.3) \cdot 81 = 71$$
$$F(\text{December}) = 401 + 71 = 472$$

Period (t)	D(t)	BF(t)	T(t)	F(t)
July	108	108	–	–
August	189	108	81	189
September	287	189	81	270
October	339	275	83	358
November	325	352	81	433
December	442	401	71	472
January				543
February				614
March				685
April				756
May				827
June				898

To solve the problem of calculating trends so that the influence of random variations is not too large, it is sometimes more suitable to determine trends for product groups. These group trends will later be used in the forecast calculations for the products in each group. Trend estimates for each product group are preferably made manually since it is then possible to take into consideration future assessments of factors such as the current business cycle or planned marketing measures influencing demand.

Consideration to seasonal variations

So far the only trends described have been systematic increases or decreases in demand. Another common type of systematic changes in demand is seasonal, in which demand varies with the annual seasons. If there are seasonal variations in demand, they must be taken into consideration when forecasting. A seasonal index, which expresses the magnitude of the seasonal variation, is determined to do this. The seasonal index is calculated using the following formula:

$$S(t) = \frac{D(t)}{D(m)}$$

where $S(t)$ = seasonal index for period t
           $D(t)$ = actual demand during period t
           $D(m)$ = average demand during all periods of the year

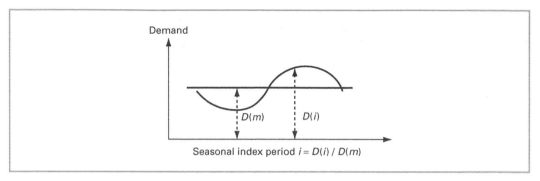

FIGURE 7.14 Calculation of seasonal indexes

The principal for calculating the seasonal index is shown in Figure 7.14. In this figure, $D(i)$ is the demand in period i and $D(m)$ the average demand during the year. After the seasonal indexes have been calculated they are normalised, i.e. rounded off and adjusted so that the sum of all seasonal indexes is equal to the number of periods during the year.

The starting point for calculating seasonal indexes is actual seasonal variations over one or more previous years. If only one previous year is used in the calculation, this means that the seasonal index for March the following year will be equal to the calculated seasonal index for March in the current year. If the quality of these calculations needs improving, the moving average approach or similar can be used to calculate new seasonal indexes from several years' seasonal variations.

It is often difficult to calculate seasonal indexes for individual products. There is a risk that random fluctuations will have too large an influence since they may be in the same order of magnitude as the seasonal variations themselves, especially for slow-moving products. Seasonal indexes should be calculated instead for product groups. The seasonal index calculated and established for the product group will then be used for individual products within the group.

When the seasonal index has been established, consideration may be given to the influence of seasonal variations as follows. Before the latest received demand values are allowed to influence forecast calculations they must be adjusted from seasonal influence, irrespective of whether moving average, exponential smoothing or any other method is used for the forecast. Seasonal adjustment is carried out using the following formula:

$$DR(t) = \frac{D(t)}{S(t)}$$

where $DR(t)$ = seasonally cleaned demand in period t
$D(t)$ = actual demand in period t
$S(t)$ = seasonal index for period t

The seasonally adjusted demand value is then used for forecasting in one of the methods presented in the earlier sections. The calculation will result in a new basic forecast. The forecast for a certain period, n, will then be seasonally adjusted using the following formula:

$$F(n) = L(n) \cdot S(n)$$

where $F(n)$ = forecast demand in period n
$L(n)$ = basic forecast for period n
$S(n)$ = seasonal index for period n

Example 7.5

Demand per month for a certain product is shown in the time series in Figure 7.3. Calculations based on demand statistics from the last three years have shown that the seasonal index for January should be set at 1.10 and for February 1.05. If the actual demand in January was 422 pieces, a seasonally adjusted forecast for February can be calculated as follows, assuming that no consideration needs to be paid to any trends and that exponential smoothing with a smoothing factor of 0.3 is applied.

Seasonally adjusted demand for January = 422 / 1.10 = 384
Basic forecast for February = 0.3 · 384 + 0.7 · 250 = 290
Seasonally adjusted forecast for February = 1.05 · 290 = 305

Actual demand in many cases may show demand patterns that appear similar to seasonal variations. However, these variations may depend on random influences. For seasonal corrections of basic forecasts to be meaningful it is important that there really is a logical explanation for the variations being seasonal. To take one example, seasonal variations in demand for lawnmowers are logical since it is reasonable to expect more to be sold during the summer than other times of the year.

7.6 Forecast Errors and Forecast Monitoring

Forecasts are always more or less incorrect in relation to real outcome. There is always good reason to estimate any forecast errors that may occur and to continuously check that forecasts remain within acceptable margins of error. This is especially the case for automatic forecasting, such as that made by moving averages or exponential smoothing. There is a risk of losing control of such automatic forecasting methods, which may lead to serious consequences. Assume that an automatic forecasting process for one or another reason generates forecasts which completely miss an increase in demand, and are thus much lower than they should be. Material planning will not be prepared for the actual rise in demand, with ensuing shortage situations and loss of sales. To be able to completely rely on automatic forecasting, it is important to supplement forecasting methods with different types of forecast monitoring in the forecasting system so that unpleasant surprises are avoided as much as possible.

Basic to all forecast monitoring is the measurement of forecast errors, which should take place continuously as a natural part of all forecasting systems. The aim is to identify individual, random errors as well as systematic errors – in which the forecast is generally too high or too low. Forecast errors are also used as a basis for sizing safety stocks.

Forecast errors (FE) are measured per period and can be defined as the difference between the forecast for one period and the actual demand for the same period. A positive forecast error means that the forecast was too large, and a negative forecast error that the forecast was too small in relation to actual demand. For continuous monitoring of forecast errors, companies usually calculate both the **mean error (ME)** and the average value of the forecast error in absolute terms, without considering whether the forecast was higher or lower than the actual demand. This measurement is usually called **mean absolute deviation (MAD)**. The **mean square error (MSE)** is an alternative measure to ME. It is an estimate of the variance of demand when the mean demand is 0. A practical difference between ME and MSE is that MSE penalises large forecast errors. The following formulas can be used to calculate the four forecast error measures:

$$FE(t) = F(t) - D(t)$$

$$ME = \frac{1}{n} \sum_{t=1}^{n} (F(t) - D(t))$$

$$MSE = \frac{1}{n} \sum_{t=1}^{n} (F(t) - D(t))^2$$

$$MAD = \frac{1}{n} \sum_{t=1}^{n} |(F(t) - D(t)|$$

where FE = forecast error
        ME = mean error
        MSE = mean squared error
        MAD = mean absolute deviation
        $F(t)$ = forecast demand for period t
        $D(t)$ = actual demand in period t

Continuous calculation and updating of MAD is carried out most simply using exponential smoothing as in the following formula:

$$MAD(t) = \alpha \cdot |F(t) - D(t)| + (1 - \alpha) \cdot MAD(t - 1)$$

where $MAD(t)$ = MAD up until period t
        $F(t)$ = forecast demand for period t
        $D(t)$ = actual demand in period t
        α = exponential smoothing factor

With this formula, MAD can be continuously calculated at every change of period when the actual demand for the period is known. If the latest period forecast is 12, actual demand in the latest period is 15, α is equal to 0.3 and the previous MAD value was 2.4, then the new MAD value will be equal to $0.3 \cdot |12 - 15| + 0.7 \cdot 2.4 = 0.3 \cdot 3 + 0.7 \cdot 2.4 = 2.58$.

Mean absolute deviation is a measurement of the distribution of the forecast relative to actual demand. If demand is normally distributed, which may often be assumed to be the case, it is possible to calculate the standard deviation, σ, from the MAD value:

$$\sigma = 1.25 \cdot MAD$$

The standard deviation is used among other things for sizing safety stocks and setting control limits for forecast error monitoring and demand error monitoring.

The calculation of mean error (ME) may also be carried out with the aid of exponential smoothing as in the following formula:

$$ME(t) = \alpha \cdot (F(t) - D(t)) + (1 - \alpha) \cdot ME(t - 1)$$

where ME is the mean forecast error up until period t

While MAD is a measurement of the spread of the forecast relative to actual demand, ME gives an indication of whether the forecasts are systematically incorrect or not. A low MAD is not a sufficient measure of good forecast quality. A good forecast method must also give small ME values in the long-term, i.e. equally often give forecasts which are too high as forecasts which are too low, relative to actual demand. A forecast method that is working perfectly has an ME of zero.

The sizes of MAD and ME are influenced by low or high mean demand. They are difficult to compare between different forecast objects and between different lengths of planning periods.

To make such comparisons possible, relative or percentage measurements will be required. **Mean absolute percentage error (MAPE)** is a relative measure of MAD and mean percentage error (MPE) is a relative measure that corresponds to ME. These two relative measurements are defined in the following ways:

$$MAPE = \frac{1}{n} \sum_{t=1}^{n} \left| \frac{FE(t)}{D(t)} \right| \cdot 100$$

$$MPE = \frac{1}{n} \sum_{t=1}^{n} \frac{FE(t)}{D(t)} \cdot 100$$

The calculation of ME, MSE, MAD, MAPE and MPE is illustrated below:

Period (t)	D(t)	F(t)	Error (F(t) − D(t))	Square error (F(t) − D(t))^2	Absolute error |F(t) − D(t)|	Percentage error $\left(\dfrac{F(t) - D(t)}{D(t)} \right) \cdot 100$	Absolute percentage error $\left\| \dfrac{F(t) - D(t)}{D(t)} \right\| \cdot 100$
August	189	189	0	0	0	0	0
September	287	270	−17	249	17	−5.9	5.9
October	339	357.63	18.63	347.08	18.63	5.5	5.5
November	325	432.80	107.8	11 620.84	107.8	33.2	33.2
December	442	471.59	25.59	654.85	29.59	6.7	6.7
January	422	531	109	11 881	109	25.8	25.8
Total			248.02	24 752.77	282.02	65.3	77.1

ME = 248.02 / 6 = 41.34
MSE = 24 792.77 / 6 = 4125.13
MAD = 282.02 / 6 = 47.00
MAPE = 77.1 / 6 = 12.8%
MPE = 65.3 / 6 = 10.8%

TABLE 7.2 Illustration of calculating ME, MSE, MAD, MAPE and MPE

Demand error monitoring

Demand error monitoring is made to prevent incorrect or extremely deviating demand values from influencing forecast calculations. This monitoring means automatically testing how reasonable all new actual demand values are. Exceptional demand values for individual periods may occur due to unusually large one-off orders, or a demand peak in the form of extreme outbound deliveries taking place during a certain period as compensation for manufacturing disruptions in the previous period, with an ensuing inability to deliver. Such events must not be allowed to influence forecasts for normal ongoing activities, and the demand values produced should be corrected or completely eliminated. Demand error monitoring is a way of preventing forecast errors of this type arising.

By testing how reasonable demand values are before they are allowed to be used in forecast calculations, unacceptable influence on forecasts can be avoided. A common way of doing this is to compare demand values with the latest forecast, which acts as a reference value if the maximum deviation allowed from this value is a certain factor multiplied by MAD. Demand values which do not fulfil this condition are rejected and the forecast remains unchanged. The factor is selected to give a reasonably low probability of not excluding values which should be included and thereby allowed to influence the forecast. A commonly applied value for this factor is 4. This value corresponds to a 99.8 per cent degree of confidence, meaning that there is a 99.8 per cent probability that excluding the demand value in question is the correct measure to take.

Forecast error monitoring

Forecast error monitoring is aimed at automatically detecting and signalling when forecasts are systematically too high or too low, i.e. being able to verify that forecasts are close to average. This monitoring is most simply carried out by periodically following up whether the mean forecast error lies within acceptable limits in the form of established control limits, one for positive and one for negative mean forecast errors. These control limits are expressed as a factor multiplied by MAD. When the mean forecast error exceeds these limits, it is a sign that the forecast calculation is going wrong and that manual corrective measures should be taken. Selecting a suitable value for this factor is a question of balance between manual measures to correct forecasts and the extent to which systematic forecast errors can be accepted. A commonly applied value for this factor is 1. One way of solving the issue of priority is to differentiate the factor and thus the size of control limits. Less important products from the viewpoint of management and sales are given higher factor values, or wider control limits, and thereby larger tolerance towards forecast errors. Smaller factor values and thus tighter control limits are selected for products that are more significant to business activities for various reasons. Normally, a factor value between 0.5 and 1.5 is chosen.

Forecast error monitoring can be illustrated graphically by using control charts as illustrated in Figure 7.15. The diagram shows the mean forecast error relative to the two control limits. If the mean forecast error lies close to the upper control limit, this is a sign that forecasts are systematically too high. If it lies close to the lower control limit, forecasts are systematically too low. As long as the mean forecast error remains between these limits, forecast calculations can be described as under control. The lower graph in Figure 7.16 shows how forecast errors may be monitored and visualised in a demand planning software.

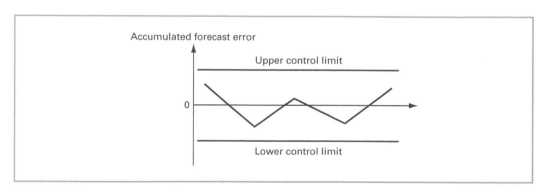

FIGURE 7.15 Illustration of the use of forecast error monitoring using a control chart

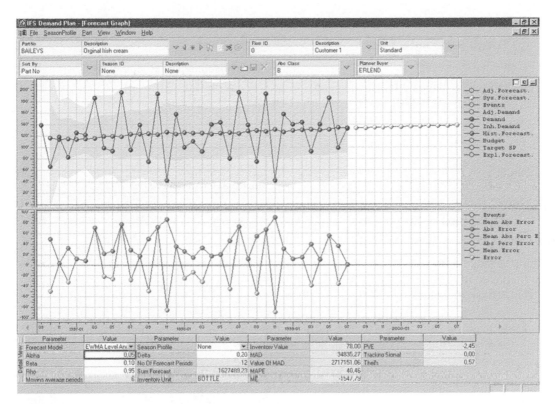

FIGURE 7.16 Example of screen of forecast data

The upper graph visualises the forecast and actual demand. The lower graph visualises the forecast error, and the detailed view under the graph shows detailed forecast parameters. The picture represents a function from the ERP system IFS Applications

7.7 Summary

Theories and methods for forecasting as a basis for company planning and decision making on issues related to manufacturing planning and control have been treated in this chapter. The basic conditions required for forecasting and the principles of designing systems and procedures for forecasting were described in the introductory section of the chapter, as well as the occurrence of demand uncertainty and its relation to forecasting. Later sections discussed three judgemental approaches to forecasting: the management group approach, the grassroots approach and pyramid forecasting. Calculation methods based on moving averages, exponential smoothing and focus forecasting were described. Approaches to considering demand trends and seasonal variations were taken up, as well as approaches and methods for verifying and monitoring forecasts in relation to real outcomes, i.e. forecast errors.

🔑 Key concepts

Bull-whip effect 105

Collaborative planning forecasting and
    replenishment (CPFR) 105

Exponential smoothing forecast 118

Focus forecasting 121

Forecast error 101

Forecast horizon 111

Forecast monitoring 102

Grassroots approach 116

Mean absolute deviation (MAD) 127

Mean absolute percentage error (MAPE) 129

Mean error (ME) 127

Mean square error (MSE) 127

Moving average forecasting 118

Pyramid forecasting 117

Sales management approach 116

Seasonal variation 109

Time series 108

Trend 109

⊡ Discussion tasks

1 What is the role of forecasting in the following types of companies: (a) an automotive manufacturing company that builds cars to order, (b) a manufacturer of fashion clothes and (c) an ice cream manufacturer?

2 What could be the problems of using historical sales data when forecasting future demand?

3 How can a manager use the MAD and MAPE information?

Problems[1]

Problem 7.1

The company Sugerman AB makes and sells two types of baking machines, BR203 and DB34. The table below contains sales figures for the last two years. Monthly forecasts are based on the moving average method.

Month	Year 1 BR203	Year 1 DB34	Month	Year 2 BR203	Year 2 DB34
Jan	74	87	Jan	79	103
Feb	71	93	Feb	99	109
Mar	82	113	Mar	85	104
Apr	83	108	Apr	102	115
May	70	102	May	90	95
Jun	73	91	Jun	113	94
Jul	92	82	Jul	103	98
Aug	90	92	Aug	110	84
Sep	79	79	Sep	106	71
Oct	94	81	Oct	115	75
Nov	93	91	Nov	111	78
Dec	99	89	Dec	120	100

---

1 Solutions to problems are in Appendix A. Because the forecasting problems contain large data sets and several similar calculations they are preferably carried out using Excel.

a) How many months should be included in the calculations for each product?

b) Forecast the sales for each product for January in year 3, using the chosen number of months from question (a).

Problem 7.2

The sales of two products at Stone Soap Ltd expressed in number of cases per month during the last two years are shown below.

Month	Year 1		Month	Year 2	
	STSP01	STVP07		STSP01	STVP07
Jan	170	144	Jan	207	205
Feb	174	150	Feb	194	194
Mar	190	168	Mar	183	185
Apr	197	177	Apr	177	181
May	206	187	May	194	200
Jun	193	177	Jun	202	210
Jul	181	167	Jul	171	181
Aug	209	197	Aug	185	197
Sep	174	164	Sep	177	191
Oct	197	189	Oct	189	205
Nov	175	169	Nov	173	191
Dec	182	178	Dec	206	226

The company has used the moving average method but will start to use the exponential smoothing method in year 3. In order to have an "old" forecast as input for the first exponential smoothing calculation, it has to calculate a forecast for December in year 2 by using the moving average method.

a) Chose an appropriate alpha factor for each of the products STSP01 and STVP07.

b) Forecast the sales of each product for January year 3 by using the alpha factor and the calculated "old" forecast for December year 2.

Problem 7.3

Stone Soap Ltd uses the forecast for making production plans for its manufacturing. It must forecast sales six months ahead. At the end of December, the demand for product STOP06 in January was forecast at 150 pieces. By analysing the previous year's actual demand, the trend was estimated at 8 pieces (increase) per month. Make forecasts for the first six months of the new year, based on the January forecast and the estimated trend.

Problem 7.4

Some of the products at Sugerman AB have seasonal demand during the year. The average monthly demand for product GMT65 during the last three years is shown in the table below. Determine the product's seasonal index for each month.

Month	Demand	Month	Demand
1	183	7	179
2	215	8	158
3	235	9	149
4	219	10	151
5	216	11	164
6	184	12	193

Problem 7.5

Woodmakers Ltd makes the product group Wooden Table Standard to finished goods inventory. The product group contains three types of tables. The amount and type of raw materials and capacity needed to make one table is about the same for all three tables. Last year's demand for the three tables is shown below.

Month	Table 1	Month	Table 2	Month	Table 3
1	160	1	110	1	450
2	195	2	120	2	375
3	170	3	150	3	440
4	210	4	140	4	380
5	185	5	155	5	365
6	155	6	130	6	410
7	167	7	140	7	440
8	195	8	155	8	450
9	210	9	110	9	395
10	220	10	120	10	385
11	195	11	160	11	350
12	240	12	150	12	420

a) Make forecasts for months 2 to 13 for each table using exponential smoothing with an alpha factor of 0.3.

b) Make aggregated forecasts for months 2 to 13 for the product group Wood Table Standard using exponential smoothing with an alpha factor of 0.3.

c) Make forecasts for quarters 2 to 5 for each table using exponential smoothing with an alpha factor of 0.3.

d) Make aggregated forecasts for quarters 2 to 5 for the product group Wood Table Standard using exponential smoothing with an alpha factor of 0.3.

e) Compare the forecast errors (MAD and MAPE) for the forecasts in a) to d). What are your conclusions?

f) What general conclusions can be drawn when forecasting individual products compared to product groups, and monthly compared to weekly forecasts?

Problem 7.6

A product that is made to stock has had the following actual demand during the last 24 periods (months):

Period	Demand	Period	Demand
1	143	13	190
2	142	14	201
3	153	15	192
4	157	16	210
5	139	17	231
6	156	18	200
7	181	19	225
8	180	20	230
9	151	21	232
10	180	22	240
11	182	23	251
12	171	24	270

a) Forecast the demand for periods 4 to 25 by using a three-period moving average.

b) Forecast the demand for periods 13 to 25 by using a 12-period moving average.

c) Forecast the demand for periods 2 to 25 by using exponential smoothing with an alpha factor of 0.1.

d) Forecast the demand for periods 2 to 25 by using exponential smoothing with an alpha factor of 0.4.

e) Forecast the demand for periods 2 to 25 by using exponential smoothing with an alpha factor of 0.8.

f) What are the reasons for using (i) the moving average method with several periods and (ii) a high alpha factor in the exponential smoothing method?

g) Forecast the demand for periods 2 to 25 using linear exponential smoothing with trend correction. Use an alpha factor of 0.1 and a beta factor of 0.4.

h) Forecast the demand for periods 2 to 25 using linear exponential smoothing with trend correction. Use an alpha factor of 0.1 and a beta factor of 0.7.

i) Forecast the demand for periods 2 to 25 using linear exponential smoothing with trend correction. Use an alpha factor of 0.3 and a beta factor of 0.4.

j) Forecast the demand for periods 2 to 25 using linear exponential smoothing with trend correction. Use an alpha factor of 0.3 and a beta factor of 0.7.

k) Compare the mean absolute deviation (MAD) during periods 10 (period 13 for the method in (b) to 24 for the nine methods in questions (a) to (j). What are your conclusions?

Problem 7.7
Conduct demand and forecast error tests for the forecasts in questions (d) and (h) in the previous problem. Calculate and compare the MAD for each period using exponential smoothing.

Further reading

APICS The association of operations management (2007) *Building competitive operations, planning and logistics*, APICS certified supply chain professional learning system Module 2. APICS, Alexandria.

Armstrong, J. (2001) *Principles of forecasting*. Kluwer Academic, Norwell, MA.

Brander, A. (1995) *Forecasting and customer service management*. Helbing & Lichtenhahn, Basel.

Helms, M., Ettkin, L. and Chapman, S. (2000) "Supply chain forecasting: collaborative forecasting supports supply chain management", *Business Process Management*, Vol. 6, No. 5, pp. 392–407.

Lapide, L. (2003) "Organizing the forecasting department", *The Journal of Business Forecasting*, Vol. 22, No. 2, pp. 20–21.

Lapide, L. (2006) "Top-down & bottom-up forecasting in S&OP", *The Journal of Business Forecasting*, Vol. 25, No. 2, pp. 14–16.

Lewis, C. (1998) *Demand forecasting and inventory control.* John Wiley & Sons, New York (NY).

Makridakis, S., Wheelwright, S. and Hyndman, R. (1997) *Forecasting methods and applications.* John Wiley & Sons, New York (NY).

Moon, M. and Mentzer, J. (1999) "Improving sales forecasting", *Journal of Business Forecasting,* Vol. 18, No. 2, pp. 7–12.

Silver, E., Pyke, D. and Peterson, R. (1998) *Inventory management and production planning and scheduling.* John Wiley & Sons, New York (NY).

Småros, J. and Hellström, M. (2004) "Using the assortment forecasting method to enable sales force involvement in forecasting", *International Journal of Physical Distribution and Logistics Management,* Vol. 34, No. 2, pp. 140–157.

Smith, B. (1991) *Focus forecasting and DRP – logistics tools of the twenty-first century.* Vantage Press, New York (NY).

Smith, I. (2006) "West Marine: a CPFR success story", *Supply Chain Management Review,* Vol. 10, No. 2, pp. 29–36.

Wacker, J. and Lummus, R. (2002) "Sales forecasting for strategic resource planning", *International Journal of Operations and Production Management,* Vol. 22, No. 9, pp. 1014–1031.

Wacker, J. and Sprague, L. (1998) "Forecasting accuracy: comparing relative effectiveness of practices between seven developed countries", *Journal of Operations Management,* Vol. 16, pp. 271–290.

Wilson, H. and Keating, B. (2002) *Business forecasting.* McGraw-Hill, New York, NY.

CASE STUDY 7.1: FORECASTING AT ALTO TOOLS LTD

Alto Tools makes profession tools used in manufacturing and construction companies. The products are organised in five different business areas, based on type of products and market. They have their own production sites (three in Europe and ten more worldwide) but they also buy tools from external manufacturers and include them in their assortment. Alto Tools sells most of its products directly to larger manufacturing and construction companies or through retailers. It has defined its market as consisting of three regions (Europe, Asia-Pacific, America). Each region contains two to five distribution centres (DCs) grouped into different DC areas with one or two DCs in each area. Each DC area supplies a number of selling areas. In Europe a selling area is normally the same as a country, and a DC area contains the countries closest to the DC.

In order to have short delivery times, Alto Tools applies an MTS strategy. All products produced by external suppliers are stored in the DCs. The in-house manufactured products are stored in the DCs but an increasing number of products are delivered directly from factories to retailers or end-customer companies.

The company is heavily dependent on forecasts. Five demand planners, one for each business area, have key roles in the forecasting process. Their responsibilities include the generation of forecasts and securing high forecast accuracy. They participate in the demand planning process in terms of pricing, media/activities, product change and merchandising effects to secure the sales forecast on product level. They also participate in developing the sales forecasting methodology and planning of the product change process (phasing in and out of products).

Sales frame as a forecast constraint

The forecasts are constrained by a so-called sales frame. The corporate management group at Alto Tools develops this "sales frame". It is a long-term sales plan, expressed in total sales volumes in monetary units for each of the three regions. This frame expresses the expected and anticipated total sales, is related to the strategic business plan and includes the remaining part of the current fiscal year plus five years into the future. It is updated three times per year.

These sales frames are exploded into business area sales and compared with forecasts for each business area provided by the demand planners, taking into account the business areas' growth plans and ambitions for the future. When a difference is identified between the two plans, the sales frame is broken down into sales frames per region, and further down into sales frames per product area within the respective region. This comparison and adaptation between plans is conducted three times a year, at the same time as the frame is developed. If there is a difference between the forecast and frame, the demand planners should adjust the forecast accordingly. This is normally done by adjusting the forecast in a specific country. If this is not appropriate, the forecast is proportionally adjusted in all countries.

Forecasts on three levels

Alto Tools generates forecasts on three levels as described in the figure. The forecast on selling unit level is input to the distribution requirements planning, used for generating stock replenishment orders and input to the master production scheduling and capacity booking at production sites and suppliers. Forecasts on DC group level are for example used for safety stock calculations, decisions about what DC must prioritise to supply in shortage situations but also as a complement to the plans generated from the distribution requirements planning and used as input to the master production scheduling at its own production sites and capacity booking at external suppliers. Forecasts on regional level are mainly used to check with the sales frames as discussed above.

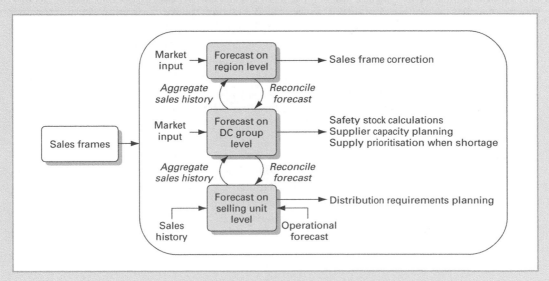

New forecasts are generated on product and selling unit level, i.e. for each product at each separate retailer (see the bottom square in the figure). The planning bucket and planning frequency are weeks and the planning horizon is 84 weeks, i.e. forecasts are generated on a rolling 84 weeks' planning horizon, with new historical sales data loaded once a week. The statistical forecast is based on an exponential smoothing method, with adjustments for trend, seasonality, product life cycle and temporary events. The statistically generated forecast is combined with a manually generated operational forecast on store level. The operational forecast is a forecast registered by the respective selling unit (e.g. store) for the coming three weeks. The sales history and the operational forecast are used to create a forecast for each product on the selling unit level. The forecast used for weeks 1 to 3 is the operational forecast and the forecast for weeks 4 to 84 is the statistical forecast based on historical sales.

Thereafter the store forecasts are aggregated and reconciled with the sales frames and market input on the DC group level. The market input on the two upper forecast levels (see the two upper squares in the figure) concerns every country's activity plan (campaigns, etc.) and estimated price changes on product level. Alto Tools encourages individual countries to have local campaigns and activities. Country-specific activity plans must, however, be decided at least six months in advance.

Demand planners review the forecasts on all levels each week in order to identify those products for which the forecast deviates considerably from actual sales. In case sales deviate considerably from forecasted figures, the demand planner looks for the reason and adjusts the sales figures or forecast model accordingly. Possible reasons and adjustments could be that the forecast model is not considered sufficient, for example, due an increased trend in a specific country or that it takes a longer or shorter time for a product to move into a new product life cycle phase. It could also be that the sales figures are higher than normal, because of a temporary campaign, and therefore should be adjusted so future statistically generated forecasts are not being overestimated. There is no formal communication between demand planners and the sales representatives who register the operational forecasts in the respective selling unit area. Therefore, it is sometimes difficult for the demand planners to understand the reasons for inaccurate forecasts.

A monthly operational meeting takes place between product developers, purchasing managers, sales managers and demand planners. The issues discussed during the meeting are, for example, overstocked products, low sale products, shortages and product quality. A result from the meeting is a central product action report communicated to the respective sales representatives, with information about overstocked and low sale products. The sales representatives at the respective selling unit are encouraged to take actions according to the report. Their actions are communicated back as registered market activities.

Forecast accuracy
The forecast accuracy differs between products and especially between business areas. Some products and business areas have shorter product life cycles and more fluctuating demand than others. Still, the forecast process is considered appropriate for all products and business areas. On average, the weekly operational forecast accuracy is about 70 per cent and the weekly product forecast three months into the future is between 40 and 60 per cent.

Demand planner and selling unit relationship

The demand planners feel that the sales representatives in the different selling units are not taking the forecasting seriously. These are examples of demand planners' opinions about the selling units:

- "25 per cent of all declared market activities registered by sales leaders are not real activities. The selling units abuse the activity routine in order to secure deliveries in situations of supply shortages."
- "The selling unit representatives don't understand what a forecast is. They see it as a best case scenario or ambitious goal. Therefore, the sum of all selling unit forecasts is too high."
- "I do not pay very much attention to the operational forecast registered by the selling units. I take a more global view on the forecast and can therefore omit the individual selling unit figures."
- "The selling units take low responsibility for their input to the forecast since they are concentrating on selling products."

On the other hand, the selling units also have concerns about the demand planners. These are example of selling units' opinions about demand planners:

- "It would be easier for us to plan our market activities if we got earlier notice about shortage and overstock products."
- "Why are there so many shortage occasions when we sell according to the registered forecast? Production and supply must be messing up the capacity and supply."
- "The statistical forecast is very wrong. It is obvious that the demand planner software generating the 84 weeks' forecast is not working."
- "Why can they not deliver according to our forecast? How difficult can it be? We are ordering more and more and they are delivering less and less!"

Discussion questions

1   Alto Tools uses a combination of quantitative (statistical) and qualitative (manual) forecast methods. In what way do the two types of forecasts complement each other in the process?
2   Forecasts are generated on three levels (four levels if including the sales frame level). Is it really necessary for Alto Tools to forecast on so many levels?
3   There are obvious conflicts between demand planners and selling unit representatives. Why do you think that is so? What impact do you think it has on the forecasting process?

CHAPTER

08

Customer Order Management

Logistics in manufacturing companies is primarily concerned with controlling the flow of materials and the consumption of resources so that market demand for the company's products can be satisfied in the most efficient manner possible in order to achieve the company's business goals. To achieve this there must be a balance between demand on the one hand, which drives material flows and resource consumption, and resources on the other hand, in the form of materials, production capacity and available capital. This in turn presupposes that demand is known or can be predicted to a reasonable degree.

The part of demand known in the company is from customer orders placed, which represent undertakings from customers to purchase the company's products. Although there may be changes in wanted delivery dates for these orders, or even they may be cancelled, in this context and for practical purposes they may be considered and treated as known information about future demand. Customer orders in this context are interpreted in a broad sense, and include **call-offs** from agreements made and delivery schedules obtained from customers. Demand apart from orders in this sense is partially or fully unknown and must therefore be estimated in different ways. In companies that manufacture and sell order-specific products, estimates of demand can be based to some extent on the backlog of open offers. The outcome of offers made is assessed for probability and this assessment is used as a guide for estimating future demand. In other areas, the unknown component of demand must be estimated by means of different forms of forecast.

Forecasting as a means of estimating future demand was treated in the previous chapter. This chapter treats the **customer order process** from the perspective of demand and logistics. It describes how customer orders and the demand information that they represent can be used in planning and controlling material flows and the use of resources. In addition it describes the specific characteristics of order processes in companies that **make to stock**, companies that **make to delivery schedule**, companies that **assemble to order** and companies that engineer and **make to order**.

LEARNING OBJECTIVES

After studying this chapter, you should be able to:

◆ Outline a general customer order process
◆ Describe and compare customer order processes in different types of companies
◆ Explain how to conduct available-to-promise calculations

8.1 The General Customer Order Process from a Logistics Perspective

The connection between the order process and other planning processes is shown in overall terms in Figure 8.1. Together with forecasts, customer orders constitute the information required for the master production schedule process. The role played by orders in relation to forecasts is dependent on the extent to which the company's products are made to stock or to order. The order process often has a direct relationship to material planning, especially in the case of stocked purchased components and manufactured semi-finished items that are sold directly to customers as well as being used in production. This is the case for spare parts, for example.

The approach to working with orders varies considerably between different types of companies. From a logistics perspective and at a superficial and overall level of description, it is still possible to talk about a general customer order process with common activities for all types of companies. This general customer order process is illustrated in Figure 8.2. It may be considered as a conceptual and simplified model of information flows and material flows in the process

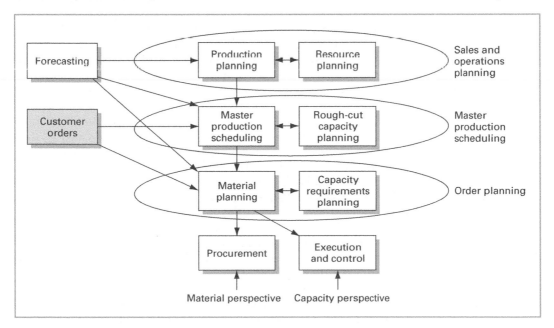

FIGURE 8.1 The relationship between the order process and other planning processes

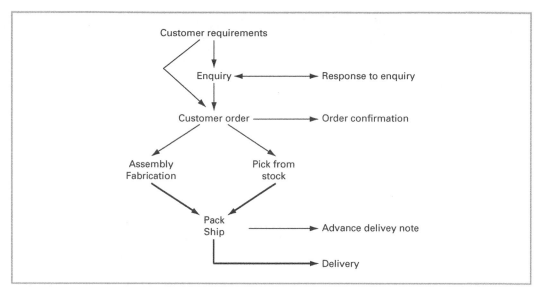

FIGURE 8.2 Model of the general customer order process from a logistics perspective

from need to delivery. The aim of the model is to obtain a framework as a starting point for later descriptions and comparisons of the structure of these processes in different types of companies.

In a manufacturing company, the order process starts with a need that a customer has, or perceives that he or she has, and finishes when the need has been satisfied through the delivery of manufactured products. Between these two points there are a number of sub-processes that initiate different flows of information and materials.

Depending on what the material need is for, there are two main alternatives. If what the customer wishes to purchase is a known and standardised product and if he or she has a developed channel to the supplier, the need may directly result in an order to the supplier. This may be the case when delivery schedules are sent to subcontractors or when the customer himself or herself enters the order directly in the supplier's order system, via a web shop for example. In other cases the customer, in addition to evaluating and selecting the supplier, must also make enquiries to the supplier before an order can be placed. These enquiries may be of a simple type, such as for prices and possible delivery dates. They may also be very complicated in cases where the product does not exist at the time of making the enquiry other than in the form of a rough specification of functions and performance. The enquiry may then be in the form of a **request for quotation**, and for the supplier this means making a general design, calculating a price to offer, investigating the options of getting raw materials and making an assessment of capacity requirements so that information on possible delivery dates can be given. Between these two extremes is a spectrum of more or less complex questions which the supplier must be able to answer for an order to be possible. The majority of these questions relate to material flows and the utilisation of resources, and obviously this means that they have a strong connection with the company's manufacturing planning and control.

When a supplier has received an order, it is often acknowledged by sending an **order confirmation** to the customer. In straightforward cases, this will mean that the supplier confirms he or she undertakes to deliver a certain quantity of a certain product at a certain price and at a certain delivery date. In more complicated cases, an order confirmation may also contain drawings, specifications, promised performance and so on for the product intended to be delivered. In simpler types of orders there may be no order confirmation as such; a procedure known as tacit acceptance is used. This means that the order is automatically considered as confirmed if the supplier has not contacted the customer within a certain period of time, such as one week.

If the order is for standard items in stock, a **picking** list is printed when delivery is to take place and the stated quantities are picked. If the order is for products to be made or assembled to order, a manufacturing order is released. This order must be started early enough for the delivery to take place according to the agreement. Whether products are manufactured or picked from stock, they will be packed and dispatched to the customer. In some cases the supplier will send advance shipment notice at the time of dispatch, informing the customer that the delivery is on the way. This is especially common among companies that make to delivery schedule.

The following sections in this chapter describe in more detail different elements of the customer order process in the five types of company named above. Due to their similarities, make to order and engineer to order are treated as one category. Only those elements of process in which orders have a direct connection with the control of material flows and utilisation of resources are included.

8.2 The Order Process for Make-to-Stock Companies

In companies which make to stock, there are standard products which are fully known at the time of order and which are manufactured without requiring any orders from customers. Quantities manufactured are always delivered to stock and withdrawn from stock in conjunction with delivery to customer order. Picking and delivery often take place immediately when an order is received, so delivery excluding transportation time is in principle instantaneous. There are also cases when withdrawal from stock and delivery take place with an amount of **delivery lead time**. In this situation there will be an order backlog in the form of **allocations** on hand. Delivery with delivery lead time also occurs when there are shortages in stock and for this reason immediate deliveries are not possible.

Orders in MTS companies may be of two types: traditional orders and call-offs. The traditional order, in addition to information on the ordering party, contains details of delivery conditions, terms of payment and so on (see Figure 8.3). It also consists of one order line per product ordered, with information about item number, price, quantity and delivery date. A call-off is a simplified ordering procedure which is carried out in cases of general agreements or blanket orders. Under these agreements, commercial conditions including prices are specified and call-offs in most straightforward cases need only contain information on the item number, quantity and delivery date. From the material planning viewpoint, there is no significant difference between these two types of order.

Enquiries about delivery time

When the customer enters an order directly in a supplier's order processing system, there is no need for any special enquiry to the supplier regarding availability in stock and delivery lead time. This information is directly accessible in the system, and so no reply is made to the customer. Whether it is the customer who accesses information on current delivery options or if it is the supplier's own personnel, **available to promise (ATP)** at the desired date is calculated as a basis for providing information on delivery lead times. Most **enterprise resource planning (ERP)** systems include qualified user support to make such estimates.

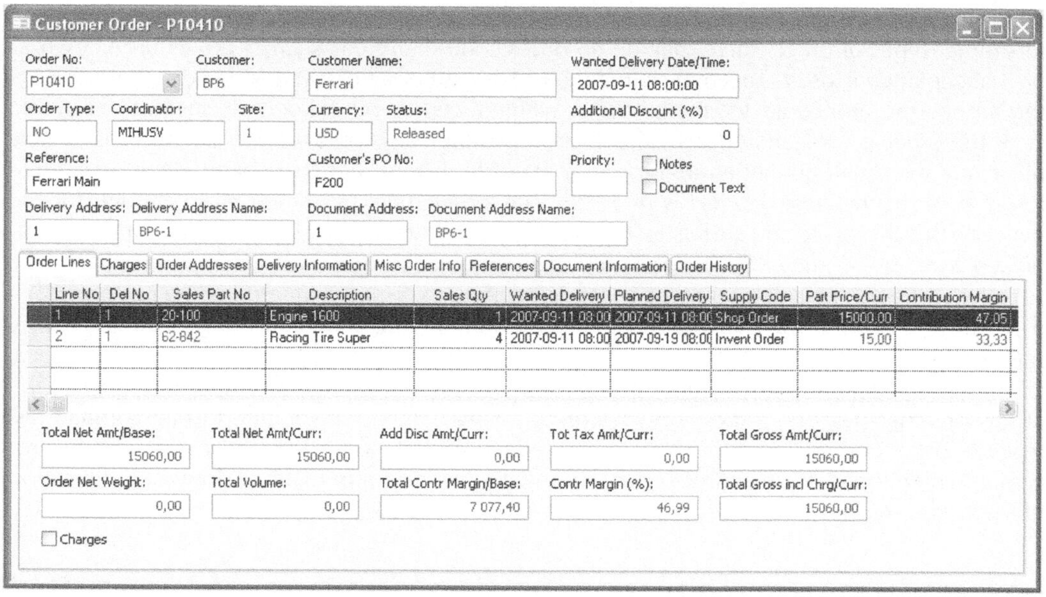

FIGURE 8.3 Example of screen of customer order registration

The picture illustrates the registration of two order lines, the first for one piece of item 20–100 and the second for four pieces of item 62–842, on the order number P10410 to customer BP6. The picture represents a function for the ERP system IFS Applications.

The term "available to promise" refers to the quantity at a given point in time that can be delivered from stock without affecting any other customers' deliveries. The fact that there are seven items of a product in stock is not the same as having seven that are available to promise. A number of them may be allocated for another customer. Neither is available to promise equal to the estimated inventory balance period for periods in the future.

The calculation of available to promise is illustrated with the aid of an example in Figure 8.4. In this example, the length of the time period is one week in order to give more clarity. Under normal conditions, delivery precision will be in terms of days. The first row in the figure refers to the total allocated quantities from orders on hand. An allocation constitutes a booking for a

Week		1	2	3	4	5	6
Allocated quantities		60	25	40	15	5	0
Scheduled receipts		–	–	100	–	–	100
Projected inventory	90	30	5	65	50	45	145
Available to promise		5		40			100
Cumulative available to promise		5	5	45	45	45	145

FIGURE 8.4 Example of available-to-promise calculation

certain quantity for delivery at a later date to a specific customer order. The second row refers to **scheduled receipts** from open or planned manufacturing orders, or purchase orders in the case of products purchased. The opening stock is 90 pieces and row 3 illustrates the projected inventory estimated as outbound deliveries to orders on hand and scheduled receipts from manufacturing orders or purchase orders take place over time. The fourth row shows the amount of the scheduled receipts that is available to promise. The final and final row shows the size of cumulative quantities remaining in stock available to promise for a new customer.

At the beginning of week 1, the available to promise amount includes the opening stock on hand (90 pieces) plus items to arrive during the first week (none in this example) minus allocated quantities made prior to the next scheduled receipt, i.e. week 3 (60 + 25 pieces). 5 pieces (90 + 0 − 60 − 25) are consequently available to promise during week 1. No quantities will arrive during week 2 and therefore no additional quantities are available to promise in week 2. 100 new items will arrive in week 3. The additional amount of items to be available to promise during week 3 is the 100 quantities minus the allocations made prior to the next scheduled receipt, i.e. week 6 (40 + 15 + 5). 40 pieces (100 − 40 − 15 − 5) are consequently added to the cumulative available to promise in week 3. The cumulative available to promise during weeks 3 to 5 is thus 45 pieces (5 + 40).

Available-to-promise calculations are conducted in the first period and in periods with scheduled receipts. The amounts of available to promise identified in periods with scheduled receipts are added to the cumulative available to promise. The following formulas are used when making the available to promise estimates:

First period available to promise = Opening stock on hand + Scheduled receipt during first period − Allocations made prior to the next inbound delivery

Following periods available to promise = Scheduled receipt during the period − Allocations made prior to the next scheduled receipt

Note that available to promise in week 1 is not 30 pieces, even though there are 30 pieces remaining in stock at the end of the week. If these 30 pieces were promised to a new customer, the customers already promised delivery of 25 pieces in the next week would be affected. Consequently, if the company receives an enquiry from a customer for the delivery date of 10 pieces, the answer must be week 3. The alternative is to divide the order into two split deliveries.

Order registration

When an agreement between a customer and a supplier has been reached, the order is entered in the supplier's ERP system either by the supplier's own personnel or by the customer. The order can also be transferred in an electronically readable form with the help of **electronic data interchange** (**EDI**). At the same time as the registration is made, the products on the different order lines are allocated in those quantities and at the delivery dates stated on the order. In this way, information about demand – which the customer order represents – is stored in the system and consideration can be taken to the equivalent outbound delivery requirement in material planning.

The entered order can be confirmed to the customer through an automatic printout from the customer order file, which is then sent by mail or fax. It may also be sent in the form of an email or an EDI transaction. The extent to which this takes place is dependent on tradition and on the agreement reached between the parties involved. If orders are entered in the supplier's system directly by the customer, there is no need to send an order confirmation.

Picking and delivery

When the delivery date for an order comes up and products ordered are in stock as planned, a picking list is printed. The products are picked from stock, packed and sent to the customer. When the products are picked, the inventory balance is decreased by the quantity picked for the order. At the same time, allocations must be removed for the available quantities in stock to be correct. If the quantities in stock at the time of picking are insufficient, the delivery must either be delayed until the full quantity can be delivered or the order must become a split delivery, meaning that some of the promised quantities are back-ordered and delivered on a later occasion while the part of the order available in stock is delivered. In the case of a back order, allocations for quantities back-ordered are retained.

From the accounting perspective, withdrawals from stock must also be booked. This takes place at the same time as entering revenues, expenses, goods sold and accounts receivable. Stock deductions, removal of equivalent allocations and accounting of financial transactions associated with a delivery to the customer are often processed in conjunction with invoicing. This procedure may then mean that the stock balance shown in the ERP system is higher than the physical balance in stock if time elapses between the time of withdrawal and time of invoicing. The available quantity will still be correct, however, since allocations remain in the system as long as the inventory balance has not been reduced by the equivalent quantity. One way of avoiding the problem with temporary inconsistencies between physical and accounted inventory balance is to transfer the stock value picked at the time of picking from the inventory account to an account for products to be delivered. The balance can then be reduced and the allocations withdrawn at the same moment in time as picking is reported in the ERP system.

8.3 The Order Process for Make-to-Delivery-Schedule Companies

In companies that make to delivery schedule, standard products are fully known at the time of ordering. The difference is that manufacturing takes place to a **delivery schedule**, which constitutes a type of order. Quantities manufactured can be delivered directly to the customer in accordance with the delivery schedule, or to stock and then delivered from stock at the rate stated on the delivery schedule. The choice between these two alternatives depends on whether it is financially feasible for the supplying company to use manufacturing order quantities that are as small as the quantities in the delivery schedule. If this is not the case, the manufacturing order quantities must correspond to some form of economic order quantity, and delivery then takes place from stock. Even if manufacturing is made in order quantities that directly correspond to the delivery schedule quantities, it is common that deliveries take place from stock in terms of handling, administration and accounting. The difference is that the delivery quantities cannot be consolidated into one manufacturing order quantity and that manufacturing orders are not initiated by the size of stock being sufficiently large to cover future needs or not. The manufacturing order is initiated directly by the delivery schedule.

The delivery schedule may be considered as a special type of order. It often consists of three parts, as illustrated in Figure 8.5. The first part is a call-off for delivery and may be considered as a pure order from the customer. Normally, this part may cover anything from a couple of days to a number of weeks. The time period depends on current delivery lead times and the delivery agreements reached between the parties which regulate how the delivery schedules are to be interpreted and processed. This first part also constitutes the frozen part of the delivery schedule, i.e. it may not be changed by the customer from one delivery schedule to the next without good reason. The agreement reached between the parties regulates the extent to which changes may

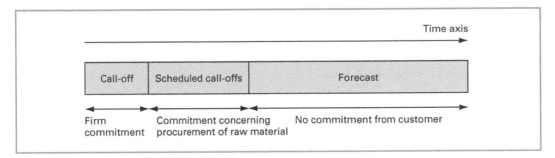

FIGURE 8.5 Parts of the delivery schedule

take place in the frozen part of the schedule. The call-off part of the delivery schedule may be considered as allocations and consequently constitute a form of order backlog.

The next part of the delivery schedule is made up of the quantities which are expected to be called off. These expected call-offs are not completely firm undertakings from the customer to definitely call off. However, there is often some form of commitment from the customer, for example in the form of paying for costs of raw materials that the supplier must order in advance to have time to manufacture before the call-off takes place. The most far-reaching part of the delivery schedule is only a forecast of probable needs of future deliveries. It is used for the purposes of master production scheduling and is treated in a later chapter.

Reception and processing of delivery schedules

Enquiries of a routine nature do not exist in this type of company in connection with delivery schedules being sent out. Material flows are of a repetitive character and controlled to a large extent by long-term agreements and the exchange of demand information with long planning horizons.

Since delivery schedules contain much information and are sent to suppliers rather frequently, in many cases more than once a week, from the administration point of view it is cumbersome to handle them in the form of paper documents which must be registered in an ERP system. For this reason, delivery schedules are generally communicated by EDI or via the **Internet** in an electronically readable form for the receiving system.

There is no order confirmation in the true sense in this type of company. Relations and procedures between customer and supplier are completely governed by the delivery agreements between them. The closest corresponding sub-process to order confirmation is the supplier's check that changes in the frozen part of the delivery schedule made by the customer lie within the limits of what has been agreed. If this is not the case, there may be reason for the supplier to make contact with the customer to indicate they wish to make changes. It may also be necessary to inform the customer that due to changes in the frozen part of the agreement it is no longer possible to deliver the quantities desired. If the new delivery schedule complies with the agreement that has been made, and it is possible to execute, it will be automatically considered as confirmed without the necessity of any special information being exchanged.

Picking and delivery

The sub-process of picking and delivery for this type of company is practically identical to the previous type. Instead of removing allocations in the traditional sense, delivery schedule quantities are removed in a corresponding fashion. In other aspects the procedure is only different in that it can be made simpler and more routine since in this type of company deliveries are often repetitive and frequent. More advanced system support is often used, such as goods flags equipped with bar codes for marking delivered goods so that identification at the customer's **goods reception** is facilitated. It is above all in this type of company that **advance shipment notices** are used and sent to the customer, such as in the form of EDI transactions.

8.4 The Order Process for Assemble-to-Order Companies

In companies of the type assemble to order, specific product variants are not completely known until the orders arrive. They are specified in conjunction with the order and they are assembled or finally manufactured to order. On the other hand, the raw material and the components that the products are made of are standardised manufactured semi-finished or purchased items that are either made or purchased to stock.

Since assembly/final manufacture takes place to order, delivery cannot be done directly when receiving the order, but is always done with a delivery lead time. This delivery lead time must be at least as long as the lead time in assembly/final manufacture. As a result of this relationship, there is always an order backlog in this type of company. Assemble to order also means that allocations do not exist in the same sense as in companies that make to stock. In addition, every order received from a customer results directly in manufacturing.

Specification of orders and enquiries for prices

Since products delivered from this type of company are not completely defined in advance, they must be specified in conjunction with order entry. To enable this to take place in an efficient manner and to be able to create different product variants from a standardised range of manufactured semi-finished items and purchased components, products are based on modules. This means that every product variant can be built by combining different variants from each of the modules that it consists of. In many cases, in addition to the different modules, there are common items which are included in all product variants irrespective of which variant is selected. When specifying a desired product variant, the starting point is a so-called product model, i.e. the variant group that represents all the different product variants which can be produced by combining different module variants. For example, a car model may consist of an engine module and a gearbox module. The engine module may be a six-cylinder or a four-cylinder engine, and the gearbox module may be manual or automatic. A product composition of this type is illustrated in principle in the left part of Figure 8.6.

A large number of different product variants can be created from the product model, which consists of a number of modules and which in turn may be made from a number of variants. The specification of the desired product variant takes place when an order is placed. This specification procedure means that the customer specifies exactly which variant is desired for every module. The approach is illustrated in the right-hand part of Figure 8.6. The car variant X which has been chosen by the customer in this case has a four-cylinder engine and an automatic gearbox.

Enterprise resource planning systems often contain a tool called a product **configurator**, which supports work with configuring product variants. Using this system support, available module alternatives can be crossed off on a screen. When all the alternatives have been selected and the configuration program is closed down, a bill of material for the product variant selected by the customer is automatically created. Product configurators generally have

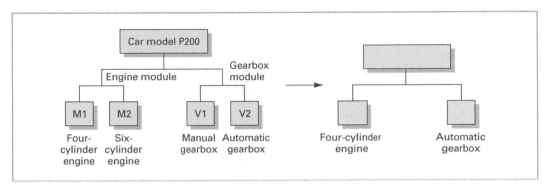

FIGURE 8.6 Illustration of configurations of a product based on modules

functions that ensure non-valid combinations of module variants cannot be chosen. Answers to enquiries about prices for the product variant that has been configured may also be provided by the product configurator.

Enquiries about delivery time

The question of when it is possible to deliver a certain quantity of a product variant is somewhat more complicated in this case than for companies that manufacture standard products to stock. Available to promise must cover both capacity and access to materials. The fact that consideration must be paid to **available capacity** depends on that some manufacturing is taking place to order, for example the final assembly operation. Available to promise must therefore also include available capacity to promise. Two different methods may be used to calculate possible delivery dates with respect to available capacity in assembly/final manufacture. The first method involves current workload being continuously updated as orders are placed. When an order is received, a calculation is made of when the number of hours required for the order is earliest available in the work centres used. The method is described in more detail in the next section. The second method is somewhat simpler to work with. It is based on an overall master production schedule for the product models manufactured by the company. This master production schedule concerns quantities that can be delivered per time period for each product model, irrespective of any variants of the model. The master production schedule is checked off from the capacity aspect during master production scheduling and therefore represents possible capacity in quantities per time period. Refer to Chapter 10 for more details on master production scheduling.

As orders are received for different product variants, these quantities are ticked off the master production schedule for the corresponding product models. Quantities that are not ticked off are those quantities for which there is capacity remaining to manufacture and deliver. The procedure is illustrated with the aid of an example in Figure 8.7. As shown in the example, delivery of 10 pieces can be promised in week 4 of a product variant whose model is planned to be manufactured according to the master production schedule in the figure. The method is above all useful in cases where different product variants of a certain model have reasonably similar capacity requirements.

To be sure of being able to promise a delivery date which can be kept, the availability of materials required for the product variants must also be checked. This is because it cannot be assumed that the distribution between different module variants will always be exactly as planned. Available to promise must therefore be calculated for all module variants selected for

Week	1	2	3	4	5	6
Customer order allocations	20	20	18	15	5	0
Master production schedule	20	20	20	25	25	25
Available to promise	0	0	2	12	32	57

FIGURE 8.7 Example of available-to-promise calculation regarding capacity for assembly/final manufacture to order

configuration. The same approach as described above for the company type make to stock may be used. The module variant that results in the latest delivery date will be decisive for when the order can be delivered with respect to availability of materials.

As an alternative to determining possible delivery dates for a customer order by using available-to-promise calculations as described above, many companies use a more or less fixed policy delivery lead time, generally based on experience. The drawback with fixed delivery lead times is that they must always be sufficiently long to work in situations when workload is temporarily higher than normal. The advantage is that the principle is easy to apply and that delivery information can be given to potential customers very quickly.

Order registration

When the desired product variant has been specified and a possible delivery date established, the order can be entered into the customer order file. The specification, often in the form of a code string, is stored with the order. Since product variants typically do not have an item number, they are often assigned an identification number when the order is received. This identification number may be the customer order number plus an order line number.

Since the product to be delivered is specific for every order, there is more reason to confirm the order than in the case of companies that make to stock or make to delivery schedule. This may take place through an automatic printout from the order file which is then sent by mail or fax to the customer. The order confirmation may also be sent in the form of an email or EDI transaction.

To manufacture the product variant specified in the order, a manufacturing order must be created. This may take place automatically when the customer order is entered or separately by manual registration. At the same time as the manufacturing order is entered, the semi-finished items and purchased components required for the product variant will be allocated in the ERP system.

Manufacturing and delivery

At the time when the manufacturing order is planned to start, the semi-finished items and purchased components required for the product variant are withdrawn from stock. At the same time as they are withdrawn, stock on hand is updated and data about the value of the materials withdrawn is transferred from a stock account to a work-in-process account in the accounting system.

When the order is reported as completed the balance is not updated since the manufactured product variants are not kept in stock. Neither are allocations updated. The value of the quantity manufactured is transferred from the work-in-process account to an account for goods to deliver. When the quantity manufactured has been packed and delivered, invoicing and accounting takes place in the same way as in the case of make and deliver from stock.

8.5 The Order Process for Engineer-to-Order and Make-to-Order Companies

The most complicated order process is in types of company that make to order and engineer to order. In these cases, it is not only the final product which is unspecified before the order. The required semi-finished items and purchased components may also be unknown to some degree. In addition there is often engineering work and preparation of products for manufacturing to be carried out. This work is directly connected with individual orders and must be performed before manufacturing can start. This work also influences the order process, for obvious reasons. In both of these types of company there is always a certain delivery lead time required, the length of which depends on the extent of the work required in connection with the order.

Specification of order and price enquiries

Since products manufactured and delivered in these types of company are to a large extent undefined in advance, they must be at least partially specified before an order can be prepared. This takes place during a **quotation** process. The customer submits a request for quotation to possible suppliers. This request for quotation may involve cases where only overall preferences for functions and product performance are included, cases where a complete engineering description is included, and cases where only small modifications of existing product types are stated. Product type in this context refers to a basic design from which different product variants can be created through additions, removals, or modification of components and accessories included in the product.

Irrespective of the type, work with preparing quotes in this context involves the specification of products demanded to such an extent that it is possible to assess capacity requirements, estimate delivery dates for materials with critical lead times and calculate what it may cost to produce, and thus what price is required. The extent of work with quotes will depend on the degree of deviation from existing basic designs. In extreme cases there may be no basic design to start from, and assessments must be made on experience from similar previous orders.

The results of this work will be an offer which the customer can consider. It will include specifications with respect to functionality and performance, as well as price and other commercial conditions. The offer will also include information on possible delivery dates.

Enquiries about delivery time

In the types of company discussed here, material requirements and capacity requirements will be unique to each order to different degrees. The available-to-promise analysis must thus include both access to materials and access to capacity. Consideration must be given to available capacity, since manufacturing takes place partly or wholly to order. If removals and modifications of the basic design required for the product type are of a reasonable extent, product types can be used as a basis for calculating a possible delivery date with regard to available capacity. The same procedure as described above for companies assembling to order can then be used, i.e. quantities of corresponding product type master production schedules can be successively checked off for every new order so that remaining capacities represent quantities per period that remain available for any new orders received, as illustrated in the example in Figure 8.7. The smaller the differences between capacity requirements for different orders of the same product type, the more advantageous this approach will be.

Another alternative, which is also applicable in extreme cases including a lot of new engineering work and manufacturing preparation, involves continuously updating capacity requirements in hours or other unit of measurement as orders are received and delivered. It is appropriate to do this with the aid of so-called **capacity bills**. A capacity bill expresses the total capacity requirements per item of a product in those resources required for its manufacture. Since in this context products are not engineered and ready for production when the order is received, information on capacity requirements cannot be retrieved from the basic data files. Instead it must be derived from experience-based estimates of expected capacity required for entire manufacturing departments at a time, and occasionally for individual work centres which are known to be critical, such as bottleneck sections. A maximum of some 10 resource groups are usually sufficient to include in capacity bills.

The time estimates which are made in conjunction with work on **quotations** can be used as a basis for producing capacity bills. At a later stage, when the order has been received and more detailed information is available, they may be refined to allow greater accuracy in the capacity planning that is carried out in this context.

When analysing and assessing a possible delivery date from the capacity viewpoint, the estimated capacity requirements per resource group are compared with available capacity. The calculation is carried out for every resource group separately as illustrated in Figure 8.8. In this figure, allocated capacity refers to the capacity planned to be used for orders on hand. In week 3 there are 40 available hours, and in week 4 there are 60 additional available hours. If a new order requires 30 hours, it can be manufactured and delivered as early as week 3 by the resource group in question. If it required 50 hours, it could not be completed and delivered before week 4. Out of those resources in the capacity bill, it will be the resource group which gives the longest delivery lead time which determines the possible delivery date which can be promised to the customer.

For the same reasons as in the case of assemble-to-order companies, it is not only available capacity which is decisive for a delivery date. Access to materials will also influence possible delivery dates. Since products of this type have material contents that are generally little known in advance, it is not possible to do this completely. It is often necessary to accept manual control of individual items which are known to be critical for the delivery lead time. If the items in products are customer specific, it may be necessary to contact suppliers before a possible delivery date can be established. In this type of environment it is generally either access to capacity or delivery lead time for individual purchased components with critical delivery lead times that determines when it is possible to promise delivery to a customer.

Week	1	2	3	4	5	6
Capacity in hours	200	200	200	200	240	240
Allocated capacity	200	200	160	140	230	–
Available capacity	0	0	40	60	10	240

FIGURE 8.8 Example of available-to-promise calculation regarding capacity for engineer-to-order and make-to-order companies

In these types of company, too, delivery lead times are established through policy used as an alternative to determining the earliest possible delivery date with the aid of capacity bills. Policy-determined delivery lead times are especially common in cases when there are small amounts of additions and modifications to existing basic designs. In the types of manufacturing discussed above, the use of fixed delivery lead times tends to produce later delivery dates to customers than is necessary.

Order registration

Orders from customers in this type of manufacturing environment have another character than those for delivery of standard products from stock and products configured from standard modules and assembled to order. The products do not usually have identities in the form of item numbers. Instead they are described in the form of running text in the customer order document, and the details are defined in the form of drawings, functional requirements, performance requirements and other types of specification which are attached to the order. Since the products are specific for each order, there is good reason to confirm what has been agreed. It is more common in these types of company that orders and order confirmation documents are sent by mail than in other types of company.

In order to manufacture the product variant specified in the order, a manufacturing order must be created. The same order often contains several sub-manufacturing orders. It is also common that there are order-specific items in the product which must be purchased as a direct consequence of the order received. Release and registration of manufacturing orders and purchase orders will usually take place manually, especially in those cases where there are no bills of materials or routings already entered in the basic data files. Items incorporated in the product are allocated when entering the manufacturing order.

Manufacturing and delivery

When a manufacturing order is planned to start, the semi-finished items and purchased components are withdrawn from stock and at the same time inventory balances are reduced and the value of the withdrawn material is transferred from a stock account to a work-in-process account in the accounting system in the same way as in companies that assemble to order.

Since manufactured products are not kept in stock, inventory balances are not updated when orders are reported as completed. The value of the manufactured product is transferred from a work-in-process account to an account for goods to deliver. When the manufactured quantity has been packed and delivered, invoicing and accounting take place in the same way as in the case of make and deliver from stock. Advance shipment notice is less common in this type of environment.

8.6 Summary

This chapter has treated the customer order process from the perspective of demand and logistics. It has described how orders and the demand information they represent can be used in planning and to control the flows of materials and use of resources. The appearance and contents of the order process have been described specifically for companies that make to stock, companies that make to delivery schedule, companies that assemble to order and companies that make and engineer to order. The principles for delivery schedules and available to promise were treated, among others.

🔎 Key concepts

Advance shipment notice 148

Allocation 143

Assemble to order 140

Available capacity 149

Available to promise (ATP) 143

Call-off 140

Capacity bill 152

Configurator 148

Customer order process 140

Delivery lead time 143

Delivery schedule 146

Electronic data interchange (EDI) 145

Enterprise resource planning (ERP) 143

Goods reception 148

Internet 147

Make to delivery schedule 140

Make to order 140

Make to stock 140

Order confirmation 142

Picking 143

Quotation 151, 152

Request for quotation 142

Scheduled receipt 145

Discussion tasks

1 Consider the general customer order process. How will it differ in the following sales situations: (a) a CNC machine that is made to order, (b) a specific metal component that is made to schedule and (c) selling copy paper to wholesalers?

2 An important objective of the customer order process is to allow for delivery lead time enquiries, i.e. conducting so-called "available to promise" (ATP) calculation. This calculation is conducted somewhat differently in different environments.

    Compare the ATP calculation in (a) make-to-stock (MTS), (b) assemble-to-order (ATO) and (c) engineer-to-order (ETO) environments. Emphasise what considerations to make and what to base the calculations on in the respective environment.

3 The objective of the available-to-promise logic is to give customers delivery promises where the promised delivery times depend on the present order and capacity situation. Several firms, however, work with fixed delivery times without considering the order and capacity situation. Compare the impact of working with available-to-promise calculations and fixed delivery times, respectively.

Problems[1]

Problem 8.1

The company Cutting AB makes chainsaws of various sizes and capacity. For one of the larger models, the stock on hand is 25 items. The table shows the received customer orders and scheduled receipts from manufacturing during the next six weeks.

Week	1	2	3	4	5	6
Customer order reservation	60	45	40	30	20	5
Scheduled receipts	100		100			100

a) Determine the projected inventory balance at the end of each week during the six weeks, considering that no more customer orders are received.

b) Determine the available to promise, i.e. the number of additional items the company can deliver during the six weeks.

Problem 8.2

A customer calls at the beginning of week 1 and wants to buy 50 chainsaws of the model in the table above. What alternative possibilities does the company have to fulfill this customer request?

Problem 8.3

A customer order planner at Woodmakers Ltd gets a phone call from a customer who wants to get 40 chairs of a specific model to be delivered during week 4. The planner looks at the "available-to-promise" picture in the ERP system and sees the following:

Week	1	2	3	4	5	6
Customer order reservations	80	45	60	30	70	25
Scheduled receipt	120		120		120	
Available to promise	45	45	75	75	100	100

Stock on hand at the beginning of week 1 is 50 pieces. The planner accepts the order with delivery of 40 chairs in week 4. What quantities are now possible to promise during the six weeks?

Problem 8.4

PC Amateurs AB is a company that assembles personal computers (PCs) from standard components purchased from a large number of suppliers. The PCs are configured and delivered based on customer specifications. Assembling and testing are conducted along an assembly line with a capacity of 20 PCs per day. The lead time is 4 days. In order to promise delivery dates when receiving a customer order, the inventory availabilities of the critical components are checked. All components must be available in order to start assembly.

Right now the order backlog is according to the table (figures represent number of PCs that have been promised to be delivered during the respective day):

Day	1	2	3	4	5	6	7	8
Planned customer deliveries	20	20	20	20	18	5	4	0

---

1 Solutions to problems are in Appendix A.

A request to deliver 5 PCs containing the critical components A and B is received. Each PC contains one A and one B. The following tables contain information about reservations for the planned assembly orders and planned deliveries from suppliers during the next 8 days. Stock on hand at the beginning of day 1 is 15 for component A and 17 for component B.

Component A:

Day	1	2	3	4	5	6
Reservations	8	2	3	0	0	0
Scheduled receipts		20				20

Component B:

Day	1	2	3	4	5	6
Reservations	12	3	2	0	0	0
Scheduled receipts				20		

When is the earliest possible day when the company can promise delivery to the customer?

Further reading

Bowersox, D., Closs, D. and Cooper, B. (2002) *Supply chain logistics management.* McGraw-Hill, New York (NY).

Crum, C. and Palmatier, G. (2003) *Demand management best practices: process, principles and collaboration.* J. Ross Publishing, Boca Raton (FL).

Pibernik, R. (2006) "Managing stock-outs effectively with order fulfillment systems", *Journal of Manufacturing Technology Management*, Vol. 17, No. 6, pp. 721–736.

Schönsleben, P. (2007) *Integral logistics management.* St. Lucie Press, New York (NY).

Stadtler, H. and Kilger, C. (eds) (2005) *Supply chain management and advanced planning.* Springer Verlag, Berlin.

CASE STUDY 8.1: CUSTOMER ORDER MANAGEMENT USING CAPABLE-TO-PROMISE
PLANNING AT SSAB[2]

SSAB Oxelösund is a niche producer of high-strength steel. It has about 10 000 customers spread over the world. Only two of them stand for more than 5 per cent of SSAB's total sales. Most customers are consequently fairly small. The customers are served from about 200 sales managers. The mill, where the steel is produced, is producing 24 hours a day, 7 days a week, but does still not have enough production capacity to fulfil the customer demand. There is consequently a heavy focus on capacity utilisation and prioritising the right products and customers. The objective of the master production scheduling is thus to determine the optimum product mix in order to reach high utilisation, customer service and contribution margin.

The sales function at SSAB Oxelösund is organised in different business areas, which correspond to different geographical areas. Every business area has a business area manager. The business areas are divided into smaller geographical sales areas, which are divided into even smaller sales regions, each with its own manager. The sales function is responsible for forecasting its respective market's future demand, promising delivery dates and accepting orders.

SSAB's customer order management process is part of an integrated tactical planning process consisting of demand forecasting, master production scheduling and customer order management (available-to-promise/capable-to-promise calculation).

The planning software
The following six planning software modules support the process;

♦ *Jeeves Order Entry* is a module used to register customer orders.

♦ *Lawson M3 Advanced Production Planner* (APP) is a module used for the detailed production activity control.

♦ *Beppe* is an in-house developed module used for shop floor routing. It supports the process with routing data.

♦ *i2 Demand Planner* (DP) is a forecasting module allowing quantitative forecasting, aggregation and disaggregation of forecasts in terms of product groups, geographical regions and time periods in multi-user environments.

♦ *i2 Supply Chain Planner* (SCP) is a module using data about forecasted future demand, available capacity and related prices, costs and constraints as input to a calculation, based on linear programming (LP) or mixed-integer linear programming (MILP) models, that generates an optimum production plan. The optimisation can consider infinite or finite capacity and generate the minimum cost or maximum profit solution. Due to the huge amount and detail of data required in the model, only constrained (or near-constrained) resources are normally modelled in detail.

♦ *i2 Demand Fulfilment* (DF) is customer order management software, including available-to-promise (ATP)/capable-to-promise (CTP) functionality.

---

2 Based on Jonsson, P., Kjellsdotter, L. and Rudberg, M. (2007) "Applying advanced planning systems for supply chain planning: three case studies", *International Journal of Physical Distribution and Logistics Management*, Vol. 37, No. 10, pp. 816–834; and Cederborg, O. and Rudberg, M. (2008) "Advanced planning systems: master planning in the process industry", *Proceedings of the EurOMA Conference 2008*, Groningen, The Netherlands.

▶

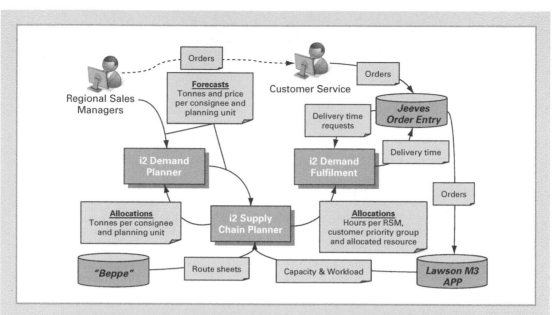

The figure illustrates how the different software modules are linked in the planning process at SSAB.

Forecasting
Forty per cent of the demand is made to stock (MTS) and stored at several warehouses around the world. The aim is to have close to 100 per cent service level of all stored products. The other 60 per cent of the demand is made to order (MTO) with a delivery time of 12 weeks plus transportation time. In total, there are 370 stock-keeping unit (SKUs) and 160 different MTO products forecasted in tons and prices for every customer, specified per month, with the first three months disaggregated to weeks.

Master production scheduling
The forecast information in the DP is imported to the SCP module, where the first three months are disaggregated from monthly to weekly buckets. The SCP uses route sheets from Beppe to calculate the needed capacity for the forecasted demand. In the SCP, each individual forecast (regional sales manager, customer, product, price and time bucket) is kept intact and matched with current production capacities, which is imported from the APP module, in order to develop a master production schedule with optimum product mix.

The optimisation is based on the following four main types of business rules:

1 Supply the most important customers, based on a four-level ABC customer classification.
2 Supply all MTS products.
3 Supply all business areas up to their respective specified minimum levels.
4 Supply the most profitable forecast.

The fourth profit optimisation rule ensures that SSAB produces and sells the product mix to the customers that results in highest profit contribution. This is possible, because both volumes and prices are forecasted. Based on the business rules and planning data, constrained and optimised sales plans and master production schedules are generated. The SCP optimisation is a time-consuming process and takes between six and ten hours. The constrained delivery plan expressed in tons is exported from the SCP to the DP, in order for salespeople to check their current allocations in the DP. The same problem as in the old planning approach exists with tons as planning units. It does not relate to the needed production capacity, but it is a well known unit that everyone understands and is therefore used. The constrained master production schedule is exported to the APP and there used as input for releasing production orders.

Customer order management

The constrained master production schedule expressed in hours per allocated resource is exported from the SCP to the DF. There are 118 different production resources defined in the production system, but only 15–20 of them are used as allocated resources. When a salesperson receives a customer order, it is sent by email or fax to the customer service department. Here, the order is entered into the order entry system (Jeeves), which connects to the DF and looks for available capacity for the salesperson in the needed resources and time buckets. The allocated resources normally have workloads between 98 and 99 per cent because they are the bottlenecks, but they are also the ones that it is important not to overbook, since that unconditionally will cause delays.

In the capable-to-promise (CTP) calculation, the DF seeks available capacity based on the following rules. Based on the four-level ABC classification, highest priority is given to the most important customer group. First the DF seeks to find available capacity for the regional sales manager in the actual customer's ABC group. If capacity does not exist, DF seeks capacity in lower ABC customer groups for the actual regional sales manager. As a third step, and only for customers in the three most prioritised ABC groups, DF seeks available capacity first in less important customer groups of other regional sales managers in the same sales area and then in other sales areas' least important customer groups. This method is used to level out the fluctuations in individual sales and the above procedure is the standard setting. Every business area manager can decide exactly how and which rules should be applied in their own business area. If the DF finds available capacity, a delivery promise is returned to Jeeves, and communicated to the salesperson. If not, the order is not accepted.

Some planning experiences

The use of profit maximisation as the object function in the SCP is uncommon. However, due to the business rules used, there are only a few per cent of the total demand that are actually affected by the profit maximising objective. The use of profit optimisation is not uncontroversial at the company. There are several opinions on how to calculate sales costs for different sales offices in different countries, and there are managers claiming that profit optimisation is a short-term approach that makes the company more vulnerable in the long run.

SSAB also struggles with long run times in the SCP. This is partly because all forecasts for each regional sales manager and product group is kept as an individual "order" in the planning process. This makes the planning cumbersome and slow, but the use of profit optimisation requires every order to be considered individually and SSAB wants to communicate the constrained delivery plan from the SCP to the forecasters, which also makes it impossible to consolidate the forecasts.

Discussion questions

1 Several sub-processes are integrated in SSAB's tactical planning approach. Discuss the interface to other planning processes of the customer order process.

2 The planning approach at SSAB is heavily dependent on the implemented software support. Discuss the importance of software support when conducting capable-to-promise planning.

CHAPTER

09

Sales and Operations Planning

Planning in a manufacturing company must be carried out both in the short term and the long term. However, planning far ahead in the future is not a goal in itself. Uncertainties in plans made will increase in proportion to the length of the **planning horizon**, and plans will have lower value as they stretch further into the future. Long-term plans may also unnecessarily limit options for various measures and result in less **capacity** to adapt to changing production technologies and conditions on the market. One principally decisive factor for how far in the future plans should be made is the time it takes to adapt a company's capacity to the production and distribution of its products in relation to existing market demand. If advance-planning time is shorter than this reaction time and **demand** is on the increase, opportunities on the market will not be able to be fully exploited. Customers and market shares will be lost. If, on the other hand, demand is decreasing in conditions of short advance-planning time, the company's resources will not be able to be fully exploited. There will be a low utilisation rate in production, and stocks of goods at various stages of refinement will increase.

The length of the planning horizon is related to the required level of detail in planning, as shown in Chapter 3. For this reason companies work at different planning levels. **Sales and operations planning (SOP)** is the most long-term planning level and at the same time the least detailed. At this overall level, issues involve interactions between a number of units at the company. Planning is influenced by available production capacity, capacity to manufacture with respect to supplies of raw materials, purchased components and semi-finished products, the capacity to distribute, the capacity to sell and to receive orders, as well as financial capacity. This chapter looks at the aims of sales and operations planning and the relationship between the limitations and opportunities that exist in different units of a company. The design of systems, procedures and processes for sales and operations planning are also discussed. Finally, examples are given of sales and operations planning in different types of companies. The main emphasis in these descriptions is on companies' production units.

After studying this chapter, you should be able to:

- Explain how to design systems and processes for sales and operations planning
- Describe the steps of a general sales and operations planning process
- Describe and compare sales and operations planning calculations for different company types

9.1 Processes and Relationships

Sales and operations planning can be defined as a process at the top management level of a company that involves working out and establishing overall plans for sales and production operations. The relationship between these and other planning processes can be described as in Figure 9.1. The process is aimed at achieving a balance between supplies and demand to optimise the company's efficiency and competitiveness. To accomplish this, the process also aims to co-ordinate goals and plans in different departments and units, and to co-ordinate all operations in the company that influence or are affected by the flows of materials and the utilisation of resources.

Achieving a reasonable balance between **supply** and demand (Figure 9.2) from month to month is one of the most basic requirements for running successful operations in a manufacturing company. Supplies consist of current production capacity and stocks of raw materials, purchased components, internally manufactured semi-finished items and finished products, while demand, used as the basis for sales and operations planning, consists of **forecasts** of

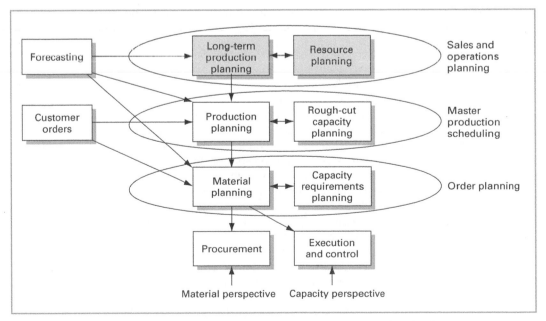

FIGURE 9.1 The relationship between the sales and operations planning process and other planning processes

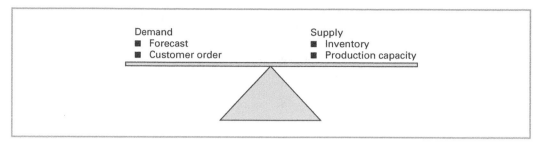

FIGURE 9.2 The balance between supply and demand

expected future sales. In engineer-to-order type of companies considered demand may also consist of customer orders.

If demand exceeds supplies, consequences will include the following:

- Loss of sales, meaning lower revenues and loss of market share
- Poorer delivery capacity to customers, both in the form of a lower service level if the company makes to stock and in the form of longer delivery lead times if the company makes to order
- Higher costs, caused by overtime costs in production and higher freight costs as a result of back orders and split deliveries, or due to express transportation

If the opposite relationship holds true and capacity is higher than market demand, consequences may include the following:

- Increased stocks and consequently unnecessary tied-up capital and higher storage costs
- Higher unit production costs due to low utilisation of machines and labour
- Lower revenues as a result of price cuts or discount offers necessary in order to increase demand to maintain a reasonable level of capacity utilisation

Consequences of the above types affect most departments in the company. Similarly, all departments in the company may constitute a limitation in some respect for what is possible to achieve in terms of balancing supply and demand. For example, financial limitations may mean that increased stocks to maintain capacity utilisation at a high level are not possible. Limitations in marketing and sales resources may likewise cause difficulties in utilising existing production capacity. In a corresponding fashion, limitations to opportunities of procuring more raw materials and purchased components or limitations to the option of expanding production capacity may prevent the manufacture of quantities that can be sold on the market. It is primarily for these reasons that sales and operations planning is an issue for company management. It concerns processes in which key persons in the company's various departments should be involved, not least those responsible for marketing/sales, production, material supply and finance. Sales and operations planning involves drawing up plans for each of these departments and ensuring that the various plans are reconciled with each other.

The starting point for drawing up these plans has its basis in the overall goals set by the company and the strategies formulated for operations. Issues include strategies for delivery capacity, which markets are to be given priority and which products will be phased in and out of different markets. There may also be strategies for storage and capacity expansion. Common overall goals that influence the sales and operations planning process include profitability goals such as profit margins, or growth goals such as percentage increase in sales of certain products on certain markets.

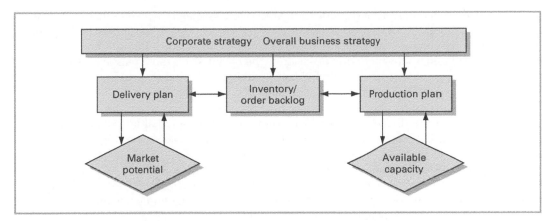

FIGURE 9.3 The relationship between business goals, delivery plans and production plans

Those plans that are primarily a result of a sales and operations planning process are **production plans** and **delivery plans**, sometimes also called **sales plans**. Production plans relate to volumes to be manufactured per time period, while delivery plans refer to volumes to be delivered to customers. In this context, the delivery plan is not the same as delivery schedules from customers as described in the previous chapter. Here it is a question of roughly planned outbound delivery volumes, often at the product group level. There is generally a rather simple relationship between these two plans. For companies that make to stock and deliver from this stock, the relationship during a time period is as follows:

Closing inventory = Production volume + Opening inventory – Volumes delivered

This formula is sometimes called the basic equation of logistics. There is an equivalent formula for the case of make to order. In this formula, outbound stock and inbound stock are replaced by outbound **order backlog** and inbound order backlog.

The delivery plan is related to the company's sales plans and sales budgets, while the production plan is related to production budgets, long-term material planning and capacity planning in the company (Figure 9.3). Delivery plans and material procurement plans together influence the company's cash flow planning.

9.2 Designing Systems and Procedures for Sales and Operations Planning

To achieve an efficient process for sales and operations planning, a number of parameters must first be established. It may involve the objects of delivery plans and production plans, what units they will be expressed in, the length of the planning horizon and so on. Some of these parameters will be treated in this section.

Planning objects

Sales and operations planning refers to the overall and long-term planning of a company's future operations regarding procurement, production, sales and deliveries to customers. The level of detail should consequently be low with respect to the unit of expressed demand in the sales and delivery plan as well as production volumes in the production plan.

When expressing delivery and production volumes, it is generally preferable to work at the product group level, i.e. with groups of products rather than individual products. Exceptions may occur in companies that engineer to order. In such companies, the number of products may be so small and unique from one occasion to the next that it is more reasonable to describe them in terms of product types than individual products, or basic designs of products that are adapted to each customer order. Forecasting and planning is carried out on the basis of these product types as planning objects.

When products are divided into product groups, it is important that all products belonging to a certain group have as similar a demand behaviour as possible, for example the variation of demand between seasons or business cycles, since it is their aggregate demand which will be forecast. Products belonging to a common product group should also be as uniform as possible with respect to the production resources required for their manufacture and the start-up materials used. The reason is that capacity requirements planning and calculations of start-up materials are based on the product group as a whole.

Units of capacity

For the same reasons as for planning objects, the level of detail should also be low for the unit of capacity and **available capacity**. There may be cases when an entire workshop, assembly line or even departments within a factory are used as units for planning. For units of this type, capacity requirements are expressed in terms of machine resources, such as machine-hours, or in the form of human resource requirements such as number of man-hours or number of employees. In companies that engineer to order, engineering and production preparation work is also carried out to customer orders. Accordingly, in these types of companies it may be desirable that sales and operations planning also include operations in the engineering department which will then constitute special capacity units.

In companies with extremely uniform production, capacity may be expressed in production volume, i.e. in the same unit as in the production plan. Capacity units such as euros, number of items, square metres or kilograms may be used in such contexts. To provide some examples, companies that manufacture flooring products often use square metres as units and companies that assemble white goods on assembly lines use pieces as units. If it is not possible to use the same unit for production volume as for capacity requirements and capacity, volumes in production plans must be converted into capacity requirements units such as machine-hours or man-hours as above. Conversion of production volumes to capacity requirements is achieved with the aid of **capacity bills** described in Chapter 14.

Planning horizon

Planning as far into the future as possible is not a goal in itself. The longer the forecasts made, the more uncertain the forecasts become and the less meaningful will be the allocation of resources and time to carry out such forecasts.

Since there is a strong relationship between sales and operations planning and budgeting, it is not appropriate to plan with less than one year's horizon. If demand varies during the seasons of the year, it is also necessary to have a planning horizon of at least one year. How much further in the future plans should extend is largely a question of how far in advance planning is required to be able to adapt production capacity. Such adaptation may mean investing in new manufacturing equipment, employing or laying off production personnel or establishing new subcontractor relations. The length of the planning horizon must also be adapted to the accumulated product lead time, i.e. the time from ordering of raw materials and other start-up materials for manufacturing until the point in time when the product can be delivered. To obtain information as a basis for assessing the availability of delivery-critical materials in sales and operations planning, the planning horizon must clearly exceed this accumulated production lead time. Time required for the development and introduction of new products may also influence the length of the planning horizon.

The majority of all companies use a planning horizon of between one and two years, although there are some cases in which planning horizons extend to more than three years.

Time fences for changes in plans

When a new production plan is drawn up there is always an old plan in use that has governed the sizing of production capacity and the procurement of materials. Typically sales and operations plans are rolling on a monthly or quarterly basis. Since it takes time to make changes, both of access to capacity and procurement of materials, it is difficult in practice to change production volumes at short notice. The next weeks' and sometimes months' volumes can seldom be changed to any large degree. How far into the future the current production plan must be tied to subsequent planning occasions will depend on the length of the lead times for the procurement of materials and manufacturing. It also depends on the flexibility and adaptability of operations.

To be able to handle this type of issue in practical terms, a company may apply simple regulations to the extent to which changes to current production plans may be allowed. Such regulations may state **time fences** for when changes may be carried out and percentages for how much may be changed within these time fences. The following guidelines provide an example of such regulations:

Within the coming month	Only changes to volume sanctioned by the chief executive officer (CEO) are allowed
During month 2	Maximum 20 per cent change in volume allowed
During month 3	Maximum 40 per cent change in volume allowed
From month 4	No limitations on basis of current plan

The size of the time fences and percentage changes allowed are always dependent on the conditions prevailing at each individual company, especially with regard to the products' accumulated lead times and the flexibility with which the production facility has been structured. The competitive situation may also influence how much extra cost a more rapid adaptation may be worth. Examples of time fences and permissible plan changes are shown in Figure 9.4.

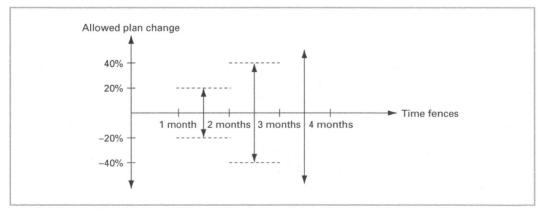

FIGURE 9.4 Examples of time fences for permissible plan changes

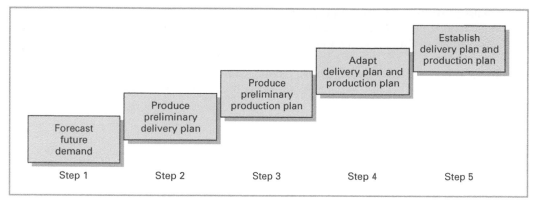

FIGURE 9.5 The planning process in sales and operations planning

9.3 The Planning Process

Sales and operations planning is not a question of running a computer program that will calculate quantities to be produced and quantities to be delivered. Neither is it a question of one individual's decision on how much will be produced, or another individual's decision on how much can be delivered. Sales and operations planning is a process carried out in consecutive steps with a number of key persons from different departments involved. It is also a process in the sense that it is executed repetitively, either monthly or otherwise. The following consecutive steps are normally included in the sales and operations planning process. The different steps are illustrated in Figure 9.5.

Step 1: Forecasting future demand

The first step in the sales and operations planning process is when the marketing department produces a forecast of the coming planning period's expected demand. This type of forecast refers to product groups, is often expressed in monetary values and extends over a relatively long time in the future, normally one to two years. It is crucial for the continued process that the assessments of future sales made at this stage are not influenced by wishful thinking; for example, influenced by the need to sell more because current utilisation of capacity is insufficient, or the need to sell more because stocks are too large. Assessments should not either be affected by production capacity being at full utilisation, thus preventing increases in delivery volumes. Judgements should only be made of sales volumes demanded by the market, based on the best possible evaluations. Demand assessments should include the effects of marketing activities that are planned to be executed and which are expected to have an effect within the planning horizon.

Step 2: Preparing a preliminary delivery plan and setting up goals for inventory levels or customer order backlogs

In step 2 of the sales and operations planning process, the marketing department prepares a preliminary plan for future sales and delivery volumes. As a first stage in this work, the previous sales and delivery plan should be compared with volumes actually delivered. The aim of this comparison is primarily to learn from previous experience, but also to follow up the marketing department's responsibility for its part of the process, i.e. the responsibility to meet the sales and delivery volumes that were established at the previous planning process.

The delivery plan is based on the produced and established forecast of future demand. This may be identical to the sales plan if delivery normally takes place from stock as orders are received. It may also deviate considerably in time when order-specific products are involved, especially if delivery times are long and there is thus a long interval between the time of sale and the time of delivery. To be able to achieve a balance in material flows through sales and operations planning, it is necessary to plan from the marketing department for delivery volumes since production must adapt its activities to delivery dates. As in the case of forecasts, delivery plans are expressed in volumes per period, for example in euros per month.

While the above forecasts refer to volumes demanded in the future by the markets on which the company operates, delivery plans refer to volumes that the company wishes to sell and deliver per period. Volumes in delivery plans are often equal to forecasted volumes, but not necessarily in all cases. One reason for differences occurring between forecasts and delivery plans may be when a company decides to decrease sales of a product being phased out in order to increase sales of a newly introduced product. Another reason may be that forecast volumes are not felt to be acceptably high compared with sales budgets established. With the aid of marketing activities and price offers, a company may plan to sell and deliver more than the volumes originally forecast.

In step 2, goals are also established for inventory size or order backlog, depending on whether the company delivers from stock or to order with delivery lead times. The starting point is the inventory or order backlog that exists at the time of planning. In simple cases, the goal may be limited to the size of inventory at the end of the planning horizon. Inventory levels in the interim periods are then calculated automatically. It is also possible to state period for period how the inventory should develop. This may be particularly suitable when it is necessary to increase the inventory to cope with seasonal variations in demand or to manage demand during a campaign. The goal with regard to order backlogs is to increase or decrease the time involved, i.e. how many days' or weeks' capacity current order backlogs are equivalent to. A company may wish to reduce order backlogs from three to five weeks' production in order to achieve competitive delivery lead times to customers.

Step 3: Preparing a preliminary production plan

In step 3 of the sales and operations planning process, production departments and those departments responsible for the procurement of start-up materials for manufacturing will prepare a preliminary production plan. This production plan relates to volumes to be produced and delivered from production for each period during the planning horizon. For the same reasons as in step 2, step 3 starts by making a comparison between the previous production plan and the actual outcome of production. Deviations are analysed for future knowledge, and are part of the follow-up of responsibility for producing volumes promised according to the previous plan.

The starting point for drawing up the preliminary production plan is the preliminary delivery plan. If the company delivers its products from inventory, the inventory constitutes a supply that can be exploited to cover delivery needs. Consideration must also be given to the size of the inventory desired at the end of the planning horizon. Those volumes that cannot be covered by the current inventory must be manufactured in order that the delivery plan can be fulfilled. It thus represents production needs. The relationship that exists between delivery volume, production volume and inventory changes as defined by the basic equation of logistics means that only two of these three variables can be chosen freely. The third will be a result of the two selected. If, for example, the delivery plan and desired inventory changes are established, the production plan will be determined as a direct consequence of the delivery plan and inventory changes. If instead the delivery plan and production plan are established, this will result in certain changes to inventory.

The same reasoning applies to companies that make and deliver directly to order. Opening and closing inventory then corresponds to opening order backlogs and closing order backlogs, with the difference that order backlogs can be considered as a negative inventory and therefore represent demand, not supply. To be able to decrease order backlogs during the planning horizon and therefore decrease delivery lead times, a company must have a larger capacity than that which corresponds to the delivery plan. If it is desired to increase the order backlog, the reverse relationship is true; capacity needs will be less than those in the delivery plan.

The volumes that must be manufactured after reconciliation with desired inventory changes or changes in the size of the order backlog will constitute the first preliminary production plan, which expresses what volumes must be manufactured per period for the delivery plan to be fulfilled. Before this first preliminary production plan can be accepted it must be reconciled with the available supply of start-up materials that is required to achieve planned production volumes. It must also be reconciled with the available supply of production capacity to manufacture the volumes in question. This work is a part of the **resource planning** at the sales and operations planning level. Methods and approaches for executing resource planning are described in Chapter 14.

If the supply of start-up materials is judged to be sufficient and the capacity that is available is assessed to be large enough for the production volumes required to fulfil the delivery plan, the preliminary production plan can be accepted. If this is not the case, the production and logistics managers must adjust the plan so that it corresponds to what is both possible and desirable from their points of view. Such an adjustment may be in the form of an increase in volume compared with the first preliminary estimated plan, if that would result in an unacceptably low level of utilisation. It may also be in the form of a volume decrease if the preliminary plan would result in an unfeasible production capacity requirement, or require impossibly large amounts of start-up materials from the procurement viewpoint.

Step 4: Adjusting the delivery plan and production plan

The fourth step involves a reconciliation meeting between the managers of the company's marketing, production, procurement and logistics departments. Since delivery plans and production plans have a great influence on the capital tied up in the company and its cash flow, it is also appropriate that representatives from the financial department are involved. The preliminary delivery and production plans drawn up are presented at this meeting, and important consequences and issues are discussed. Any discrepancies that may occur are also taken up.

The aim of the meeting is to make any adjustments to the production plan, delivery plan or goals for changes in inventory or order backlogs in order to achieve a balance between supply and demand in such a way that financial requirements are met. The consequences of any adjustments made are analysed and evaluated. If there are large differences between various preferences expressed at the meeting, more thorough adaptations and impact assessments of the plans drawn up may need to be carried out. It may then be necessary to refer the work back to each department and to resume the reconciliation meeting at a later date.

When consensus has been reached, a proposal is established and a recommendation made for a final delivery plan and production plan for the coming planning periods.

Step 5: Settling the delivery plan and production plan

The proposal for a delivery plan and production plan, including any other preliminary decisions taken as a result of plans drawn up, are put forward to the company's top management group. Any remaining unresolved issues are taken up. When an agreement has been reached, the members of the management group meeting will settle the delivery plan and the production plan, and these plans will constitute a target for continued operations at the company. Expressed in straightforward terms, we can say that this decision at the management level signifies an undertaking from the marketing department to sell quantities such that the delivery plan can be realised. The decision also involves an undertaking from the production and material procurement departments to achieve production volumes that are large enough to fulfil the production plan.

How often such a planning process should be carried out varies from case to case, depending on the type of business, current delivery lead times, how rapidly the market changes and the frequency of product renewal. It also depends on how often it is necessary to check off operations with their budgets and to make new budget forecasts. In most cases, sales and operations planning processes are carried out quarterly or monthly.

9.4 Sales and Operations Planning Calculations

In most ERP systems there is support for drawing up and working with delivery plans and production plans. Simpler tools such as tables or spreadsheets can often be used too. To be effective, such tools should contribute to providing an overview of current plans so that a total solution can be produced. They should also support any calculations that are necessary, primarily those related to the above-mentioned fundamental equation of logistics, which establishes whether there is a balance between supply and demand. There should also be calculation tools for converting production plans into capacity requirements for long-term **capacity planning**.

Two examples are shown in this section of sales and operations planning. The first example illustrates a case of making to stock, and the second case is of making or assembling to order.

Planning in make-to-stock companies

In companies that make to stock and deliver to customers from stock, inventory changes are the differences between volumes in the delivery plans and production plans, in accordance with the fundamental equation of logistics. Since delivery is normally intended to take place directly from stock, delivery time is negligible – which means that planned sales volumes for each period are equivalent to planned delivery volumes for the same periods. One example each for producing a delivery plan, a production plan and a plan for the resulting inventory changes for a certain product group is shown in Figure 9.6.

The left-hand side in Figure 9.6 shows results for past periods with respect to real delivered volumes, real produced volumes and changes in inventory. The right-hand side shows plans for the following eight periods. The plans may be interpreted in such a way that the expected rise in demand as expressed by the increased volumes in the delivery plan cannot be completely covered by new production. It will take a number of periods before production capacity can be adapted to the higher delivery volumes. During this time, demand is satisfied through decreasing inventory. The plans in Figure 9.6 may also be interpreted in such a way that the company plans to decrease capital tied up in inventory, and for this reason chooses not to produce at the same rate as deliveries take place.

	−3	−2	−1	−	1	2	3	4	5	6	7	8
Delivery plan												
Plan	70	70	70		70	70	90	90	90	90	90	90
Actual	54	75	87									
Deviation	−16	5	17									
Production plan												
Plan	80	80	80		80	80	80	80	80	90	90	90
Actual	59	72	78									
Deviation	−21	−8	−2									
Inventory												
Plan	55	65	75		48	58	48	38	28	28	28	28
Actual 45	50	47	38									
Deviation	−5	−18	−37									

FIGURE 9.6 Examples of a delivery plan and a production plan when making to stock

Planning in make-to-order companies

In the case of a make-to-order company, whether it is assemble to order, make to order or engineer to order, it is changes in the size of the order backlog and thus delivery lead times to customers that constitute the difference between volumes in delivery plans for each period and production plans. Since deliveries are made with delivery lead times, in this case there is a delay between sales plans and delivery plans that must be taken into consideration during planning. Figure 9.7 shows an example for drawing up a delivery plan, a production plan and a plan for the resulting changes in order backlog for a certain product group.

In the same way as for Figure 9.6, the left-hand side of Figure 9.7 shows the outcome for past periods with respect to real delivered volumes, real produced volumes and changes in the order backlog. The right-hand side shows plans for the following eight periods. The plans in the figure may be interpreted in such a way that a decrease in demand is expected, as expressed by the reduced volumes in the delivery plan. Production volumes are cut back as a result of this. In order to reduce the order backlog and thus delivery lead times to customers, production volumes are not decreased as much as delivery volumes are expected to decline. The plans in Figure 9.7 may also be interpreted in such a way that it is not financially justifiable to decrease production volumes at the same rate as outbound deliveries decline, and that a reduction in the order backlog must be accepted.

	−3	−2	−1	–	1	2	3	4	5	6	7	8
Delivery plan												
Plan	60	60	60		60	60	60	50	50	50	50	50
Actual	64	77	54									
Deviation	4	17	−6									
Production plan												
Plan	65	65	65		60	60	60	55	55	55	60	55
Actual	71	69	62									
Deviation	6	4	−3									
Order stock												
Plan	92	87	82		90	90	90	85	80	75	65	60
Actual 97	90	98	90									
Deviation	−2	11	8									

FIGURE 9.7 Example of a delivery plan and a production plan in the case of make to order

9.5 Summary

Sales and operations planning may be defined as a process at the management level of a company that involves preparing and settling overall plans for sales and production operations. The process is aimed at achieving a balance between supply and demand in order to promote the company's efficiency and competitiveness. This chapter has described sales and operations planning and the relationship between the limitations and opportunities that exist in different departments in a company. The principles of the system and processes for sales and operations planning, such as consideration of the selection of planning objects, planning units, capacity units, planning horizons and time fences for changes in plans have been discussed.

In addition, sales and operations planning processes were described in five steps: forecasting future demand, preparing preliminary delivery plans, preparing preliminary production plans, adjusting delivery plans and production plans, and settling delivery plans and production plans. The principles involved in sales and operations planning are generally similar, irrespective of the type of company. However, there are certain differences. At the end of the chapter delivery plans and production plans for make-to-stock companies and make-to-order companies respectively were illustrated.

 Key concepts

Available capacity 164	Planning horizon 160
Capacity 160	Production plan 163
Capacity bill 164	Resource planning 168
Capacity planning 169	Sales and operations planning (SOP) 160
Delivery plan 163	Sales plan 163
Demand 160	Supply 161
Forecast 161	Time fence 165
Order backlog 163	

Discussion tasks

1 Sales and operations planning is related to both budgeting and master production scheduling. What are the differences between sales and operations planning and budgeting? What are the differences between sales and operations planning and master production scheduling?

2 The sales and operations planning process is sometimes considered the most important "supply chain process" of a company. Discuss why that is so.

3 The sales and operations planning process can have several objectives. Outline and discuss the different objectives.

 ## Problems[1]

Problem 9.1

Olsson and Sons Inc. is working on the sales and operations plans for the next year. The products are made to order, because they are to some extent customer specific. The demand has a seasonal variation but the company has still decided to have a levelled production plan, with a constant number of operators and capacity need throughout the year, which equals the average demand.

The quarterly forecasts (= estimated sales) made at the marketing department are 9000, 12 000, 7000 and 9000 units, respectively, for quarters 1, 2, 3 and 4. At the beginning of the year, the order backlog was 2200 units and the goal was to have an order backlog of 600 at the end of the year. The numbers of working days for the four quarters are 56, 62, 42 and 58, respectively; the production capacity per operator is 4 units per day.

a) What production capacity (number of units) per day is necessary to fulfil the sales estimates and the order backlog goal for the end of the year? How many operators are needed?

---

1   Solutions to problems are in Appendix A.

b) How large is the order backlog and the delivery time at the end of quarter 2 if the forecasted demand is correct and the customers accept the actual delivery times?

Problem 9.2

Velab AB makes products to stock. The manufacturing capacity is 10 units per operator and day, with 20 working days per month during 12 months a year. In order to minimise costly capacity changes the management has only accepted capacity changes during half-year shifts, i.e. between June–July and December–January, and works with levelled production plans with fixed production quantities during the rest of the year. The capacity strategy is also to not allow for planned over-capacity in order to temporarily increase the capacity. The finished goods inventory at the beginning of the year contained 1000 units. To decrease the tied-up capital the company plans to lower the inventory to 500 units at the end of the year. The marketing manager has delivered the following sales forecasts for the year: 5000 for quarter 1, 5000 for quarter 2, 6000 for quarter 3 and 7000 for quarter 4.

a) Develop a production plan (in units per day) that matches the agreed capacity strategy.

b) How many operators are needed?

c) What would the consequences be if the actual demand during quarter 1 was 1000 larger than the forecast?

Problem 9.3

Robinson Products Inc. is about to conduct sales and operations planning for next year. The demand per month (in € million) for the next year was forecasted according to the table:

Jan	Feb	Mar	Apr	May	Jun	Jul	Aug	Sep	Oct	Nov	Dec
8	10	7	11	12	7	8	10	9	11	12	15

The aggregated sales volume is consequently estimated to be €120 million, calculated as cost of goods sold. No products were available for delivery from stock at the beginning of the year and no products were expected to be kept in stock at the end of the year. In order to not lose customers, the company has decided to make and deliver every customer order received. At the end of December this year, the production capacity was equivalent to delivery volume of €11 million per month.

The costs associated with the production plan are mainly related to the inventory carrying costs and the costs for changing capacity between months. The inventory holding costs are estimated to 25 per cent of the inventory value and the costs for changing capacity are estimated at €100 000 per changing occasion.

a) Develop two alternative production plans, one based on a levelled strategy (i.e. the same production rate per period) and one based on a chase strategy (i.e. the same production and demand rates per period).

b) Calculate the average tied-up capital in inventory for the respective strategy. Base the calculations on the inventory balances at the end of the respective month.

c) How large is the sum of inventory carrying costs and capacity change costs for the respective strategy?

d) Develop an own suggestion of a production plan based on a hybrid strategy. Calculate the sum of the inventory carrying cost and the capacity change cost for this production plan.

Further reading

De Kok, T., Janssen, F., Doremalen, L., Wachem, E., Clerkx, M. and Peeters, W. (2005) "Philips Electronics synchronizes its supply chain to end the bullwhip effect", *Interfaces,* Vol. 35, No. 1, pp. 37–48.

Fogarty, D., Blackstone, J. and Hoffmann, T. (1991) *Production and inventory management.* South-Western Publishing, Cincinnati (OH).

Grimson, A. and Pyke, D. (2007) "Sales and operations planning: an exploratory study and framework", *International Journal of Logistics Management,* Vol. 18, No. 3, pp. 322–346.

Jonsson, P., Kjellsdotter, L. and Rudberg, M. (2007) "Applying advanced planning systems for supply chain planning: three case studies", *International Journal of Physical Distribution and Logistics Management,* Vol. 37, No. 10, pp. 816–834.

Lapide, L. (2004) "Sales and operations planning part I: the process", *The Journal of Business Forecasting,* Vol. 23, No. 3, pp. 17–19.

Lapide, L. (2004) "Sales and operations planning part II: enabling technology", *The Journal of Business Forecasting,* Vol. 23, No. 4, pp. 18–20.

Lapide, L. (2005) "Sales and operations planning part III: a diagnostic model", *The Journal of Business Forecasting,* Vol. 24, No. 1, pp. 13–16.

Ling, R. and Goddard, W. (1988) *Orchestrating success.* John Wiley & Sons, New York (NY).

Mentzer, J. and Moon, M. (2004) "Understanding demand", *Supply Chain Management Review.* May/June, pp. 38–45.

Proud, J. (1999) *Master scheduling.* John Wiley & Sons, New York (NY).

Sheldon, D. (2006) *World class sales & operations planning: a guide to successful implementation and robust execution.* J. Ross Publishing, Ft Lauderdale (FL).

Wallace, T. (2004) *Sales and operations planning: the how-to handbook.* T.F. Wallace and Co, Cincinnati (OH).

CASE STUDY 9.1: SOP/MPS AT AKZO NOBEL SURFACTANTS EUROPE[2]

Akzo Nobel Surfactants makes specialty chemicals. It is divided into three regional organizations – America, Asia and Europe – of which the latter is the focus here. Europe contains three production plants, makes 1050 products and has an annual production volume of 110 000 tons. Many of the products are produced in more than one process step, often involving more than one plant, which means that there is a large flow of intermediate products between the plants. Every month, Europe buys a certain amount of products produced at plants in America and Asia. Europe also undergoes contract manufacturing from 15 production plants. In case of

---

2 Based on Kjellsdotter, L. and Jonsson, P. (2008) "Prerequisites for using APS in S&OP and MPS processes", *Proceedings of the EurOMA 2008 Conference,* Groningen, The Netherlands.

capacity shortage, it is to some extent possible to use idle capacity at other plants or capacity could be bought from contract manufacturers. Europe has about 1400 external customers, who are divided into different priorities with respect to annual sales value, growth potential and strategic value, and between 70 and 80 suppliers located in different countries. Most products are made to stock in order to have acceptable delivery times to customers. There are capacity constraints in production resources, storage, packaging and some raw materials.

Planning process, organisation and software support

In order to co-ordinate the material flows between the production plants, Akzo Nobel Surfactants Europe has established a centralised planning organisation and a centralised process that correspond to a combination of sales and operations planning (SOP) and master production scheduling (MPS). The software tools used in the planning process contain advanced planning and scheduling modules, for example Zemeter Supply Planner, an ERP system and Excel. Zemeter's supply planner calculates an optimised production plan on product level using a linear programming (LP) model for the three plants. The target function of the LP model is to fulfil every demand and target inventory at highest possible contribution margin. Production volume is measured in tons and in the planned production month at the actual plant. The principle, "one preferred site" is used in the model, meaning that it is predetermined in which plant the products will be produced. The reason is that only about 10 per cent of the products are transferable between the production plants, and that customers, for quality reasons, require that products are produced in particular plants. Contract manufacturers are not included in the LP model, nor are the other regional organisations. This is because it is difficult to receive the information needed from contract manufacturers and the dependencies between the regional organisations are not considered high enough in relation to the expected time and effort needed to make the planning model work.

The actors involved in the planning process are the centralised planning organisation (consisting of a supply chain manager and two supply planners), business managers, sales managers, customer service representatives, production managers, site schedulers and contract manufacturing representatives. The centralised planning organisation is formally responsible for setting up and running the Zemeter supply planner model and it is supposed to control the entire process.

The common SOP and MPS process

The conducted planning process consists of a demand planning process, a production planning process and a SOP meeting (see figure).

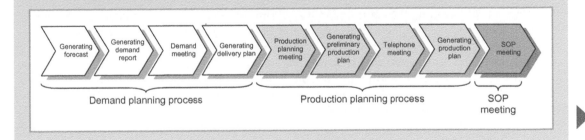

The aim of the *demand planning process* is to generate a delivery plan. It starts with generating the forecast. The sales manager registers sales forecast figures per customer/product/ton/month. Every time the system rolls forward one month, the statistical forecast on customer/product combination is updated and the forecast for the new month, 18 months ahead, is made visible. The 18-months forecasts are remained until changed by the sales manager. The statistical forecast is based on three years' sales history and the demand planning software uses different quantitative forecast methods to produce the statistical forecast. At the beginning of each month, the sales manager needs to update the three coming months of forecast. When this is done, the closest month's forecast is frozen and it is no longer possible for the sales manager to change it.

The supply chain manager, who is part of the central planning organisation, is now taking over the demand planning process. She puts together a demand report from the demand planning software and distributes it to all business managers. The demand report consists of forecast figures and information about existing forecast accuracy for the coming three months.

The business managers should review the demand report so that they are well prepared for the monthly demand meeting. Here, the central planning organisation meets each of the four business managers at different occasions to discuss every individual market segment (a business area consists of four to six market segments). The forecast accuracy is discussed as well as if it is possible to increase sales, if a specific customer is profitable or not, etc. During this meeting a delivery plan, expressing the expected monthly future sales per product, is established.

After the meeting, the central planning organisation makes possible updates of the delivery plan and the established delivery plan is then automatically sent to the supply planner module and the production planning process begins.

The aim of the *production planning process* is to generate a production plan that meets the established delivery plan. The production planning process starts with the central planning organisation, exporting the delivery plan from the sales module software, and finalising specific Excel-based delivery plans with the information of interest to each of the three production plants and contract manufacturers, respectively. This plan is distributed to the production managers at the respective plant and to the contract manufacturers.

Then, Zemeter's supply planner module is updated with master data (bill of material, intermediates, production times, transportation time, costs, etc.), the delivery plan, actual inventory balances and capacity figures. The LP model in the supply planner module is run to create a preliminary production plan. In general, the central planning organisation runs a series of scenarios to analyse how the optimum plan changes if the actual demand is more or less than the forecast. The identified capacity constrains and the over-capacity is analysed.

The output of these scenario analyses is discussed during the monthly production planning meeting. At the meeting, the central planning organisation, the operations manager, the purchasing manager, the production managers and the contract manufacturing representatives participate and discuss changes, capacity utilisation and bottlenecks, and if it is possible to meet the expected sales expressed in the delivery plans.

After the production planning meeting the centralised planning organisation updates the supply planner module with accurate data and the LP model is run to create a final production plan. The production plan is made on product level and specifies the volumes per product to produce during the next 18 months. The first month's plan is more detailed and specifies what to produce at each reactor. Before finalising and releasing the production plan, the centralised planning organisation also has a telephone meeting with Surfactants America and Surfactants Asia to discuss common material flows. Some minor corrections to the preliminary plan could be conducted after these meetings. The final production plan is then distributed to the production plants, where it is used as input to the detailed production planning activities.

Every second month the CEO, marketing managers, purchasing manager and the central planning organisation meet in a *SOP meeting*. At this meeting a report of the present production and delivery situation is given by the participators and the focus is to identify risks and solve problems that have not been solved before. The support for this meeting is the delivery plan and the production plan; however, the focus is on fill rate at each production plant, and common questions concern capacity shortage, should Surfactants Europe buy capacity or say no to customer requests, etc.

Discussion questions

1 The planning process at Akzo Nobel is described as a combination of SOP and MPS. In what respects is it a SOP process and in what respects is it an MPS process?

2 What would you consider the main objectives of the process to be for Akzo Nobel?

3 What is the role of the centralised planning organisation in Akzo Nobel's process?

Master Production Scheduling

The aim of sales and operations planning as described in the previous chapter is to achieve a balance at an overall level between demand and supply of the company's products and in such a way that will support the company's primary business goals. In more concrete terms this means planning to achieve consistency between planned deliveries of the company's products and the **capacity** available for producing the volumes stated in the **delivery plan**. **Sales and operations planning (SOP)** is generally carried out for whole product groups and is more oriented towards volumes to be delivered and produced than towards the mix of different products within each product group.

A greater degree of planning detail is required to operatively control material flows into, through and out of the company. The plans must then also include individual products and access to capacity must be ensured for smaller manufacturing units than entire workshops or production departments. The process required to achieve this is called master production scheduling and involves a greater level of detail in comparison with planning for sales and operations, and at the same time it has a shorter planning horizon. Since the level of detail is higher, differences in how the planning process can be carried out will be larger between companies that make to stock and those that make to order. The overall relationships between master production scheduling and other planning processes are described in Figure 10.1.

There are large general similarities between sales and operations planning and master production scheduling. This means that it is not always relevant to plan separately at both of these planning levels. In many companies the two planning levels are merged together and are then most often called master planning. For example, if there is no need to plan at the product group level because the number of products is small and a suitable planning horizon is not longer than one year, both planning levels may be carried out together to some advantage. It may also be appropriate to merge the two planning levels if a company has long delivery lead times to customers or if incoming orders are relatively stable over time. Under these conditions, the planning process may be carried out relatively infrequently, such as once a month. In this case it is still possible to involve company management in the planning process.

This chapter describes aspects of what master production scheduling involves, what functions it fulfils and what relationships it has with sales and operations planning. Different parameters are discussed in the context of designing master production scheduling systems. The structure

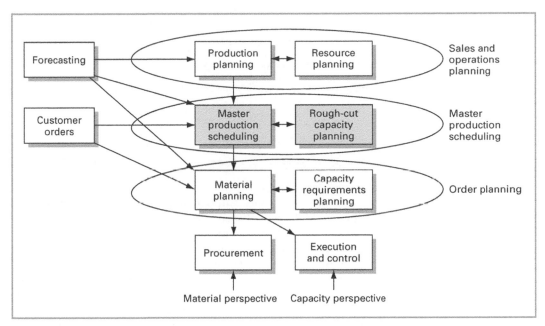

FIGURE 10.1 Relationships between master production scheduling and other planning processes

of the master production scheduling process is explained for companies that make to stock, companies that assemble products to order and those that engineer and make to order. Finally, **multi-site master production scheduling** and scheduling final assembly operations are discussed.

LEARNING OBJECTIVES

After studying this chapter, you should be able to:

- Explain how to design systems and procedures for master production scheduling
- Describe the steps of a general master production scheduling process
- Describe and compare the master production scheduling process in different types of companies
- Explain the similarities and differences between the **master production schedule (MPS)** and the **final assembly schedule (FAS)**

10.1 Functions and Relationships

Master production scheduling can be defined as a process that involves developing and establishing plans for a company's sales and production operations. From the point of view of logistics, these are delivery plans and master production schedules showing what quantities of the company's products are planned to be manufactured and delivered in each period. An example of a master production schedule for three different products over a period of six weeks

Week	1	2	3	4	5	6
AX – 102 456	200	200	200	200	240	240
AX – 345 679	120	120	120	120	120	120
BD – 365 911	40	40	40	40	40	40

FIGURE 10.2 An example of a master production schedule for three different products over a period of six weeks

is shown in Figure 10.2. In the same way as in sales and operations planning, master production scheduling is aimed at achieving a balance between supplies and demand in such a way that a company's efficiency and competitiveness are promoted. Supplies are made up of current production capacity and inventories of raw materials, purchased components, internally manufactured semi-finished items and finished products. Likewise, the demand that is the basis of master production scheduling consists of orders on hand and/or **forecasts** of expected future sales, in the same way as with sales and operations planning. The principal difference is the length and degree of detail in the plans made. Another important difference is that the master production scheduling process is carried out more frequently than the sales and operations planning process, often more than once a month and in some make-to-order and engineer-to-order companies every time a new customer order arrives.

Master production scheduling is somewhat less of a process and more of an automated calculation procedure, especially for companies that manufacture standardised products to stock or assemble variants of products from standardised components to order. In these contexts it has more the character of material planning of end products. However, in principle the same steps are carried out as in sales and operations planning. Fewer people from fewer departments in the company are involved and master production scheduling does not require the same involvement of top management personnel if there is a sales and operations planning level.

The starting point for drawing up delivery plans and master production schedules in the master production scheduling process are those plans that have been established in sales and operations planning. These plans make up a framework within whose limits master production scheduling must remain. This means in principle that the sum of volumes in master production schedules for different products belonging to a certain product group must be equal to the volume in the overall master production schedule established for the product group at the level of sales and operations planning, as illustrated in Figure 3.5. The same conditions apply to delivery plans in master production scheduling. For practical reasons it is appropriate to establish policy regulations for how much deviation is allowed before changes must be taken up and decided at management level. A policy rule may be to allow deviations of up to 10 per cent per month from volumes established in sales and operations planning.

The same relationship exists between delivery plans and master production schedules as at the level of sales and operations planning, and the fundamental equation of logistics applies here too, as shown in Figure 10.3. The options regarding modification of the master production schedule are also limited by **available capacity**. At this level the limitations are even greater, since the planning horizon is shorter and measures such as overtime and reallocation of personnel within the workshop are among the few possibilities to make adjustments to capacity. For the type of **capacity planning** in question, the same methods and tools may be used as in sales and operations planning, such as **rough-cut capacity planning** with the aid of **capacity**

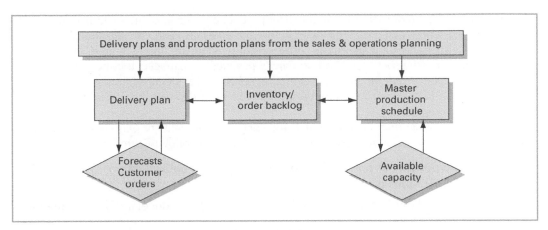

FIGURE 10.3 Important relationships in master production scheduling

bills. It is also possible to use **capacity requirements planning (CRP)** in companies that make to stock and in companies that assemble to order. These methods and tools for capacity planning are discussed in more detail in Chapter 14.

Compared with the level of sales and operations planning, delivery plans at the master production scheduling level must be more dynamic since in many cases inflowing orders must be continuously considered, and in some cases cancellation or **rescheduling** of orders occurs. Orders received must also be deducted from the current forecast so that quantities are not counted twice. This is called **forecast consumption** and is further analysed in Sections 10.3 and 10.4.

The master production scheduling process consists in principle of the same steps as the process of sales and operations planning, i.e. as follows:

1 Forecast future demand.

2 Generate a preliminary delivery plan on the basis of forecasts and orders on hand.

3 Generate a preliminary master production schedule on the basis of the preliminary delivery plan and current and targeted inventory levels or sizes of **order backlogs**.

4 Reconcile plans drawn up and the conditions for realising them, for example with respect to access to materials and capacity. Adapt plans as necessary.

5 Settle the prepared plans.

With the exception that the forecasting object is somewhat different in different types of companies, step 1 in the master production scheduling process is basically the same in all types of companies. Forecasts produced as a basis for making delivery plans must cover the same future time periods as the planning horizon for the master production schedule. For this reason it is necessary to take into consideration any trends, seasonal variations, business cycle impact and so on when forecasts are made. The effects of planned marketing measures of various types must also be considered. Steps 2 and 3 are however somewhat different and are further described in Sections 10.3 to 10.5.

In step 4, reconciliation is made with the current overall delivery plan and **production plan** as established by sales and operations planning. In the case of any deviations greater than the tolerance limits as determined by policy, the quantities are adjusted. Tolerance limits of +/– 20 per cent may for instance be suitable. Adjustments are made if there are any remaining discrepancies between delivery plan and master production schedule. An analysis is also made of other consequences and measures taken before the plans are ready to be established.

Both the delivery plan and the master production schedule are finally established. This may take place at a meeting between the company's production manager, logistics manager and marketing manager. When the master production schedule has been established, it is used as input for material planning and thus for controlling and co-ordinating the flow of materials and manufacturing operations.

The master production schedule, which together with the delivery plan is the result of an executed master production scheduling process, includes the following functions:

- It is an instrument for putting into practice the plans and intentions determined by sales and operations planning.

- It is an undertaking from the production department to produce and deliver quantities planned.

- It is a link between the marketing department and the production department.

- It is intended to co-ordinate and control the entire flow of materials into, through and out of the company.

- It must be able to operate as a tool for facilitating sales and order entry to answer questions regarding the timing and quantities of deliveries.

10.2 Designing Systems and Procedures for Master Production Scheduling

The approach to master production scheduling is influenced by a number of factors. When the planning procedures and the master production scheduling system are designed, a number of parameters must be selected and established for the system to function well. As in the case of sales and operations planning, issues involved may be what the delivery plans and master production schedules refer to, how long the **planning horizon** will be, the frequency with which planning should take place, etc. Some of these important parameters are taken up in this section.

Planning objects

The selection of planning objects, i.e. what the delivery plan and the master production schedule will refer to, depends to a certain extent on what type of company in question. If the company makes to stock, then the product will be the most natural planning object. This will apply to both delivery plans and master production schedules, even if a product group is used as an object of forecasting. If product groups are used in forecasting, the forecast is exploded with the aid of a forecast bill of material for each individual product. The bill of material quantities are set equal to each product's sales in relation to the total sales of the group over a number of months. The product is also the natural planning object in companies that make to delivery schedules from customers.

In companies that assemble to order, products are specified when customer orders are received as described in Chapter 8. This means that from the customer order point of view, the planning object may be a product. However, the forecast cannot refer to a specific product variant since this is not known. Since the delivery plan is the sum of quantities from orders and forecasts in this type of company, the delivery plan cannot use a product as a planning object.

In order to solve this dilemma, forecasts and delivery plans are made at the level of models. When deducting forecast quantities as orders arrive, the equivalent model in the forecast is adjusted downwards by the relevant order quantity.

A similar procedure can be applied in companies that make to order and engineer to order. In these cases forecasts are made of **product types**. As orders are received for certain variants of this product type in a certain quantity, the quantity is deducted from the forecast of the equivalent product type.

The planning unit

Delivery plans and master production schedules at the level of master production scheduling have a more direct link to material flows than plans at the sales and operations planning level. It is therefore generally appropriate in companies that make to stock and make to delivery schedules from customers to use stocked units as the planning unit. In companies that assemble to order, make to order and engineer to order, there are no stocks of end products. In these cases, sales units make a suitable planning unit. If the products are sold as numbers of items, delivery plans and master production schedules will also refer to numbers of a **product model** or product type.

What do the plans express?

The delivery plan is often a combination of a number of orders and summarised unspecified quantities that have been forecast. In companies that make to stock, the delivery plan can be totally based on forecast quantities per time period, especially if delivery generally takes place directly from stock. The opposite situation may prevail in companies that make to order. In these companies, the delivery plan may consist entirely of orders in hand. This is particularly the case when products are so different from one order to another that forecasting in the normal sense is not possible, and in cases when the delivery time to customers is so long that the order backlog alone can give a sufficiently long planning horizon.

The quantities stated in the master production schedule in Figure 10.2 could refer to a quantity which must be available for delivery in the current week, or a manufacturing order quantity for delivery in the current week. In companies that make to stock and make to delivery schedules from customers, it is most reasonable to have the master production schedule as a number of manufacturing orders. In this environment, delivery plan quantities must be reconciled with the current stock-on-hand balance and manufacturing is generally in the form of batches, i.e. one manufacturing order includes more than one customer order requirement. Calculations of capacity requirements are based on released and planned manufacturing orders.

For companies that assemble to order, make to order and engineer to order, the master production schedule will be a combination of manufacturing orders and quantities for delivery in the current week. These environments are characterised by manufacturing orders that directly correspond to customer orders. In the short term, or as long as the order backlog extends into the future, the master production schedule consists of manufacturing orders, generally of quantities that can be delivered. In companies that make to order and engineer to order, it is possible to have the master production schedule consisting only of manufacturing orders if the backlog of orders is long enough. If the level of detail provided by sales and operations planning is considered to be sufficient beyond the order backlog, this approach may also be applied to cases with relatively short order backlogs. If the master production schedule consists of quantities to be delivered, capacity planning for final assembly is based on capacity bills, since capacity requirement planning is based on manufacturing orders. Refer also to Chapter 15.

Order types

Master production scheduling, in contrast to sales and operations planning, is characterised by more automatic calculations and other types of system support. In order to verify its partly automated procedures it is necessary to distinguish between different types of order to be able to treat them in different ways when planning. This is particularly relevant for companies that make to stock and to delivery schedules from customers, where the master production schedule represents manufacturing orders that have been generated on the basis of stock balance and lot sizing. How these manufacturing orders are generated is described in Section 10.3. In this type of planning environment, the order types **planned orders**, **firm planned orders**, **released orders** and orders released to the shop floor are common.

The manufacturing orders that are automatically generated by the ERP system are called planned orders. These orders are created automatically and are rescheduled automatically both with respect to order quantities and delivery dates. In order to limit options in the system for rescheduling, such as to avoid so-called system nervousness, but to still allow much of the planning work to be carried out by the ERP system, it is possible to use firm planned orders. A firm planned order is characterised by certain limitations in how and when the ERP system is permitted to reschedule it. One common limitation is to not allow changes in quantities for planned manufacturing orders within a certain planning horizon, since such rescheduling could have considerable repercussions in the form of bringing forward or postponing already outstanding manufacturing orders of incorporated items in subsequent material requirements planning. Firm planned orders are also used to manually adjust automatically generated manufacturing orders and their processing by the system, for example when it is desired to select an order quantity that deviates from what is normally intended to be used in accordance with the lot sizing rules of the ERP system.

Planned orders and firm planned orders are transferred to released orders at a certain advance time interval. Releasing means that the components included in the product to be manufactured are allocated for the manufacturing order. Material requirements are created for planned and firm planned orders at the lower level through requirements explosion. Order release can take place automatically with a manually established advance time margin, or completely manually. After release, manufacturing orders are checked manually and any necessary rescheduling must take place manually. The ERP system only contributes by creating action messages to the planning manager with proposals for rescheduling or other types of measure. Such measures may include bringing forward an order, postponing an order or cancelling an order. The characteristics of these three types of manufacturing orders are summarised in Figure 10.4.

Released orders are checked by the planning department in the company. They are released to the workshop in advance for picking of start-up materials and preparation for manufacture. They are then called released orders to the shop floor and are mainly checked by workshop personnel.

Damping functions

If a company chooses to allow an ERP system to automatically identify the need to reschedule and to send this information via action messages to the planning manager, there is always a risk that the system will become oversensitive. It generates more information than is possible to handle in practice. **Damping functions** can be introduced to avoid this situation and to limit the number of action messages produced, meaning that action messages are only sent under certain conditions. For example, it may be undesirable to reschedule manufacturing orders that

	Released order	Firm planned order	Planned order
Deviation message	Yes	Yes	No
Automatic replanning	No	Only partially	Yes
Ready to release to shop floor	Yes	No	No
Input material reserved	Yes	No	No

FIGURE 10.4 Characteristics of different manufacturing order types

are estimated to be delayed by less than two days, or to postpone manufacturing orders that have already been started and that have low order values and thus limited influence on capital tied up in inventory.

Planning horizon

As regards the length of the planning horizon, the same reasoning applies generally as to sales and operations planning. One of the main purposes of master production scheduling is to co-ordinate and control material flows into, through and out of the company. To achieve this, the planning horizon of the master production schedule must be longer than the accumulated lead time for the product with the longest accumulated lead time. This requirement is illustrated in Figure 10.5. The figure shows a product P with its product structure drawn in such a way that the length of the lines for the different component items expresses the lead times for manu-facturing and procurement. The critical line, i.e. the sequence of orders from procurement to

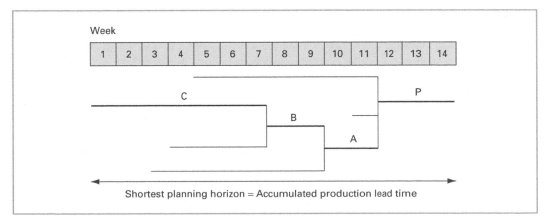

FIGURE 10.5 The planning horizon and accumulated product lead time

deliverable product which has the longest accumulated lead time, is marked in the form of thick lines. As shown in the figure, the longest **accumulated production lead time** in this case is 14 weeks. To be able to calculate material requirements of purchase item C, this means that the master production schedule must be at least 14 weeks long. The reason is that it takes 14 weeks from the placing of a purchase order for item C, the manufacturing of items B and A, and the assembly of the end product P.

It is often an advantage when purchasing to have not only an idea of when the first need for C occurs but also an idea of the material requirements for a number of weeks ahead. In other words, a planning window is desirable. In order to achieve this, the master production schedule must be made longer than 14 weeks. If a planning window of 10 weeks is desired, i.e. a possibility of estimating material requirements for 10 weeks ahead when a new purchase order is to be released, the length of the master production schedule must be 24 weeks.

Master production scheduling also covers capacity planning. This function can also put demands on the length of the master production schedule. The longer the advance time required to make adjustments of available capacity that are both possible and acceptable within the framework of the master production scheduling level, the further in the future the master production schedule must extend, and thus the delivery plan. The length of the planning horizon for master production scheduling is normally around six months to one year.

Period length

For master production schedules that are wholly or partly made up of quantities that must be available for delivery, one week is often a suitable period length. This length of time is usually long enough to give reasonably large quantities to work with and to match a forecast-based delivery plan. It is not too long with respect to desirable delivery precision. If the delivery plan is based on manufacturing orders, their delivery dates should be expressed per day to give an acceptable delivery precision, and the master production schedule will then also have one day as a period length.

The time precision with which quantities to be delivered are represented in the ERP system does not necessarily need to be the same as the length of the period used when master production schedules are presented to the user. Even if all quantities and manufacturing orders are stated per day, the master production schedule may be presented as quantities per week by adding together planned quantities per day and manufacturing orders to be delivered within the same week. It will then be easier to have an overview of the whole planning horizon. It is often appropriate to use a shorter period in the near future and a longer period in the more distant future. For example, a period length of one day could be used during the first month of the schedule, a period length of one week in months two to six, and a period length of one month in months seven to twelve in a master production schedule with a twelve-month planning horizon.

Time fences

To gain control over the automatic generation of new master production schedules and how and when rescheduling is supposed to be executed, it is possible to use a number of **time fences** and a system of regulations for how the system behaves within these time fences – in a similar way to that used for sales and operations planning.

Commonly used time fences for master production scheduling are illustrated in Figure 10.6. The **demand time fence** and the **forecast time fence** are used to balance order backlog and forecast when calculating delivery plans. The demand time fence is equivalent to the time horizon within which new customer orders are not normally received, i.e. the normally applied delivery

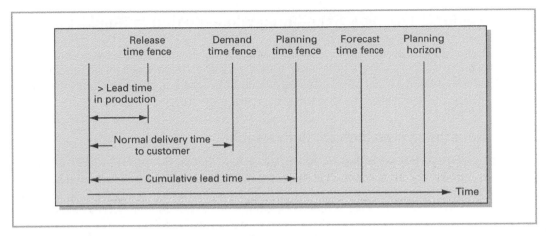

FIGURE 10.6 Time fences in master production scheduling

lead time to customers. The forecast time fence is made equivalent to the point in time when the order backlog is normally so small as to be negligible compared with the forecast. The method used to balance the forecast and orders is accounted for in more detail in Section 10.4.

The other two time fences are used to control the processing of manufacturing orders. The **release time fence** will determine how long before start-up the manufacturing order will be released. The manufacturing order with a delivery lead time this side of the release time fence is automatically released by the ERP system. Alternatively, an action message is sent to the planning manager to manually release the order. The time up to this release time fence must be at least as long as the lead time for manufacturing a product, otherwise the order will not be ready in time. The further ahead time fences lie, the more preparation time is available before start of manufacture and the earlier materials and capacity can be allocated for the order.

Between the release time fence and the **planning time fence**, manufacturing orders are of the firm planned type. The transfer from the status of planned to firm planned is generally done automatically by the ERP system when the delivery date for the planned manufacturing order approaches the planning time fence. In theory, the planning time fence should be set at the end of the accumulated lead time for the product. The automatic rescheduling which influences released orders for incorporated components in the product can then be kept under control. The planning time fence can also be used as a method of distributing planning work between the ERP system and manual planners. One way of achieving this is to set tight planning time fences for low volume/value items which are of less significance in terms of planning. These items will then be planned automatically to a greater extent. High volume/value items are instead given planning time fences further ahead, resulting in their being planned manually to a higher degree.

The capacity unit

The capacity unit in master production scheduling should be smaller than in sales and operations planning, thereby enabling a more detailed reconciliation between the capacity requirements stemming from the master production schedule and the capacity available. However, this does not mean that capacity requirement calculations should be carried out for each individual work centre. It is generally sufficient to plan capacities for whole departments or foreman areas. When necessary such coarse resource groups can be supplemented by individual work centres which are known to be critical or which often cause bottlenecks in operations.

10.3 Master Production Scheduling when Making to Stock

The different general steps in the master production scheduling process were accounted for in Section 10.1. The content and approach in steps 2 and 3 is somewhat different depending on the type of company involved. This section describes the steps for companies that make to stock. The steps are similar for companies that make to delivery schedules.

Step 2: Generating a preliminary delivery plan

A delivery plan can be generated on the basis of forecasts produced and orders on hand. Enterprise resource planning systems generally provide support for carrying this out automatically, but there is of course nothing to prevent it from being carried out manually. The selection of a suitable approach is to a large extent a question of the number of products to be processed and the degree to which delivery will take place with delivery lead times.

Since delivery when making to stock normally takes place from stock directly for incoming orders, the fusion of forecasts and orders does not cause problems. The forecast can be transferred directly to the delivery plan. In companies that make to delivery schedules, this fusion is not a problem either since delivery schedules are obtained from customers and there are no forecasts to reconcile.

In addition to generating a delivery plan, goals are also established for stock development within the planning horizon during this step in the process.

Step 3: Generating a preliminary master production schedule

The preliminary delivery plan and current and targeted stock levels constitute the basis for generating a preliminary master production schedule. This can normally also take place in the ERP system. The method of calculation is the same as that used in material requirements planning as described in the next chapter. One example of the calculations that are carried out is shown in Figure 10.7. In this example, stock on hand at the start of the planning period is 42 pcs and the order quantities for manufacturing orders are 50 pcs. In the example there will be a shortage in stock of 20 pcs in period 3. A new manufacturing order of 50 pcs with an inbound delivery in the same period is generated. To avoid future stock shortages a new manufacturing order is generated for the same reasons in week 5. No consideration has been taken of the need for safety stocks to safeguard against forecast uncertainties and other unforeseeable events, which is normally the case in reality.

In this simplified case consideration has only been paid to stock quantities at the start of the planning period. Stocks during the planning horizon have been permitted to vary without

Week		1	2	3	4	5	6
Delivery plan		18	20	24	20	20	20
Stock on hand	42	24	4	30	10	40	20
Master production schedule				50		50	

FIGURE 10.7 Generation of a preliminary master production schedule

Week		1	2	3	4	5	6
Delivery plan		18	20	24	20	20	20
Planned extra stock					10	10	10
Stock on hand	42	24	4	30	50	20	40
Master production schedule				50	50		50

FIGURE 10.8 Generating a preliminary master production schedule when planning stock build-up

restrictions. If it is desired to control stock development, more advanced calculation procedures must be used. If the intention is to increase stocks, it is possible to add one extra requirement row parallel to the delivery plan which contains the quantities to be allocated in order to build up the stock. This is illustrated in Figure 10.8 in the case where a company wishes to build up a campaign stock in weeks 4, 5 and 6 having a total of 30 pcs of the product in question. The row "planned extra stock" in this table refers to "demand" related to building up the extra stock. It is of course possible to make a completely manual calculation and adjustment of stock.

This step also includes the calculation of capacity requirements caused by the master production schedule. Since master production schedules in this type of company are expressed in the form of manufacturing orders, the methods which utilise capacity bills as well as capacity requirements planning may be implemented.

10.4 Master Production Scheduling for Assemble to Order

In the case of companies that assemble to order steps 2 and 3 are in principle carried out in the following way.

Step 2: Generating a preliminary delivery plan

For companies that assemble product variants to order, forecasts will relate to product models, whereas customer orders will relate to product variants that are configured from these product models. This problem is handled by expressing both the order part and the forecast part of the delivery plan as quantities of models, and when making calculations each order variant is converted into its equivalent model. The same procedure as in the case of making to stock can then be used. Since deliveries in this type of company are made with delivery lead times, a combination of forecast quantities and customer order quantities of models through addition would result in demand being counted twice to some extent. To avoid this, forecasts must be subtracted from those orders received; in other words, forecast consumption is applied.

A simple and rather common method of deducting forecast quantities from orders is to use the time fences that were described above in Section 10.2, i.e. a demand time fence and a forecast time fence as in Figure 10.6. Prior to the demand time fence, only orders are included in the delivery plan quantities since new orders are not normally received during the delivery lead time that corresponds to the demand time fence. The delivery plan quantity for the periods between the demand time fence and the forecast time fence is calculated as the larger of the

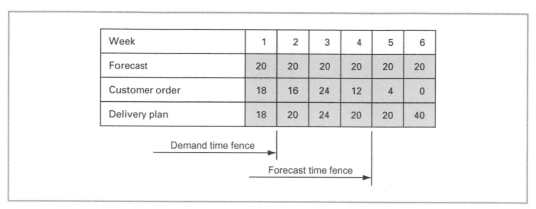

Week	1	2	3	4	5	6
Forecast	20	20	20	20	20	20
Customer order	18	16	24	12	4	0
Delivery plan	18	20	24	20	20	40

Demand time fence

Forecast time fence

FIGURE 10.9 Illustration of forecast consumption

forecast quantity and the order backlog quantity. Beyond the forecast time fence, the delivery plan is set equal to the forecast quantity. The procedure is illustrated by an example in Figure 10.9.

Step 3: Generating a preliminary master production schedule

In companies that assemble to order, the preliminary delivery plan is the starting point for generating a preliminary master production schedule, which can normally take place in the ERP system. The calculation process is mainly the same as in the previous case, with the exception that there is no stock deduction.

Since every customer order corresponds to a manufacturing order for a special product variant, in the short term the master production schedule will consist of manufacturing orders. Further ahead the delivery plan consists of forecast quantities. The master production schedule will then also consist of quantities promised for delivery. At one stage between the demand time fences and the forecast time fences there may be an element of both manufacturing orders and pure quantity information, i.e. manufacturing orders from customer orders in hand as well as forecast quantities which have not yet been consumed by orders.

Using such a procedure the generated master production schedule will thus consist of both manufacturing orders and quantities to be delivered of unspecified product variants. If desired, the quantities in the manufacturing orders may be aggregated for each model that their product variants belong to. Irrespective of the alternative selected, the capacity requirement from the master production schedule must be estimated using capacity bills since the plans in both cases are wholly or partly expressed in terms of delivery quantities and not in manufacturing order quantities. Not adding the quantities on manufacturing orders may result in slightly higher precision in cases where different product variants belonging to the same model have different capacity requirements.

10.5 Master Production Scheduling when Engineering and Making to Order

In companies that make to order or engineer to order, products are only partly specified on receipt of order, and in certain cases are unspecified. With respect to how master production scheduling may be executed, these types of company differ to a considerable extent in steps 2 and 3 from both of the previously described types.

Step 2: Generating a preliminary delivery plan

In companies where the proportion of order-based manufacture is so limited that it is possible to use product types for planning purposes, the work of producing a preliminary delivery plan can be carried out in generally the same way as for companies that assemble variant products to order. This means that the forecast is for product types, while orders are for product variants which are manufactured on the basis of product types. Both the order part and the forecast part of the delivery plan are expressed as quantities of product types and for calculations every order of a variant is converted into the equivalent product type. Since deliveries in this type of company take place with delivery lead times, forecasts must be deducted from the orders which have come in; in other words, forecast consumption must also be used in this type of company.

Under certain circumstances and when the proportion of order-based manufacturing is limited, it happens that product types are made to stock while waiting for orders to be received. This takes place especially in periods of low rate of incoming orders and is used as a method of maintaining a reasonably high utilisation of capacity. When the finished product types are retrieved from stock they are adapted to relevant customer specifications by adding, removing or modifying items or functions, or by supplementing with accessories. Producing a delivery plan then involves reconciliation with stocks to some extent, in a similar fashion to companies that make to stock.

If the proportion of order-based manufacturing is large and there is much new engineering and production preparation on each order, generating delivery plans will be different and more company specific. It may be necessary to plan at a wider level such as in sales and operations planning, or to accept a planning horizon dictated by the backlog of unfilled orders.

Step 3: Generating a preliminary master production schedule

The preliminary delivery plan is in this case, too, the starting point for generating a preliminary master production schedule. In cases with a reasonable proportion of order-based manufacturing, the approach is relatively similar to that used in companies that assemble to order. If, on the other hand, the proportion of order-based manufacturing is large, conditions will be different.

Since every customer order corresponds to a manufacturing order of special variants of a product type, in the short term the master production schedule will consist of manufacturing orders. Further ahead the delivery plan will consist of forecast quantities of such product types.

Using this approach, the master production schedule generated will consist of manufacturing orders and quantities of unspecified product variants. If so wished, the quantities in the manufacturing orders can be aggregated for the product type which each product variant corresponds to. Irrespective of which alternative is chosen the capacity requirement from the master production schedule is calculated by using some kind of capacity bills, since the plans in both cases are wholly or partly expressed in quantities. The former case may give somewhat higher precision in cases where different product variants corresponding to the same product type have different capacity requirements.

In situations where the proportion of order-based manufacturing is large, it is not so easy to forecast and plan for product types and then deduct the planned quantities of different product types as orders are received. Since the products in these cases are to a large extent unknown, information is lacking about the material content of them as well as the manufacturing steps which must be carried out. When planning capacity at the master production scheduling level, it must be accepted that the advance planning margin stems from orders in hand and capacity requirement is estimated by using capacity bills. In general, the rough type of capacity planning that can be made by sales and operations planning must be relied upon.

Since orders in this type of company may include engineering and production preparation work, consideration should also be taken of those resources required for this when consequences of the first master production schedules produced are being analysed. This may be done by including estimated requirements of the type of resource when using capacity bills.

10.6 Multi-site Master Production Scheduling

In multi-site master production scheduling, master plans for more than one production site are developed at the same time. The aim is to balance demand and supply and synchronise the material flow along a supply chain by developing master plans which are optimal for the supply chain as a whole. Multi-site master production scheduling not only balances demand with available capacities but also assigns demands (production and distribution amounts) to sites in order to avoid bottlenecks, wherefore it has to cover one full seasonal cycle, or at least 12 months in terms of weekly or monthly time buckets. The optimisation models used are defined as linear programming (LP) or mixed integer programming (MILP) models considering the total supply chain costs, forecasted demand and sales prices and defined constraints and restrictions of the entities included. These models are included in advanced planning and scheduling (APS) software.

The first step in multi-site master production scheduling is to define the nodes and links included in the supply chain to be modelled. If for example a supply chain contains three production sites which make products to stock and store the products in two distribution centres, the production sites and distribution centres are the supply chain nodes. The flow of materials and products between production sites and between sites and distribution centres are the links. Bottleneck resources are defined as constrained resources. The costs of each resource in a node and link, as well as available capacities for each constrained resource (e.g. work centre, distribution centre, transport) need to be determined. Restrictions can also be added to the resources, for example, minimum resource utilisation, safety stock levels, minimum delivery quantities, maximum production quantities, etc. These data, together with the forecasted demand and sometimes sales prices per product and market are included in the optimisation. Different objective functions can be used in the optimisation, for example, minimise costs or optimise profitability while meeting all constraints and restrictions.

Due to the complexity and detail required in the model, only constrained (or near-constrained) resources are modelled in detail. To increase the solvability of the model, most software distinguishes between hard and soft constraints. While hard constraints have to be fulfilled, the violation of soft constraints only renders a penalty in the objective function. Late delivery, unfulfilled demand, overtime and excess inventory could in some situations be defined as soft constraint. Due to model complexity and solvability the model uses aggregation, and stochastic features are typically not included in APS software. Depending on the planning issues to address and to achieve reasonable solution times a great deal of technical expertise is required to determine and limit the model size.

Multi-site master production scheduling is a centralised approach to master production scheduling. Therefore, companies conducting this type of planning normally have established centralised master planning functions. The main feature of multi-site master production scheduling is the ability to co-ordinate sourcing, production, distribution and seasonal stock decisions, etc. on a multi-site basis. Besides the positive planning effects, the approach can also enhance visibility and co-ordination throughout the supply chain. A similar approach can be used in sales and operations planning as described in Chapter 9. Optimisation is then used for scenario and what-if analysis when preparing preliminary production plans and adjusting the delivery and production plans. Cost and profit impact of changes in demand, price, utilisation, etc. could then be evaluated.

10.7 Master Production Scheduling in Relation to Final Manufacturing and Assembly Scheduling

As described above, the master production scheduling function means planning at the product level. The master production schedule drawn up is closely connected to the plans made for manufacturing end products. Schedules for final manufacturing, or final assembly schedules as they are often called in the mechanical industry, are not identical to master production schedules, however. Master production schedules are based on orders and forecasts, while final assembly schedules are only based on customer orders or stock replenishment orders. In the latter case, final assembly schedules are more a part of material planning or inventory control, as material planning is often called in this context. The final assembly schedules are also more short-term and always relate to specific products in contrast to product models, product types and similar terms used in master production schedules. Lastly, final assembly schedules always consist of manufacturing orders, while a master production schedule may refer to both quantities to be delivered and manufacturing orders.

Final assembly schedules and master production schedules are distinguishable in different ways depending on the type of company involved. In companies that make to stock and make to delivery schedules, both the master production schedule and final assembly schedules concern unique and defined products. The master production schedule consists of planned manufacturing orders and the final assembly schedule is thus made up of manufacturing orders that are released from these planned manufacturing orders. In those cases where the master production scheduling function is not fully introduced or not used at all, manufacturing orders are generated in the final assembly schedules directly from the material planning unit, often with the aid of time-phased order points or other similar planning methods. This process is described in more detail in the next chapter. This approach is particularly common in small companies. It is also widely used in controlling stocks of components and semi-finished items that are delivered to customers, such as is the case with spare parts.

In companies that assemble to order, the differences between master production schedules and final assembly schedules are greater. Final assembly schedules refer to released manufacturing orders for specific product variants, while master production schedules also consist of quantities of product models for delivery, irrespective of which specific product variants will be made. There is always an order from a customer behind manufacturing orders in the final assembly schedules, and release takes place at the same rate as orders are received.

When making to order and engineering to order, the end manufacturing plan always includes released manufacturing orders of specific product variants, and they are always tied to customer orders. Order release takes place as order-based variants are specified and production preparations are executed. For these types of company, master production schedules consist of quantities of different product types planned to be available for delivery, i.e. basic designs of products. Exceptions occur in those cases where manufacturing of product types is to stock, to be later adapted to customers as orders are received. Order release may also take place on the initiative of the planning unit at the company. In more extreme cases of engineering to order, i.e. where products are more or less completely engineered and made to order, the end manufacturing plan may be identical to the master production schedule since in this context it is not possible to predict what products will be in demand, and consequently it is not possible to draw up a master production schedule that extends longer than unfilled orders.

10.8 Summary

Master production scheduling has much in common with sales and operations planning, and for this reason it is not always relevant to plan separately at these two levels. In many companies the two planning levels are merged and are then typically called master planning. This chapter illustrated the various aspects of master production scheduling, what functions it fulfils and what links it has with sales and operations planning. The selection and design of different parameters in determining a master production scheduling system were accounted for, including planning objects, planning units, order types, damping functions, planning horizons, length of planning periods, time fences and capacity units.

The master production scheduling process consists in principle of the same steps as in the process for sales and operations planning, i.e. (1) forecasting future demand, (2) generating a preliminary delivery plan on the basis of a forecast and customer orders in hand, (3) generating a preliminary master production schedule on the basis of the preliminary delivery plan as well as current and targeted stock levels or sizes of customer order backlogs, (4) reconciling the plans produced and the prospects of realizing them, and (5) establishing the plans drawn up. The content and approach in each step varies to some extent depending on the type of company. In this chapter, master production scheduling was described and compared when making to stock, making to delivery plan, making to order and engineering and making to order. Master production scheduling in the case of engineering and manufacturing to order differs especially from other company types. The master production schedule that is generated in these cases has more the function of ensuring the medium-term access to capacity rather than controlling the flows of material since the material contents of products are only known to a limited extent before an order has been received.

In multi-site master production scheduling, plans for multiple sites are generated simultaneously with the aim of optimising a supply chain or production network as a whole. Master production scheduling involves planning at the product level. Final assembly schedules, however, are not identical with master production schedules. Master production schedules are based on orders as well as forecasts, while final assembly schedules are only based on customer orders and stock replenishment orders. Multi-site master production scheduling and design and use of final assembly schedules were described at the end of the chapter.

🔓 Key concepts

Accumulated production lead time 186	Delivery plan 178
Available capacity 180	Demand time fence 186
Capacity 178	Final assembly schedule (FAS) 179
Capacity bill 180	Firm planned order 184
Capacity planning 180	Forecast 180
Capacity requirements planning (CRP) 181	Forecast consumption 181
Damping functions 184	Forecast time fence 186

Discussion tasks

1 Which are the key functions of master production scheduling (MPS) in each of the following environments: make to stock, make to order and assemble to order? Characterise and compare MPS in the three environments.

2 It could be argued that appropriate time fences are necessary for carrying out planning in practice. But why is that so? (a) Explain the main objectives of using time fences in planning; (b) Time fences are important in several planning processes. What are the main objectives and time fence strategies of using time fences in master production scheduling?

3 What are the similarities and differences between a master production schedule and a forecast?

Problems[1]

Problem 10.1

Industry Gate Inc. makes customer-specific factory gates to customer order. One of these gates has, in addition to customer-specific items, the bill of material in the figure. Numbers in brackets show lead times in weeks. Items A, C, F, G, H, J, K and L are purchase items and the other are manufacturing items.

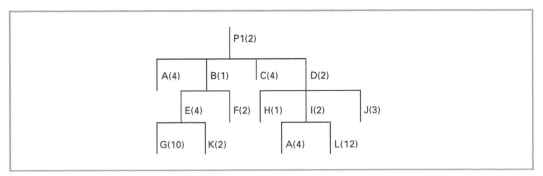

---

1   Solutions to problems are in Appendix A.

a) How long should the planning horizon at least be in order to use the material requirements plan for planning new orders?

b) How long should the planning horizon for this product be if it takes at least 20 weeks to adjust the manufacturing capacity to actual demand?

c) No matter different customer orders, the products require about the same manufacturing capacity. But the material requirements are very different between orders, especially regarding customer-specific measures. Therefore, the material requirements planning has to be based on actual customer orders. The maximum acceptable delivery time is 10 weeks. What problem will this lead to and what could be done to solve the problem?

Problem 10.2

The company Scania Mechanical makes a product with the bill of material in the figure. Numbers within brackets show required lead times in weeks to make and purchase the respective item. Items A, C, F, G, H, J, L and M are purchase items.

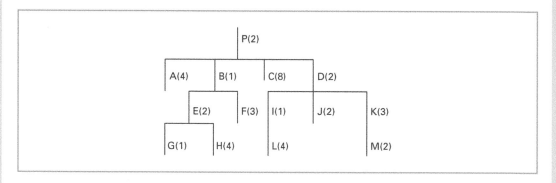

a) How long should the planning horizon for master production scheduling at least be if using material requirements planning to plan manufacturing and purchase orders?

b) How long should the planning horizon at least be if actual future material requirements should be used as the basis for determining purchase order quantities? For policy reasons, order quantities corresponding to more than eight weeks' cover time are not accepted.

c) How long should the release time fence be for product P?

d) How long should the planning time fence be for item E?

Problem 10.3

Company Worker Closet makes customer-specific lockers for working-clothes to customer order. The delivery time is never shorter than two weeks and normally not longer than six weeks. Now, a new master production schedule is to be developed. The table shows the forecasted demand and actual customer orders

Week	1	2	3	4	5	6	7	8
Forecasted demand	10	10	10	10	10	10	10	10
Actual customer orders	8	7	12	6	15	13	0	0

a) How long are the appropriate demand time fence and forecast time fence for the product?

b) Choose a policy to merge forecasts and actual orders. Then develop a delivery plan for the following eight weeks based on this policy.

c) What should especially be considered when developing such a policy?

Problem 10.4

Belgium Plastic makes plastic pallets of different sizes to stock. The products are sold to wholesalers throughout Europe. The aim is to continuously adjust the manufacturing to changing market demands and at the same time have as level a load as possible in the factory. This is partly accomplished by allowing most products to be made on most machines. For one of the products, PTV 2X, the marketing department has developed an eight-week delivery forecast according to the table. Forty-seven pieces are available in stock at the beginning of week 17. Of these 47 pieces, 30 are a safety stock to guard against forecast errors. The lot size in production is 150 pieces.

Week	17	18	19	20	21	22	23	24
Forecasted delivery	60	60	70	70	80	80	70	70

The company is working with its master production scheduling for weeks 17–24. Because it is getting closer to the holiday period when the factory is closing for some weeks, Belgium Plastic wants to build up a stock of 200 pieces, excluding safety stock. In order not to get too unlevelled a load they have decided to build up the holiday stock during a four-week period.

Consider these conditions when developing a master production schedule for PTV 2X for weeks 17 to 24. The holiday period starts in week 25.

Problem 10.5

The company SeaFood & Stuff makes ready-cooked fish dishes to stock. The products are delivered in large packages to retail distributors. The table shows the forecasted demand for one of the products, crayfish in dill, during weeks 29 to 36. The demand for this product is considerably higher during a couple of weeks in August compared to the rest of the year.

Week	29	30	31	32	33	34	35	36
Forecasted delivery	250	250	300	350	650	600	250	250

The crayfish dish is made together with other products in a cyclical pattern every second week. In week 28, a master scheduling process for the next eight weeks is conducted. Despite the peak demand during some weeks, the company strives to achieve as even a load as possible on the personnel making the fish dishes. The possibility of levelling the load is, however, constrained because the products have a limited durability. Therefore, products should not be kept in stock for longer than two weeks.

Develop a master production schedule for the product crayfish in dill, based on these conditions.

Further reading

Fogarty, D., Blackstone, J. and Hoffmann, T. (1991) *Production and inventory management.* South-Western Publishing, Cincinnati (OH).

Landvater, D. (1993) *World class production and inventory management.* The Oliver Wight Companies, Essex Junction (VT).

Proud, J. (1999) *Master scheduling: a practical guide to competitive manufacturing.* John Wiley & Sons, New York (NY).

Wallace, T. (2004) *Sales and operations planning: the how-to handbook.* T.F. Wallace & Co., Cincinnati (OH).

CASE STUDY 10.1: MASTER PRODUCTION SCHEDULING AT TOYS & GAMES LTD

Toys & Games Ltd is a European manufacturer and distributor of toys and games specifically made for children. The company develops and makes most of the products itself but distributes products from other manufacturers as well. Toys & Games has sales companies with stocked products in 11 countries in Europe and North America. The demand for most of the products is rather seasonal with peaks in the autumn close to Christmas-time. This case study only deals with the manufactured toys products.

The company applies a make-to-stock manufacturing strategy and uses a batch type of production with economic order quantities. The set-up times in the production are quite high. As a consequence, manufacturing order quantities are comparatively big, typically corresponding to more than one month's demand. Due to the big order quantities and due to functionally organised manufacturing, lead times are long, typically between three and five weeks. To store and distribute the manufactured batch quantities a central warehouse close to the production site is used.

Replenishment orders from the sales companies

Replenishment orders from the 11 sales companies are received at the distribution department once a week and two weeks in advance of expected delivery from the central warehouse. This means that the company always has a backlog of two weeks' deliveries. These orders are entered into the ERP system and the availability is checked in batch mode. The availability checking is made by means of an available-to-promise technique considering stock on hand, already-made allocations and scheduled receipts from manufacturing orders. If it is not possible to deliver replenishment orders as expected, the sales companies concerned are notified. They must then add supplementing replenishment orders next week instead. Orders possible to deliver are confirmed to respective sales company.

Forecasting

Forecasts are received from the various sales companies once every four weeks. They comprise estimated weekly demand per product for the coming 52 weeks, starting from the two weeks covered by replenishment orders. Some of the sales companies make these forecasts available through a web server application while others send them to the master planning department as Excel documents attached to emails.

The received forecasts are entered into the demand planning module in the ERP system and processed to get an overview of the total forecasts per product. The figures are checked by the forecast responsible at the marketing department and changes are made if needed. Manual adjustments are also made in cases where a product is planned to be phased out from the assortment and in cases when new products are planned to be introduced on the market within the planning horizon.

Master planning of production

Toys & Games makes no difference between sales and operations planning and master production scheduling. The two planning processes are integrated in a common process called master planning. The master planning process starts when the delivery plan consisting of the manually adjusted aggregated forecasts and the two-week backlog per product is available from the marketing department.

Based on the weekly quantities in the delivery plan, capacity requirements are calculated in the ERP system by using capacity bills. These capacity bills express the total capacity requirement per product in terms of man-hours needed per 100 pieces in manufacturing as illustrated in the table for one of the wooden products.

Manufacturing department	Required man-hours per 100 pieces
Cutting	12
CNC-milling	140
Grinding	55
Painting	40

The results of the calculations are weekly requirements of capacity in the various manufacturing departments. It constitutes a preliminary capacity requirements plan and represents capacity needed to manufacture what has to be delivered according to the delivery plan.

When this calculation is carried out, a master planning meeting is held. Here, representatives from the planning and production departments review the capacity requirement and compare it with the capacity planned to be available. They also analyse the need for adjusting the capacity in order to produce the volumes requested by marketing. Examples of strategies used at the company for balancing capacity and capacity requirements, are to add another shift in critical resources or to produce in advance in order to even out periods with peak load, especially in the autumn before the Christmas season. In the latter case they move the quantities in the requirements plan. If these capacity and requirements adjustments are not enough for meeting the demand, reductions have to be made in the delivery plan. This step in the process is carried out together with representatives at the marketing department and includes decisions concerning allocation of limited quantities to prioritised markets. The outcome of the process is a master production schedule stating weekly quantities during the coming 52 weeks.

Based on the master production schedule, material requirements planning is used to generate manufacturing order for end products, semi-finished items as well as orders for purchased components. The total annual demand in the delivery plan is also used for updating economic order quantities for manufacturing orders in the ERP system.

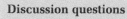

Discussion questions

1 Rather simple capacity bills are used at Toys & Games in the sense that the capacity requirement for a product is put in the same week as it is supposed to be delivered. What does this mean concerning the accuracy of the capacity planning at the company?

2 How is stock on hand of finished products, semi-finished and purchased components considered in the planning process at the company? What effect will this have on the reliability of the master production schedule?

3 What might the effect of calculating capacity requirements based on estimated demand per week instead of calculating them based on generated manufacturing orders be from a capacity planning point of view?

PART 4
Detailed Planning and Execution

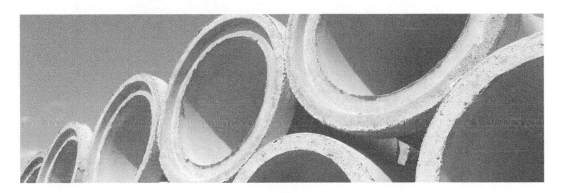

The fourth part (Chapters 11–18) treats detailed planning and execution. Chapter 11 presents the fundamental starting points for material planning and the order processes and planning processes which initiate flows of materials. The most commonly occurring material planning methods are described and compared. Lot sizing methods and various mechanisms for buffering against uncertainties are central parts of materials management. Chapter 12 treats lot sizing and Chapter 13 describes safety stocks. Chapter 14 takes up capacity planning at different levels of planning. Different action variables for achieving a balance between available capacity and capacity requirements in the medium- and short-term perspectives are reviewed. A summary is made of a number of commonly applied methods for capacity planning. Execution and control in pull environments are described in Chapter 15. Here, it is described how kanban and material planning without orders can be used as a means of execution. Chapter 16 concerns the main activities of production activity control, i.e. execution and control in traditional environments. The most commonly used methods for managing order release and material availability check are also described. The procurement process for purchased items is described in a similar way in Chapter 17. The book ends with Chapter 18. Here, inventory accounting including the reporting of inbound deliveries and withdrawals from stock as well as physical inventory are described.

Part Contents

Material Planning

From the perspective of logistics, a manufacturing company is characterised by the inflow of materials from external suppliers and the refinement at the company of these materials into end products. These products in turn flow outwards from the company to markets and to customers. Flows into the company are made up of raw materials, components and other purchased semi-finished items used as start-up materials for the value-adding processes within the company. During the process of refinement a number of flows are created: partly through raw materials and purchased components being refined into parts and semi-finished items of different types, and partly through purchased components, internally manufactured parts and semi-finished items being refined into finished products. Flows out from the company consist of end products which are then shipped to distribution warehouses or directly to customers. The administrative activities in a company required to control these flows on an operative level within a given production and distribution structure are called order planning. These activities are carried out within the framework of the overall master production scheduling as in the planning hierarchy discussed in Chapter 3. Order planning may thus be considered as the function within a company that executes plans established at the strategic and tactical level at the company.

Every flow of material is defined by the quantity to be transferred from a supplying unit to a consuming unit in a supply chain, and by the points in time when the quantities must be available at the consuming unit. Somewhat simplified, we can say that the main task of order planning is to establish, for every product and item, these quantities and points in time as efficiently as possible with respect to capital tied up in material flows, delivery service achieved to customers and the utilisation of resources when activities are carried out.

In order planning, consideration must be taken to current requirements of materials and capacity in relation to supplies of materials and capacity, i.e. order planning must be executed from the perspective of both materials and capacity. Order planning from the materials perspective is called material planning and will be further treated in this chapter. Its relationship to other planning processes is illustrated in Figure 11.1.

This chapter describes the basic starting points of material planning and the order processes and planning processes that initiate flows of materials. The balance between access to materials and requirements of materials that must be maintained in order to achieve cost-effective material

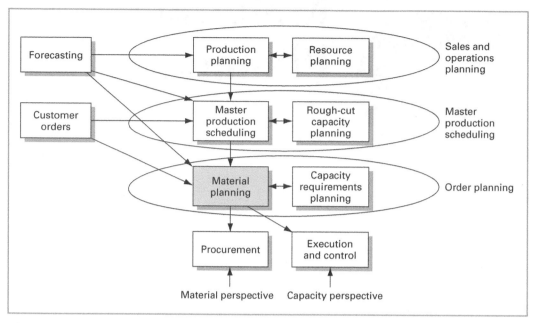

FIGURE 11.1 General relationship between material planning and other planning processes

flows and high delivery service is also discussed. The different material planning methods that are used to synchronise the flows of materials in achieving such a balance are accounted for. There are descriptions of definitions, material planning calculations, method variants and characteristic properties of the most common methods used. Their characteristic properties and serviceability are discussed for different planning situations. Methods for lot sizing and **safety stock** calculations that are used within the framework of material planning are discussed in Chapters 12 and 13.

LEARNING OBJECTIVES

After studying this chapter, you should be able to:

- Describe what material planning is in the context of production and materials management
- Describe the relationship between common business processes in a manufacturing company and material flows
- Explain what synchronisation of material flows means
- Describe a number of commonly used material planning methods and their characterising properties
- Calculate re-order points, **order-up-to levels** and run-out times
- Carry out a **material requirements planning (MRP)** calculation including **explosions** in a multi-level bills of material
- Explain in what environments the different planning methods primarily can be used

11.1 Some Basic Starting Points for Material Planning

Irrespective of the type of company and the planning situation in general, there are a number of basic starting points and principles for material planning that are always relevant. Four of these important starting points and principles will be treated here. These are the basic issues of material planning, the division into **dependent demand** and **independent demand**, available information for balancing requirements and supply of materials, and the characterisation of material planning as **pull** based or **push** based. All four of these are important for the understanding of material planning and the methods which may be used for material planning.

The basic issues of material planning

It was noted in the introduction that the main purpose of order planning and thus material planning is, for every product and other item, to establish quantities and points in time for all the orders created with the aim of initiating material flows and satisfying existing requirements. The goal is to generate these material flows as efficiently as possible with respect to tied-up capital, delivery service to customers and the utilisation of resources in internal activities. We can say that material planning is generally about answering the following four basic questions:

1 For what items must new orders be planned? (item question)

2 How large must the quantity of each item in the order be? (quantity question)

3 When must the order for each item be delivered to stock, directly to another manufacturing department or directly to a customer? (delivery time question)

4 When must the order for each item be placed with the supplier, or when must it be started in internal manufacture? (start time question)

Methods that are used for material planning all answer the first question by making proposals in one form or another for planning new **manufacturing orders** or **purchase orders** for items to ensure material supplies in supply chains. Regarding the other three planning issues, existing material planning methods provide a basis for decisions. How these bases for decisions are generated varies between the different methods, which are explained in the method overview in Section 11.4.

Due to long delivery **lead times** or throughput times in a company's own workshop, the choice of order quantities, start times and delivery times must often take place relatively early when planning new orders. In many cases the choice represents anticipation or a forecast-based assessment of what quantities may be required and when. This is especially the case in materials supply based on planning processes. The probability of quantities and times determined continuing to correspond to current needs at the planning stage may be small. This depends not least on changes in demand, and different forms of disruptions in material flows that are almost inevitable. In many contexts, therefore, there is a need to review **planned order** quantities and inbound delivery times, i.e. reschedule. The material planning department will also cover this type of **rescheduling** and certain material planning methods may be able to provide support for decisions in this respect as well.

Independent and dependent demand

The division of demand into independent and dependent types is of fundamental importance to all material planning. This division is crucial to the understanding of how material flows are integrated and for the selection of appropriate material planning methods. Independent demand for an item is demand that has no direct relationship to demand for other items. As a general rule, items stocked for delivery to customers have independent demand. This may be the case for finished standard products in the traditional sense, but also for incorporated purchased components and internally manufactured semi-finished items, if they are sold as spare parts for example.

In contrast to independent demand, dependent demand means that the demand for an item can be traced to the demand for another item. Items that are part of start-up materials are said to have dependent demand. In these conditions demand does not need to be forecast, but can be calculated from the demand for the parent item. If, for example, there are 32 spokes in a bicycle wheel and forecast sales of bicycles are 50 per week, each with two wheels, it can be calculated that 3200 spokes per week must be manufactured.

Many items that have dependent demand may also have an element of independent demand. Such is the case if an item is used both as incorporated material in the manufacture of a product and is sold separately as a spare part. The spokes of a bicycle wheel are an example of this type of item. Independent demand may also arise for other reasons. Scrapping is one reason for independent demand, as a certain proportion of the quantity picked from stocks may not be usable in production and must be scrapped. The same is true of errors in **stock on hand**, which means that quantities in stock according to stock accounting may not be available when withdrawals are to be made.

Information on requirements and supplies

In a similar fashion to the case for sales and operations planning, material planning is all about balancing demand for materials with supply of materials in the flow as cost-effectively as possible. If the supply is smaller than the demand, the flow of materials must be increased by planning new manufacturing orders or purchase orders. If, on the other hand, the supply is larger than the demand, inbound deliveries of already planned and **released orders** must be postponed if possible, or demand must be influenced by sales campaigns of different types, for example. Imbalances between supply and demand lead to large stocks if supplies are too large and shortage situations and poor delivery capacity if demand is too large.

As described above, the primary function of stock is to decouple processes in order to avoid unacceptably costly relationships of dependence. In the short term, let us say in the coming week or month, there may be reasons for not maintaining an exact balance between supply and demand. A company may wish to manufacture or purchase quantities that are larger than its immediate requirements. In the longer term, however, there must always be a balance between supply and demand, otherwise the balance must be restored by scrapping or selling surplus stocks at reduced prices if accumulated supplies have become too large. If accumulated requirements have become larger than supply, the balance may be restored by the loss of sales and customers.

For operative material planning, information on supply consists of the stock balance according to stock accounts and of information on purchase orders placed and manufacturing orders planned to be delivered. Stock represents current supply, whereas an expected delivery represents a supply expected to be available at a certain future point in time. Information about demand consists of forecasts, exploded requirements, customer **allocations** and allocations for manufacturing orders. Refer to Figure 11.2. Of the different types of information, forecasts are the least certain and reliable information about future requirements. Allocations are the most certain since they are based on decisions made by customers or internal planning. Exploded materials requirements from the information reliability perspective lie between these two

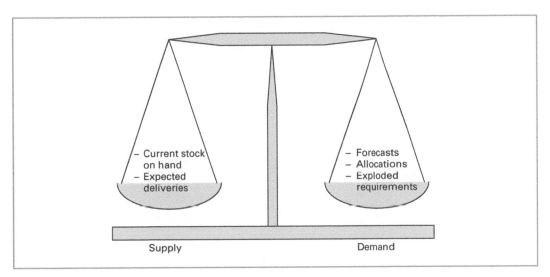

FIGURE 11.2 Information on supplies and demand for material planning

extremes. These requirements are obtained with the aid of bills of materials by calculating requirements for incorporated materials from plans of volumes of these products that the company intends to manufacture, i.e. from the master production schedules. For obvious reasons, material planning should be based as far as possible on allocations since information quality is highest for this type of demand information.

What also distinguishes the different types of information is the time horizon within which they are used. Customer order allocations are used for products in stock in the near future, i.e. the coming days or weeks, and forecasts or plans are further in the future. For items that are incorporated materials in products, conditions are somewhat different, depending on the planning environment at the time and the planning methods used. Allocations are mainly used for manufacturing orders in the near future and exploded requirements or forecasts for further in the future. If these items are sold directly to customers, for example as spare parts, allocations for customer orders are also used as information input for material planning.

Push-based and pull-based material planning

A common way of describing material planning is to distinguish between push-based and pull-based planning. There are different perceptions of what is meant by push and pull. The concepts are defined here:

> Material planning is of the pull type if manufacturing and materials movement only takes place on the initiative of and authorised by the consuming unit in the flow of materials.

> Material planning is of the push type if manufacturing and materials movement takes place without the consuming unit authorising the activities, i.e. they have been initiated by the supplying unit itself or by a central planning unit in the form of plans or direct orders.

The principles are illustrated in Figure 11.3. The decisive difference between the two types of initiating flows of materials is related to who authorises the value-adding process and/or the movement of materials. The pull principle is an expression of no such activities taking place unless the recipient and consumer of materials has ordered them. Material flows initiated by

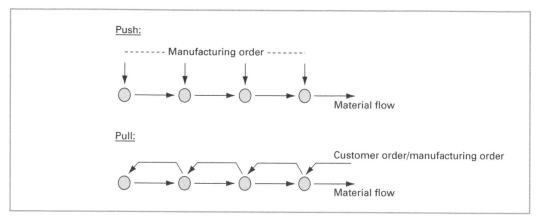

FIGURE 11.3 Illustration of push-based and pull-based planning

order processes as described above may therefore be considered as pull-based planning. The concept of pull, however, normally also means that the quantities ordered to be manufactured or moved are small and as close to the direct and immediate materials demand as possible. Material flows initiated by order processes are accordingly not the same as pull-based planning.

It would be easy to imagine that material flows initiated by planning processes are always of the push type, but this is not necessarily the case. If the consuming unit independently plans and initiates material flows it may still be of the pull type if the quantities ordered are small. The decisive factor is whether or not planning takes place centrally by a controlling unit or not.

It may be added that it is not the material planning methods as such that have push or pull characteristics. It is how they are used, i.e. their application, which determines whether material flows are push based or pull based. Most material planning methods can be used within the framework of push and pull planning. For example, a **re-order point system** may be used by a consuming unit to order material procurement, i.e. in a pull environment. However, the system may also be used by a central material planning department to order make to stock for future sales, i.e. in a push environment.

11.2 Synchronisation of Material Flows

As described above, material planning aims at balancing the supply and demand of materials in a cost-effective manner. This balancing has both a quantity dimension and a time dimension, in compliance with the basic issues described above, i.e. supply quantities must correspond to demand quantities, but supplies must also be available at the points in time when demand arises. Balancing quantities in this context is, relatively speaking, a smaller problem. Somewhat simplified, we can say that it is either a question of producing and delivering the quantities as required at each point in time, or producing and delivering a quantity equivalent to the calculated economic order quantity and which covers a number of expected future demands. Financial motives for manufacturing and delivering more than immediate requirements are described in Chapter 12.

Synchronising points in time for supply and demand represents a much larger challenge. Supplies and thus inbound delivery times for the order of an item must be made to coincide as closely as possible with demand times for the item. If delivery takes place too early, there will

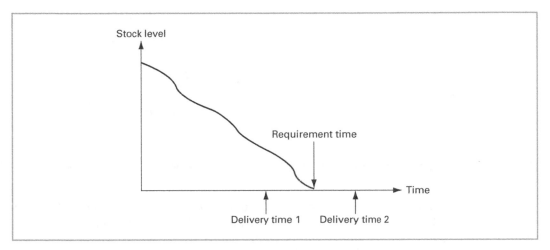

FIGURE 11.4 Illustration of synchronisation of supply and demand in material flows

be unnecessary capital tied up, and if delivery is too late there may be shortages, with production disruptions and poor delivery service to customers as a result. Figure 11.4 illustrates this relationship for stock which, due to expected withdrawals, gradually approaches zero. At the point in time at which this takes place there will be a demand to replenish stock, i.e. it is the expected requirement time for material flows into stock. If inbound delivery takes place at time 1, there will be unnecessary tied-up capital, and if delivery takes place at time 2, there will be a shortage of stock. Efficient material planning therefore depends on being able to safely and accurately predict and calculate when requirements will arise and to ensure that inbound deliveries take place as close to these requirement times as possible.

Another factor that makes synchronisation difficult to achieve is that, from a planning perspective, items can seldom be considered as isolated phenomena which can be ignored when material planning other items. In most contexts, material flows for different items are interdependent and must be planned in a holistic way. For example, if final manufacture or assembly of a product is to be carried out, the supply of several incorporated items is required simultaneously, and for a customer order with several order lines to be delivered there must be a supply of all products on the different order lines. Thus, the flows of materials must be synchronised with each other as much as possible towards nodes in material flow structures, such as those represented by the stocks of manufacturing orders and picking for customer orders.

Figure 11.5 illustrates what happens when synchronisation is lacking, as seen from the demand side for a manufacturing order for a product P, which consists of three items: A, B and C. If one of the items is scheduled to be delivered too late, such as item C in the diagram to the left in the figure, the manufacturing order start for P will be delayed and other materials will remain in stock pending delivery of the delayed item. If, on the other hand, an item is scheduled to be delivered too early – such as item A in the right-hand diagram in the figure – the item will remain in stock until the order for P is planned to start.

When ideally executed, material planning will synchronise material flows with respect to demand when planning manufacturing orders and purchase orders. To the extent to which requirements change or disruptions in material flows on the supply side occur after orders have been planned and released, synchronisation is restored if possible through rescheduling. Such ideal conditions are, however, rarely possible to achieve in practice. Stock which arises as a result of incomplete synchronisation is called control stock and was described in Chapter 2.

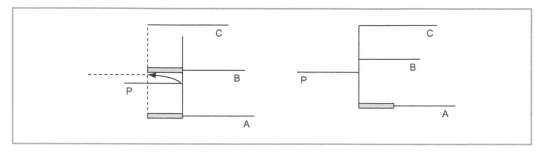

FIGURE 11.5 Two cases of defective synchronisation of requirement times at nodes in material flow structures

The lack of synchronisation that gives rise to control stocks is unintentional and is the result of functional shortcomings in material planning methods and procedures used. It is also the result of difficulties in completely predicting changes in demand and not making changes quickly enough as disruptions occur in the inbound flow of materials. Lack of synchronisation between demand and supply may sometimes also be deliberately created and there may be good reasons for doing so. This is the case when for various reasons a company wishes to link the flows of different items in time by using planned inbound deliveries of orders for co-ordination purposes. It is then possible to choose the inbound delivery times for different orders for a number of items to coincide, despite the fact that demand times for each item do not coincide.

Since the demand times for different items can vary, one of the items whose flow needs to be co-ordinated must control when new supplies should be planned. In general it is the item that needs to be replenished earliest that will dictate when delivery must take place. This relationship is illustrated in Figure 11.6. Since replenishment of stocks is calculated to take place earliest for item C, its demand time will control deliveries of items A and B. Simultaneous

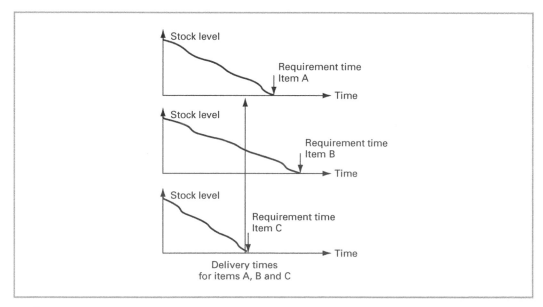

FIGURE 11.6 Illustration of deliberate lack of synchronisation in material flows in the co-ordinated delivery of several items

delivery of several items from the same supplier to decrease transport costs is an example of a situation where lack of synchronisation of supplies and demand may be motivated with respect to individual items.

Another example is the case where a company can reduce set-up times by manufacturing different items with the same or similar machine settings after each other. The stock which arises as a result of this deliberate lack of synchronisation is called **co-ordination stock**.

11.3 Methods for Planning New Orders

As a support for planning new manufacturing orders and purchase orders so that the balance between supply and demand is maintained and so that material flows are as synchronised as possible, different material planning methods can be used. These methods have different characteristics and may be more or less suitable for use in different planning environments. However, they all have the common factor that they answer the two time questions of material planning in different ways, i.e. when orders will start or will be sent to suppliers, and when inbound delivery will take place.

11.3.1 Re-order point systems

A re-order point system means a material planning method based on making a comparison between the quantity available in stock and a reference quantity, called the re-order point. When stock falls below this reference quantity, re-ordering takes place to replenish stocks, i.e. a manufacturing process or a procurement process is initiated to manufacture or procure the desired stock replenishment quantity. The basic principle behind the method is illustrated in Figure 11.7 and means that a new order is planned if the stock balance plus any planned open orders fall below the re-order point. The delivery time is set at the current state plus the lead time for the item. The vertical lines in the figure represent stock replenishment.

The re-order point quantity is the expected consumption during lead time for replenishment plus a safety stock quantity to protect against unpredictable variations in demand. Thus, in addition

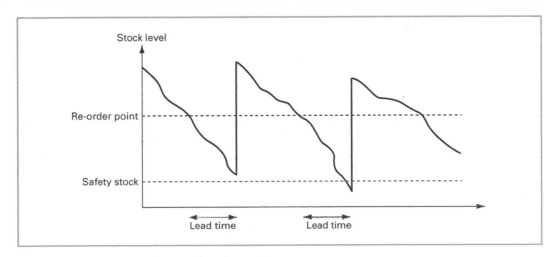

FIGURE 11.7 Basic principles of the re-order point system

to the safety stock, the re-order point represents a forecast for demand during the lead time. The following relationship is used for sizing a re-order point:

$$ROP = SS + D \cdot L$$

where ROP = re-order point
         SS = safety stocks
         D = demand per time unit
         L = lead time

Example 11.1

A re-order point system is used for material planning for a purchased item. The usage of the item during the lead time has been estimated at 20 pcs and the safety stock has been estimated at 6 pcs. At a certain point in time the stock balance is 22 pcs. Should a new order be released for the item or not?

Using the information on the item, the re-order point is equal to 20 + 6 = 26 pcs. Since the stock balance is 22, i.e. less than the re-order point, a new order should be released and placed with the supplier.

Example 11.2

A stocked item has an average consumption of 20 pcs per week. A re-order point system is used to determine when stock replenishments will take place. The lead time from released order to delivery of the item is two weeks. What re-order point quantity should be used if safety stocks should be equivalent to one week's consumption?

D = 20 pcs/week
L = 2 weeks
SS = 1 week's consumption
ROP (re-order point) = $1 \cdot 20 + 20 \cdot 2 = 60$ pcs

The theoretically correct way of determining re-order points is to add together safety stocks and demand during lead time. This procedure also creates the conditions for distinguishing the lead time-dependent and uncertainty-dependent parts of the re-order point. With changes in average values and variation in lead times and consumption, it is then possible to maintain optimal tied-up capital and delivery service by adjusting safety stock and/or demand during lead time. However, many companies do not distinguish between demand during lead time and safety stock, but instead make an assessment of the total quantity based on some kind of experience.

To use the re-order point system, information is necessary on consumption, quantities in stock and the re-order point. If the demand and the lead time are only expected to be associated with a certain random variation during future planning periods, a fixed re-order point for each item can be used. If the re-order point system is used to control items whose demand has seasonal or trend variations, the re-order point must be adjusted when demand varies, otherwise the ERP system will generate a stock shortage and/or unnecessary tied-up capital. For this reason the re-order point system is seldom the most suitable planning method to use in situations with large variations in demand.

Variants of the re-order point system

There are a number of similar material planning methods that are all variants of the traditional re-order point system. There are also a number of different applications of the traditional re-order point system. The re-order points can, for example, be continuously compared with the stock balance or at certain periodic times of review. The stock balance that is compared with the re-order point can be defined in a number of different ways. There are also a number of options when it comes to calculating re-order quantities. In addition to the estimation and calculation variants of the re-order point system, there are a number of variants in the practical application of the system. The two-bin and kanban systems which are described in Chapter 15 are such variants.

Continuous versus periodic review

Depending on when the comparison between the re-order point and the stock on hand takes place, two main types of re-order point systems can be identified: continuous review systems and periodic review systems. In practical applications of the continuous review system comparisons are made per transaction and not continuously, i.e. after every stock transaction that brings about a reduction in stock. In the periodic review system comparisons are made at given intervals. When the stock on hand falls below the re-order point, a new planned order is created, but otherwise not. In practical applications this variant is achieved by periodically, for example each week, running a re-order point system and from this system obtaining planned orders for items whose stock balance lies under their re-order points.

The safety stock that is required to achieve the desired service level in stock is dependent on demand during the so-called uncertain time, i.e. the time from the stock falling below the re-order point until a new delivery is made. Using the periodic review system, orders are not initiated directly when stock falls below the re-order point, but on the next review occasion. Half the review interval then becomes part of the uncertain time and in general larger safety stocks are required when the periodic review system is used than is the case with transactional comparisons. The review interval also influences the total lead time and thus the reaction time for covering material requirements as they arise. Transaction-oriented systems also have advantages with respect to quickly and flexibly initiating new material flows in supply chains. In transaction-based re-order point systems, planning is based on time of consumption rather than time of review.

Using the periodic review system, planning of new orders can be carried out for a large number of items together, thereby making administration more efficient. Transaction-oriented systems require that new orders are planned per item, and work takes place more frequently and sporadically.

Stock balance comparisons

In the traditional re-order point system, the re-order point is compared with stock on hand as shown by the stock accounting system. An alternative to this is to also consider open allocations and instead make the comparison with the available stock on hand, i.e. the accounted balance minus the sum allocated during the lead time. Using this procedure, consideration can be given to the completely known demand represented by allocations.

Characteristic properties of the re-order point system

The properties of the re-order point system can be characterised as follows:

- The method is easy to understand and it is generally known in industry. It is also one of the most widely used material planning methods.
- Material planning based on the re-order point system is often built on some form of accumulated future demand, such as an annual forecast or only historic consumption. Demand for the calculation of re-order points for raw materials and purchased components in manufacturing companies can also be obtained with the aid of gross

requirements planning from master schedules at the level of products. Calculated **gross requirements** are then aggregated on an annual basis and make up a forecast for future demand. Using re-order point applications based on comparisons with available stock on hand, demand in the form of current allocations can also be obtained in a simplified manner. In addition, this means that a certain amount of consideration can be taken to dependent demand.

■ Re-order point methods are not generally intended for planning materials with dependent demand, but are more suitable for items with independent demand. The theoretical conditions on which the calculation of re-order points and safety stocks are built are seldom fulfilled for items incorporated in other items and in end products.

■ A re-order point is a forecast of expected consumption during the replenishment lead time. Using re-order points for controlling materials and components incorporated in products means in principle that the entire range of items must be forecast in contrast to the method of material requirements planning, in which only final products need to be forecast. It is however possible to use re-order point methods for controlling items with dependent demand, especially in cases where the items are incorporated in so many different bills of material that the total demand tends to be smoothed out.

■ All variants of re-order point systems are usage-initiating planning methods, i.e. order initiation is based on past usage. The reason for stock becoming replenished is that stock on hand has fallen below a predetermined level due to usage. There is no direct link to future demand. However, there is an indirect link to future demand provided re-order points and order sizes are calculated on the basis of forecasts or gross material requirements.

■ When planning with the re-order point system, planned orders are generated when the re-order point is reached. This means that advance planning and the generation of **delivery schedules** is not possible within the framework of the re-order point system. When using periodic review, planning of new planned orders is carried out from the time of monitoring, while planning takes place from the time of usage when using continuous review. Depending on the variation of demand, there are different time intervals between re-ordering occasions.

Primary applications environments

The re-order point system is primarily applicable in planning environments of the following types:

■ *Finished goods stocks.* An important area of application for the re-order point system is for controlling products in finished goods stocks.

■ *Low value items.* Another primary environment is for the control of low value items such as screws and washers, for which usage is relatively even and predictable and for which the replenishment lead time is relatively short.

■ *Usage that cannot be planned.* The re-order point system is also a suitable method for controlling items whose usage cannot be planned to any great extent; i.e. it cannot be predicted and therefore cannot be handled using explosion with material requirements planning or through handling of allocations. Examples of this type are spare parts and items which also are incorporated in products that are specified to order and which must be available for immediate usage. For this type of item, the simple form of re-order point system is often the only possible alternative.

- *Frequent and continuous demand.* The relative weaknesses of the method when planning items with dependent demand become smaller with increasing independent demand, and the more frequent and continuous material requirements are. In such situations a more smoothed-out demand pattern is created, which is a precondition for the efficient use of the re-order point system. Since consideration is not paid to capacity when planning new orders, the relative advantages of the method are also greatest in environments where access to capacity does not constitute a crucial problem, for example as a result of high-volume flexibility in production.

- *Small order sizes and short lead times.* The method puts lower demands on basic data quality than material requirements planning. It also works better with smaller order sizes and shorter throughput times, since material flows then tend to be more even.

11.3.2 Periodic ordering systems

The traditional re-order point system is characterised by new planned deliveries taking place with varying intervals and delivery quantities being constant. In many contexts it may be an advantage instead to have planned deliveries with constant intervals and allow order quantities to vary and correspond to the usage that occurred during the delivery interval. This is the case when several items are procured from the same supplier. The total ordering costs and transportation costs can be reduced for each delivery by ordering a number of items. By reducing the ordering costs, conditions are also created for more frequent deliveries with decreased tied-up capital and increased flexibility as a result. The same effect can be achieved in cases when transportation can be co-ordinated into deliveries from several suppliers.

For **periodic ordering systems**, a target level or order-up-to level is specified. It is calculated by adding together the expected usage during the lead time and the replenishment interval plus a safety stock. The length of the replenishment interval is selected if possible so that the order quantity on average becomes equivalent to the economic order quantity. The order-up-to level is adjusted to changes in demand.

New orders are always planned at every periodic recurring ordering time, such as once a week and with the same periodicity as the replenishment cycle.

Choosing the difference between the order-up-to level and the current stock on hand as order quantity is identical to choosing the sum of usage since the previous ordering occasion as an order quantity. This relationship is illustrated in Figure 11.8. In this figure, S(1) plus Q(1) represent the order-up-to level. Together they also represent the total quantity available since the previous ordering occasion. Since S(2) still remains in stock, the difference between the sum S(1) and Q(1) and the remaining quantity S(2) is equivalent to the quantity used during the period, and so this quantity must be ordered.

The following formulae can be used for calculating the order-up-to level (also called target level) and order quantity:

$$T = D \cdot (R + L) + SS$$
$$Q = T - S$$

where Q = order quantity
  T = order-up-to-level (target level)
  D = demand per time unit
  R = re-ordering interval
  L = lead time
  SS = safety stock
  S = stock on hand

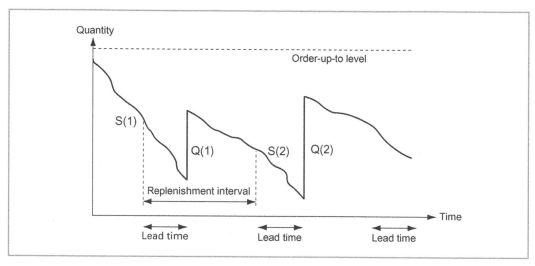

FIGURE 11.8 Basic principles of periodic ordering systems

Example 11.3

A periodic ordering system is used to replenish a stocked item. The average daily demand is 80 pcs and re-ordering of stock replenishments take place every tenth day. Lead time from order placed to delivery is two days. What should the order-up-to level be and what quantity must be ordered if stock on hand at the time of ordering is 500 pcs and the safety stock should correspond to four days' usage?

$$\text{Order-up-to level} = D(R + L) + SS = 80(10 + 2) + 320 = 1280 \text{ pcs}$$

$$\text{Order quantity} = T - S = 1280 - 500 = 780 \text{ pcs}$$

In a corresponding fashion to the re-order point system, demand information is required from usage statistics, forecasts or requirement calculations and stock on hand in order to use a periodic ordering system. The only other information required is the length of the re-ordering interval.

Variants of the periodic ordering system

Another way of managing a periodic ordering system is to allow the consumption unit on each ordering occasion to report usage since the previous order and have the material supply unit deliver this quantity. To keep the option of adjusting order-up-to levels, adjustments must be able to be reported and lead to modifications of the order quantities calculated through usage. Such adjustments may be necessary to take into consideration changes in demand and changes in safety stocks.

One further alternative variant is to report sales and corresponding allocations for customer orders or allocations for planned manufacturing orders. This also means that demand information can be provided earlier to supplying units.

Characteristic properties of periodic ordering systems

The periodic ordering system is much the same as the re-order point system, and many of their characteristic properties are similar. In contrast to re-order point methods, the interval between consecutive inbound deliveries is constant, while order quantities vary.

Primary application environments

The method's main application environments are as follows:

- *Finished goods stock.* Periodic ordering systems are primarily intended for use in environments with independent demand, i.e. for material planning of stocked items intended for sale.

- *Independent demand and frequent requirements.* The relative weakness of the method when planning items with dependent demand becomes less as the share of independent demand increases and the more frequent and continuous are material demands. Since consideration is not paid to capacity when planning new orders, the relative advantages of the method are also greatest in environments where access to capacity does not constitute a crucial problem, for example as a result of high volume flexibility in production.

- *Co-ordination of items.* Since times for ordering and deliveries can be controlled in a very efficient manner compared with most of the other material planning methods, the periodic ordering system has its greatest relative advantage in environments where the material planning co-ordination of different items is desirable. This may be the case in situations where, for transport economy reasons, it is desirable to co-ordinate inbound deliveries of a number of items from the same supplier. This may also be the case with manufacturing of items with similar machine settings. Items with similar characteristics from the viewpoint of set-up can then be ordered together. Total set-up times can be reduced by manufacturing these items simultaneously.

- *Small order sizes and short lead times.* Periodic ordering systems put lower demands on basic data quality than material requirements planning. The more the planning environment is characterised by small order sizes and short throughput times, the more efficiently the method can be made to work.

11.3.3 Run-out time planning

In a re-order point system, the re-order point excluding safety stock is calculated so that the quantity that is available in stock when it falls below this point is sufficient to cover requirements during the replenishment lead time. The requirements to cover that are calculated for a re-order point system are therefore expressed as a quantity. An alternative is to express requirements to cover as a period of time instead of as a quantity. **Run-out time planning** is a material planning method that uses this unit, and thus is closely related to the re-order point method. Run-out time planning is also called **cover time planning**.

Run-out time refers to the time that available in stock, i.e. stock on hand plus **scheduled receipts**, within lead time is expected to last. It is calculated by dividing stock on hand by expected demand per time unit. In the same way as in the re-order point system, expected demand is determined as based on usage statistics, forecasts or aggregated gross requirements through explosions of the master production schedule.

In order to safeguard against uncertainty and variations in demand during replenishment lead time, **safety lead time** is used in run-out time planning. The safety lead time multiplied by demand per time unit is equivalent to the safety stock used in the re-order point system.

The decision rule for run-out time planning can be expressed with the aid of the following formula:

Order if run-out time < Safety lead time + Replenishment lead time

---

Example 11.4

Run-out time planning is used for stock replenishment of a stocked item. The average daily demand is 80 pcs. Lead time from order placed to delivery of the item is two days. The safety lead time for the item is four days. At one review stock on hand is 320 pcs. Should ordering take place?

Since the run-out time is four days and the safety lead time plus the replenishment lead time is six days, ordering should take place. The calculation is shown below.

Run-out time = 320 / 80 = 4 days

Safety lead time + Replenishment lead time = 4 + 2 = 6 days

Run-out time < Safety lead time + Replenishment lead time → order!

Variants of run-out time planning

In the same way as for the re-order point system, run-out time comparisons can be made either when stock transactions take place or periodically in the form of batch processing of the entire range of items. Run-out time planning can, in other words, be transaction oriented or periodic. These two alternatives have the same advantages and disadvantages as for the re-order point system.

In the application of run-out time planning as described above, no consideration is given to allocations on hand. These are assumed to be included as part of the estimated demand per time unit.

Characteristic properties of run-out time planning

Run-out time planning is closely related to the re-order point system and in many respects the two systems have the same properties:

- The method is easy to apply. Since it is based on time comparisons, in contrast to the re-order point system that is based on quantity comparisons, it also provides a better understanding of the current need to order and how urgent that need is.

- Run-out time for an item represents an estimate of when stock needs replenishing. Run-out time planning may accordingly be described as proactive. The precision of the estimate is limited, however, if annual forecasts or usage statistics are used as a basis for the calculation of run-out time.

- The length of run-out time is an expression of how imminent the need for replenishment is. It can therefore be used to prioritise the manufacture of different items or their procurement from the same supplier. By dividing the replenishment lead time with the current run-out time, a priority figure is obtained which closely reflects desirable priorities.

- Run-out time planning does not contain per se any real rescheduling function. However, through the use of priority figures calculated as the relationship between the replenishment lead time and the current run-out time, a basis for rescheduling may be obtained. One prerequisite is frequent recalculations of these priority figures.

Primary application environments

Run-out time planning is primarily useful in the following environments:

- *Finished goods stock.* The method is primarily intended for use in environments with independent demand, i.e. for material planning of stocked items intended for sale. It is also in some cases an alternative to material requirements planning for items with dependent demand.

- *Independent demand and frequent requirements.* The relative weaknesses of the method when planning items with dependent demand become less as the share of independent demand increases and the more frequent and continuous are material demands. In the same way as other re-order point related methods, the more the planning environment is characterised by small order sizes and short throughput times, the more efficiently the method can operate.

- *Volume flexibility.* Since consideration is not given to capacity when planning new orders, the relative advantage of the method is largest in environments where access to capacity is not a crucial problem.

- *Low basic data quality.* Run-out time planning makes fewer demands on basic data quality than material requirements planning. Since it is less exact in comparison with material requirements planning, run-out time planning is less sensitive to poor stock on hand accuracy when compared to material requirements planning.

11.3.4 Material requirements planning

Material requirements planning (MRP) is a material planning method that is based on points in time for scheduling new deliveries being determined through the calculation of when further requirement of materials arise, i.e. when the calculated stock on hand becomes negative.

How material requirements planning is carried out is illustrated in Figure 11.9. As shown in the figure, stock is calculated to get below 0 in week 7. A new order must therefore be scheduled for delivery in that week to avoid shortages. In the example in the figure, the order quantity is 40 pcs and the lead time is 3 weeks, which means that the order has to be released in week 4.

The fundamental principle of this method is to not schedule a new order for delivery until there is a **net requirement**. The point in time for release of order is calculated as the delivery time minus lead time for the item. When a system for material requirements planning is designed, a number of parameters must be selected and established. These parameters mainly consist of the length of time periods used, the length of the **planning horizon**, the planning frequency, different types of orders, handling of rescheduling, **pegging** and planning **time fences**.

Material requirements planning can be run stock transaction by stock transaction. Material requirements can, however, also be grouped per day, week or any other time period. It is reasonable to use shorter planning periods in the short term when planning information is more exact and detailed, and longer time periods nearer to the planning horizon when requirements information is to a large extent based on forecasts.

The planning horizon for material requirements planning must as a minimum be equivalent to the longest accumulated time for manufacturing and purchasing of all items included in the end products, i.e. the lead time along the critical path of a product structure. If the planning horizon is shorter than the accumulated lead time, planned orders for purchased items at the

Week		1	2	3	4	5	6	7
Forecast/requirement		10	10	10	10	10	10	10
Stock on hand	20	10	40	30	20	10	0	–10
Planned order delivery			40					
Planned order start					40			

FIGURE 11.9 Illustration of material requirements planning

lowest product structure levels will not be planned sufficiently early. Consequences may include problems in fulfilling the current master production schedule. If the planning horizon exceeds the accumulated lead time, this problem is avoided. In most situations the planning horizon is made considerably longer than the accumulated lead time. Planned orders generated can then be used as a basis for delivery schedules to suppliers and for capacity planning purposes.

Planning frequency refers to the frequency with which new planned orders are generated and existing planned orders are rescheduled. Limitations of computer systems in the past made it impossible to carry out planning more than once a week. These days the most common ERP systems have sufficient computing capacity to carry out material requirements planning for all the items in a company several times a day. The most common procedure is to run material requirements planning daily, but companies also apply weekly planning or transactional planning.

Planned orders generated with the aid of material requirements planning cover time horizons from the present to the planning horizon. Planned orders further in the future are to a larger extent based on forecasts when compared with those in the near future. Forecast changes mean that the master production schedules are adjusted, which in turn affects material requirements planning and thus results in a need for rescheduling. Other reasons for rescheduling may be engineering changes which mean that old items are phased out and new ones phased into existing bills of material, physical inventory counting that leads to adjustments in stock on hand, scrapping production that exceeds planned levels, changed order quantities, changed delivery times and delivery quantities caused by changed orders, delivery or quality problems from subcontractors, disruptions in internal production, etc. In addition to planned orders, two further types of orders are distinguished between in material requirements planning: released orders and **firm planned orders**. These order types are described in Chapter 10.

When conditions for planning are changed, for example due to delivery delays, it is necessary to quickly identify the consequences of changes in conditions and possibly review priorities in the plans. It is then important to be able to identify how manufacturing orders and the master production schedule are affected. The process called pegging enables planners to identify the sources of gross requirements of an item. This means that requirements from material requirements explosions are provided with information on the manufacturing orders which they come from. Requirements can either be derived from or reported at one structural level at a time, called one-level pegging. Requirements may also be directly coupled to end products and show which end-product requirements are the basis for requirements for the item in question. This is called complete pegging or pegging at the end-product level. Figure 11.10 illustrates the principles of pegging. In the example, a pegging report for item B in week 3 would show that the requirement of 80 B is coupled to the production of 40 X and 20 Y.

Rescheduling an item may result in immediate consequences for other items since consideration must be given to the fact that the items are used in more than one parent item. The

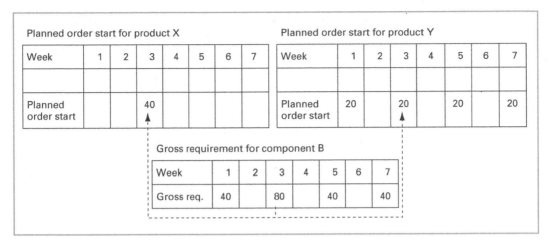

FIGURE 11.10 The principle of pegging

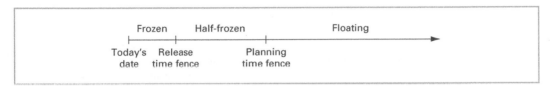

FIGURE 11.11 Planning fences and time windows

consequences of rescheduling are always more serious in the near future than when close to the planning horizon. In order to minimise the costs associated with rescheduling, planning time fences are used. Planning time fences in master production scheduling were described in Chapter 10. Commonly used time fences in material requirements planning are illustrated in Figure 11.11. The **release time fence** normally corresponds to the throughput time of an order in the workshop. During the time until the release time fence, rescheduling is normally not allowed and orders placed are considered as frozen. Since orders within the time interval are already released, changed plans may result in increased production costs and decreased production efficiency and delivery service. Sometimes one further time fence is used: the planning time fence. This is equal to the throughput time in the workshop plus the time from order placement to delivery for purchased material. Changes made in this time interval may cause problems since purchased material has already been ordered and capacity must be allocated in the workshop. This time zone is called half-frozen. Since the manufacturing order has not yet been released, the consequences of any changes will be smaller than in the frozen time zone. The period beyond the planning time fence is termed the floating time zone. Automatic rescheduling by the ERP system may be allowed in this time zone.

Application of material requirements planning

When using material requirements planning for items with dependent demand, explosion is included in the method with the help of bills of material. The starting point for material requirements planning is a master production schedule, which states in what quantities and at what times the company's end products are to be manufactured and delivered to stock or to customers. Material requirements planning consequently demands information from the master production schedule, bills of material, stock on hand, lot sizing methods and lead times.

To be able to link the master production schedule for an end product to requirements of components included, information is required on what components are incorporated in the end products. This information is obtained with the aid of the bills of material file in the ERP system.

From the master production schedule and bills of material, required quantities of components and raw materials are calculated as well as the times when these requirements occur. The existence of requirements, however, does not necessarily mean that items must be ordered for stock replenishments. Sufficient quantities of the items may be available directly from stock. Requirements that are calculated on the basis of the master production schedule and bills of material are called gross requirements. If the entire gross requirement can be covered from stock, there will be no need for further quantities. If, on the other hand, available stocks of items are insufficient, there will be a need for a new quantity equivalent to the difference between the gross requirement and stock on hand. This quantity is called a net requirement. The reason that gross requirements can sometimes be covered by items available in stock is that order quantities are often larger than the exact gross requirement.

To correctly calculate when a net requirement arises, information will be needed on current stock levels for all items. Order quantities are determined by various lot sizing methods. For internally manufactured items there is also the option of taking into consideration the degree of scrapping when sizing lots, i.e. if for example on average 2 per cent of manufactured quantities for an item is expected to be scrapped, the manufacturing order quantity will be adjusted upwards by a factor of 1 / 0.98.

In those periods when a net requirement of an item arises, there will be a need for deliveries of quantities equivalent to these net requirements. Requirements calculations will identify at what point in time and which items need to be available, but to know when items need to be ordered (from suppliers or internal manufacturing) information will also be required on each item's lead time. The lead time information will then be the basis of lead time offsetting from planned delivery time to the time for planned start. The method of calculation when running material requirements planning as described above is illustrated in Figure 11.12.

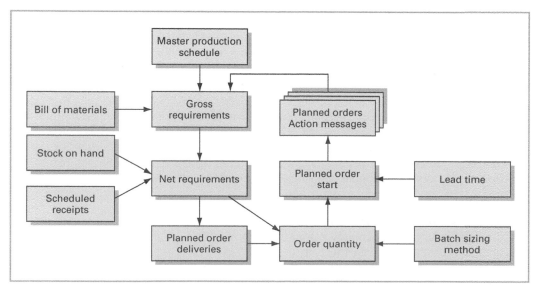

FIGURE 11.12 Calculation for material requirements planning

Example 11.5

A company manufactures two products, R and S. Table 11.1 shows master production schedules for the products during the next 12 weeks. The week numbers in the production schedule refer to planned assembly start for each product.

Week	1	2	3	4	5	6	7	8	9	10	11	12
S (pcs)	200		300	300		200	100		300		500	
R (pcs)		400		300		400		500		200		600

TABLE 11.1 Master production schedules for products R and S

Product R consists of 3 pcs of item A and 4 pcs of item B. Product S consists of 3 pcs of item B and 1 of item C. Items B and C are manufactured by the company.

Item B consists of purchased items E (1 piece) and F (4 pcs). Item C is an internally manufactured item that consists of the purchased items A (2 pcs) and G (1 piece). The bills of materials thus appear as in Figure 11.13.

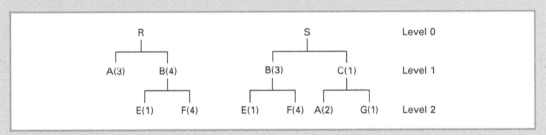

FIGURE 11.13 Bills of material for products R and S

Other information regarding order quantities, lead times and stock on hand for products and components is illustrated in Table 11.2.

Item	Order quantity	Lead time	Stock on hand
A	3000 pcs	2 weeks	3500 pcs
B	2500 pcs	1 week	1400 pcs
C	1000 pcs	2 weeks	700 pcs
E	3000 pcs	1 week	2500 pcs
F	10 000 pcs	2 weeks	10 000 pcs
G	1200 pcs	1 week	700 pcs

TABLE 11.2 Data for products and items

Material requirements planning is used to generate planned orders for items included in the end products. Below calculations of planned orders for item B, at bill of materials level 1, and for item F, at bill of materials level 2, in both products are shown. The material requirements calculations for item B are shown in Table 11.3.

Week	1	2	3	4	5	6	7	8	9	10	11	12
Part of S – 3pcs	600		900	900		600	300		900		1500	
Part of R – 4pcs		1600		1200		1600		2000		800		2400
Total gross requirements	600	1600	900	2100	0	2200	300	2000	900	800	1500	2400
Scheduled receipts												
Stock on hand	800	1700	800	1200	1200	1500	1200	1700	800	0	1000	1100
Net requirements	0	800	0	1300	0	1000	0	800	0	0	1500	1400
Planned delivery		2500		2500		2500		2500			2500	2500
Planned order start	2500		2500		2500		2500			2500	2500	

TABLE 11.3 Material requirements calculations for item B

The following calculation steps are carried out in Table 11.3:

1 *Identification of gross requirements*. The bills of material that contain the item B are identified. In this example, 3 pcs of item B are included in the final assembly of every S product and 4 pcs of B in the final assembly of every R product. This means that 3 and 4 B pcs respectively must be available when manufacturing of S and R products is started. The master production schedule states how many S and R products are planned to be started in each week. Using the bills of material and the production schedule information, it can be calculated how many B items must be available in each week for final assembly to be executed according to the master production schedule. In week 1, for example, assembly of 200 S products is started. There are 3 items of B required for final assembly of every S product. No R products are planned to be manufactured in week 1. The total requirement (gross requirement) of B items in week 1 is therefore 600 pcs. Correspondingly, the gross requirement of B items in week 4 is 2100 pcs. There are 900 pcs required for final assembly of 300 S products and 1200 pcs for the final assembly of 300 R products. The gross requirements are calculated for all weeks until the planning horizon.

2 *Netting of gross requirements*. The next step in the MRP process is to net the gross requirements, i.e. to check that the stock on hand and the scheduled receipts will be sufficient to cover gross requirements. In the example illustrated, there is a gross requirement of 600 pcs of B in week 1, but since stock on hand at the start of week 1 is 1400 pcs of B, there is no further requirement of items (net requirement). The gross requirements can be covered by withdrawing 600 pcs from stock. Stock on hand at the end of week 1 will then be 1400 − 600 = 800 pcs. In week 2 the gross requirement is 1600 pcs of B, stock on hand at the beginning of the week is 800 pcs and no scheduled receipts for that week. This means that there is a requirement for a further 800 pcs (1600 − 800). That quantity is the net requirement at the beginning of the week.

3 *Covering requirements with scheduled receipts*. The fact that a net requirement arose in week 2 means that a delivery must be planned for the start of the week. The size of the delivery (order quantity) is determined by a lot sizing rule. In this example, the order quantity is fixed and equivalent to 2500 pcs.

4 *Planning the start time in the right week (lead time offsetting)*. To make it possible to deliver a new order of 2500 pcs in week 2, it must be started the equivalent of one lead time earlier. The lead time is one week, so the order must be planned to start in week 1. This means that backward scheduling is applied, i.e. the requirement time is the starting point and the start of order is planned as late as possible to be ready exactly when the requirement arises. Planned orders generated through material requirements calculations mean that orders for B items are planned to be started at the beginning of weeks 1, 3, 5, 7, 10 and 11. If this is carried out, there will be sufficient quantities of B items available to fulfil the master production schedules for the S and R products in which B items are incorporated.

Planned order start (B)	2500		2500		2500		2500			2500	2500	
	Week											
	1	**2**	**3**	**4**	**5**	**6**	**7**	**8**	**9**	**10**	**11**	**12**
Part of B (4 pcs)	10 000		10 000		10 000		10 000			10 000	10 000	
Gross requirement	10 000		10 000		10 000		10 000			10 000	10 000	
Scheduled receipts												
Stock on hand	0	0	0	0	0	0	0	0	0	0	0	0
Net requirement			10 000		10 000		10 000			10 000	10 000	
Planned delivery			10 000		10 000		10 000			10 000	10 000	
Planned order start	10 000		10 000		10 000			10 000	10 000			

TABLE 11.4 Material requirements calculations for item F

5 *Release of orders.* As long as an order is planned but not released, it is called a planned order. When a decision is made on the order it is converted into a released order, also called a scheduled receipt. Releasing means that the order is handed over to the purchasing department for ordering from a supplier or to workshop planning department for start of manufacturing.

The corresponding steps of calculation are carried out for all items in the bills of material. The difference is that instead of using master production schedules as a basis for generating gross requirements, planned orders are used for items at higher structural levels. The material requirements calculation for item F is shown in Table 11.4.

The gross requirements of F items are obtained by calculating how many F items are part of each B item. To start manufacturing of 2500 B items, 10 000 F items are required, since 4 F items are included in every B item. The gross requirements of F items will thus be 10 000 pcs in weeks 1, 3, 5, 7, 10 and 11.

Table 11.5 shows material requirements calculations for all items included in the products R and S. The low-level code will determine in what sequence the requirements calculation for each item must be carried out. One item may be included in several products' product structures. The principal means that every item is given a "low-level code", which corresponds to its lowest structural level in the manufacturing product structures in which it is included. By first carrying out a requirements calculation for all items with the low-level code 0 and then continuing with low-level codes 1, 2, 3, etc., it is ensured that necessary input data is available when requirement calculations for each item are to be made.

Item B	Low-level code 1											
	1	**2**	**3**	**4**	**5**	**6**	**7**	**8**	**9**	**10**	**11**	**12**
Part of S (3 pcs)	600		900	900		600	300		900		1500	
Part of R (4 pcs)		1600		1200		1600		2000		800		2400
Gross requirements	600	1600	900	2100	0	2200	300	2000	900	800	1500	2400
Scheduled receipts												
Stock on hand	800	1700	800	1200	1200	1500	1200	1700	800	0	1000	1100
Net requirement	0	800	0	1300	0	1000		800			1500	1400
Planned delivery		2500		2500		2500		2500			2500	2500
Planned order start	2500		2500		2500		2500			2500	2500	

Item C						Low-level code 1						
	1	2	3	4	5	6	7	8	9	10	11	12
Part of S (1 pc)	200		300	300		200	100		300		500	
Gross requirements	200		300	300		200	100		300		500	
Scheduled receipts												
Stock on hand	500	500	200	900	900	700	600	600	300	300	800	800
Net requirement				100							200	
Planned delivery				1000							1000	
Planned order start		1000							1000			

Item A						Low-level code 2						
	1	2	3	4	5	6	7	8	9	10	11	12
Part of R (3 pcs)		1200		900		1200		1500		600		1800
Part of C (2 pcs)		2000							2000			
Gross requirements		3200		900		1200		1500	2000	600		1800
Scheduled receipts												
Stock on hand	3500	300	300	2400	2400	1200	1200	2700	700	100	100	1300
Net requirement				600				300				1700
Planned delivery				3000				3000				3000
Planned order start		3000				3000				3000		

Item E						Low-level code 2						
	1	2	3	4	5	6	7	8	9	10	11	12
Part of B (1 pc)	2500		2500		2500		2500			2500	2500	
Gross requirements	2500		2500		2500		2500			2500	2500	
Scheduled receipts												
Stock on hand	0	0	500	500	1000	1000	1500	1500	1500	2000	2500	2500
Net requirement		2500		2000		1500				1000	500	
Planned delivery			3000		3000		3000			3000	3000	
Planned order start		3000		3000		3000			3000	3000		

Item F						Low-level code 2						
	1	2	3	4	5	6	7	8	9	10	11	12
Part of B (4 pcs)	10 000		10 000		10 000		10 000			10 000	10 000	
Gross requirements	10 000		10 000		10 000		10 000			10 000	10 000	
Scheduled receipts												
Stock on hand	0	0	0	0	0	0	0	0		0	0	0
Net requirement			10 000		10 000		10 000			10 000	10 000	
Planned delivery			10 000		10 000		10 000			10 000	10 000	
Planned order start	10 000		10 000		10 000			10 000	10 000			

Item G						Low-level code 2						
	1	2	3	4	5	6	7	8	9	10	11	12
Part of C (1 pc)		1000							1000			
Gross requirements		1000							1000			
Scheduled receipts												
Stock on hand	700	900	900	900	900	900	900	900	1100	1100	1100	1100
Net requirement		300							100			
Planned delivery		1200							1200			
Planned order start	1200							1200				

TABLE 11.5 Material requirements calculations for all items

Variants of material requirements planning

Fixed, half-fixed and varying order quantities can be used within the framework of material requirements planning principles. Fixed and half-fixed order quantities are calculated and handled in the same way as in the re-order point system. Varying order quantities are achieved by adding together net requirements as successively calculated period for period ahead in time.

Planning conditions are never so stable that long-term forecasts and lead times always correspond to real outcomes. In order to create stable planning conditions and to avoid all too frequent rescheduling and adaptation to changing conditions, the option of creating buffers with the help of safety mechanisms is often utilised. Normally this is achieved through using safety stocks or safety lead times. If safety stocks are used, the opening stock quantity is reduced by the chosen safety stock quantity before the calculations of net requirements are started. If safety lead time is used, the delivery time for the planned order is set as equal to the calculated net requirement time minus the safety lead time. Methods for determining order quantities and safety mechanisms are treated in more detail in Chapters 12 and 13.

Material requirements planning is primarily designed for planning items with dependent demand. When it is used for planning independent materials requirements, the method is sometimes called **time-phased order point (TPOP)**. It may be regarded as an alternative to the traditional re-order point method. Time-phased order point uses forecasted demand by period and a time-phased MRP logic to develop planned orders.

Characteristic properties of material requirements planning

Material requirements planning is characterised by the following properties:

- In terms of applications, the method is more complex and difficult to understand and master than the three previous methods.

- Material requirements planning is, in principle, the most suitable material planning method for items with dependent demand in environments with standard products. Correct structural links between different items incorporated into products are ensured when new orders are created. The method also enables tracing the origins of material requirements resulting in new orders.

- In contrast to the previous methods, new orders are planned so that delivery can take place as late as possible.

- By its very nature, material requirements planning is oriented towards products and markets. New orders are generated only from customer order demand, or demand from established master production schedules at the end-product level.

- Available capacity is not taken into consideration when planning orders. The aim instead is to generate ideal flows of materials and then in a second step, through capacity requirements planning, to determine what capacity requirements exist to bring about these ideal flows. Making capacity adjustments is, according to the MRP approach, an issue related to the execution of plans made.

- Allocations and gross requirements are treated equally from the material requirements planning viewpoint. The difference is that allocations stem from customer orders, or released manufacturing orders, whereas gross requirements represent explosions from planned manufacturing orders.

- Advance scheduling can be achieved by generating proposals for new order plans far ahead of time. The only real limitation that exists is the horizon of the master production schedule.

■ For every material requirements plan, the correct timing of orders already planned is checked in relation to newly calculated net requirements. If such orders have been planned too late or too early, the material requirements planning system will signal the need for bringing forward or postponing them. Rescheduling may also be carried out automatically within the framework of a requirements planning run.

Primary application environments

Material requirements planning is mainly useful in the following environments:

■ *Dependent demand.* Material requirements planning is primarily useful in planning environments with dependent demand. Such environments exist, for example, when planning items incorporated in end products such as raw materials, purchased components and internally manufactured parts and semi-finished items.

■ *Allocations and seasonal variation.* Material requirements planning is also useful for items with independent demand and is an alternative to the re-order point system, periodic ordering systems or run-out time planning. The method is especially advantageous in environments where materials are allocated for customer orders and in cases with seasonal variation in demand.

■ *Bills of material.* The method assumes that there are already bills of material prepared in advance in the ERP database when it is used to control items incorporated in products.

■ *High basic data quality.* It puts high demands on basic data quality if its relative advantages are to be exploited.

11.3.5 Constraint-based material requirements planning

One of the major strengths with material requirements planning is its capability to synchronise the flow of materials at times when new manufacturing orders for items on the next higher level in the bills of material are planned to start. This is, however, only accomplished from the demand side as material requirements planning assumes infinite capacity. No consideration is given to limitations on the supply side such as late deliveries from suppliers or from the manufacturing of semi-finished items due to lack of capacity.

By carrying through the master production scheduling process before running material requirements planning, capacity considerations can be taken into account on a rough level. Still, due to the unavoidable occurrence of bottlenecks and overloaded work centres in the short term, the supply of material will be disrupted and not perfectly synchronised.

To achieve perfect synchronisation, material flows must also be synchronised from the supply side, i.e. consideration must also be given to capacity shortages and disruptions in the inbound flow of materials. The significance of including the synchronisation of material flows from the supply side can be illustrated with the aid of Figure 11.14. Assume that the requirement time for items A, B and C has been estimated at a point in time $t(1)$ by using material requirements planning. If it happens later that item C for some reason cannot be delivered from an external supplier until a point in time $t(2)$, it is no longer necessary for items A and B to be delivered at $t(1)$. These deliveries can be postponed so that they will be delivered at $t(2)$. Postponement means that tied-up capital can be reduced by avoiding deliveries that are too early, and that capacity can be freed for other manufacturing orders that may be of higher priority. Similar reasoning is valid if C is an internally manufactured item and the scheduling of its manufacturing order to finite capacity shows that it is not possible to complete it until $t(2)$.

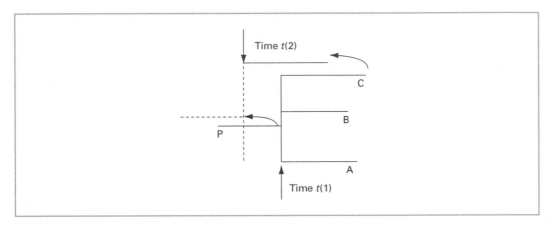

FIGURE 11.14 Illustration of the lack of synchronisation of material flows from the supply side at nodes in the material flows

Using **constraint-based material requirements planning** is a way to overcome some of these problems. Constraint-based MRP can be considered as the second generation material requirements planning and is in principle based on the **theory of constraints (TOC)** as outlined in Chapter 3. Access to some kind of **advanced planning and scheduling (APS) system**, parallel to or integrated with an ERP system, is necessary to apply the concept of constraint-based material requirements planning. Different approaches and optimisation methods are used within the concept but typically it is a stepwise procedure, in principle carried out in the following way.

When adopting a stepwise approach, it is important to identify external supply **constraints** prior to resource bottlenecks in manufacturing, for example inabilities by some suppliers to deliver as planned. The reason to start by carrying out this step is that capacity is not an issue if the material to be used in manufacturing is not going to be available. Purchase orders not being possible to reschedule to meet current requirements are then fixed and not allowed to be rescheduled by the ERP system.

A material requirements planning to infinite capacity is then run in almost the same way as traditional MRP. The difference is that all action messages concerning needs to reschedule open orders usually managed by planners are handled automatically within the system during the planning phase. This also means that manufacturing orders using materials that are not available in time due to late deliveries from suppliers are rescheduled to a later date.

From this material requirements planning run to infinite capacity it becomes possible to find out where the capacity bottlenecks are, i.e. which work centres are overloaded and constitute constraints regarding possible output from manufacturing. Having found the constraints, an appropriate next step in line with the theory of constraints is to try to elevate and exploit the capacity in the constraining work centres as much as possible, for example by using overtime or allocating resources from non-bottleneck work centres to bottleneck work centres. After having taken such capacity planning decisions, capacity levels specified in the work centre files are updated. Changes in the master production schedule to reduce load in the bottlenecks may also be a viable option.

This step could be seen as a what-if simulation analysis allowing the planner to assess consequences of various actions to resolve bottleneck problems and improve the synchronisation of the material flow. The what-if simulations can be carried out repeatedly as many times as wanted and needed until a reasonable solution has been achieved.

The next step in an appropriate working procedure when using constraint-based material requirements planning is to subordinate the utilisation of the non-bottleneck resources and to balance the material flow by carrying out a finite capacity material requirements planning run. The difference compared to traditional material requirements planning is that all manufacturing orders are planned within available capacity. This is typically done by using various rules to prioritise among the manufacturing orders competing on the same capacity in the bottleneck work centres to achieve as optimal a solution as possible. The rules used and the way they are applied are specific for each APS system.

When having found a bottleneck, a basic principle used is to not schedule the planned manufacturing order in the same way before the bottleneck as after the bottleneck. As the bottleneck work centre is limiting the material flow, it is important that the flow through work centres prior to the bottleneck is balanced against the flow through the bottleneck work centre. If not, inventories and work in process will be built up in front of the bottleneck work centre. To accomplish this, manufacturing orders carried out in non-bottleneck work centres are back scheduled from the requirements times in bottleneck work centres. Backward scheduling means basically that the last operation in a manufacturing order is scheduled first and at a time when it has to be finished to avoid shortages. After that the next last operation is scheduled backwards and so on. By doing so no order or operation starts before it is necessary.

Since the bottleneck work centres are limiting the output from the manufacturing system, it is important that the manufacturing orders and operations in work centres after the bottlenecks are carried out as early as possible. To accomplish this, these orders and operations are forward scheduled, i.e. the first operation on a manufacturing order is scheduled first and as early as possible considering available capacity as well as other manufacturing orders requiring the same manufacturing resources. After that the second operation is forward scheduled and so on. Backward and forward scheduling is further described in Chapter 14.

Like material requirements planning the constraint-based version demands information on the master production schedules, bills of material, stock on hand, lot sizing methods and lead times. In addition to this, routing information and work centre information such as available capacity is also required.

Characteristic properties of constraint-based material requirements planning

The characteristic properties of constraint-based material requirements planning are to a large extent the same as traditional material requirements planning. Some major differences are described below:

- The method is even more complex and difficult to understand and master than traditional material requirements planning.

- Traditional ERP systems are not sufficient for using constraint-based material requirements planning. An additional system of APS type is required.

- Available capacity is taken into consideration when planning new orders.

- If orders have been planned to finish too late or too early, material requirements planning systems typically generate action messages for the responsible planners to react to. With constraint-based material requirements planning necessary rescheduling is mostly carried out automatically within the system.

Primary application environments

The method is mainly useful in the following environments:

- *Dependent requirements.* Like traditional material requirements planning the constraint-based version is primarily useful in planning environments with dependent requirements.

- *Bills of material and routings.* The method assumes that bills of material and routings are prepared in advance in the ERP database.

- *High data quality.* Constraint-based material requirements planning does not just put high demands on the quality of data such as bill of materials, stock on hand and lead time. If its relative advantages are to be exploited the high demand on data quality must also be put on routings and work centre capacity.

11.4 Properties and Characterisation

Five different material planning methods have been described in this chapter. The methods have different properties and are more or less useful in different environments. The application of the methods varies between companies and certain methods are more frequently used than others. The characteristic properties of the five material planning methods treated in the chapter are summarised in Table 11.6.

11.5 Summary

This chapter treated the administrative activities in a company which are required to control material flows at an operative level within a given production and distribution structure. These activities are called material planning. The activities are carried out within the framework of master production scheduling according to the planning hierarchy discussed in Chapter 3. Material planning is considered, then, as the function in a company which executes plans established at strategic and tactical levels.

The fundamental starting points for material planning and the order processes and planning processes that initiate flows of materials were discussed. The balance between supply of materials and demand for materials that must be maintained to achieve cost-effective flows of materials and a high delivery service were also examined. The following five methods of material planning were analysed: the re-order point system, the periodic ordering system, run-out time planning, material requirements planning and constraint-based MRP. Definitions, material planning calculations, method variants and characteristic properties of the methods were reviewed. The methods are used to synchronise material flows in order to achieve a balance between supply and demand.

Property variables	Re-order point system	Periodic ordering system	Run-out time planning	Material requirements planning	Constraint-based MRP
Type of demand	Forecasts, history of usage	Forecasts, history of usage	Forecasts, history of usage	Allocations, exploded demand, forecasts	Allocations, exploded demand, forecasts
Product and component orientation	Component	Component	Component	Product	Product
Nature of demand	Independent	Independent	Independent	Dependent, independent	Dependent, independent
Initiation principle	Usage initiated	Plan initiated	Usage initiated	Demand initiated	Demand initiated
Planning principle	From time of review	From time of planning	From time of review	From time of requirement	From time of requirement or time of available capacity
Priority-basing information	No, no priority-basing information	No, no priority-basing information	Yes, gives priority-basing information	Yes, gives priority-basing information	Yes, gives priority-basing information
Replanning potential	No replanning potential	No replanning potential	Identification of replanning requirement	Identification and execution	Identification and execution
Type of material plan	Single order + call-off	Single order + call-off	Single order + call-off	Delivery plan + order	Delivery plan + order
Re-ordering interval	Varying time interval	Fixed time interval	Varying time interval	Varying time interval	Varying time interval

TABLE 11.6 Comparison of material planning methods and their properties

🔑 Key concepts

Advanced planning and scheduling (APS)
  system 229

Allocation 206

Constraint 229

Constraint-based material requirements
  planning 229

Co-ordination stock 211

Cover time planning 217

Delivery schedule 214

Dependent demand 205

Explosion 204

Firm planned order 220

Gross requirement 214

Independent demand 205

Lead time 205

Manufacturing order 205

Material requirements planning (MRP) 204

Net requirement 219

Order-up-to level 204

Pegging 219

Periodic ordering system 215

Planned order 205

Planning horizon 219

Pull 205

Purchase order 205

Push 205

Release time fence 221

Released order 206

Re-order point system 208

Rescheduling 205

Run-out time planning 217

Safety lead time 218

Safety stock 204

Scheduled receipt 217

Stock on hand 206

Theory of constraints (TOC) 229

Time fence 219

Time-phased order point (TPOP) 227

Discussion tasks

1 The manufacturing lead time is an important parameter in material planning. Discuss how the lead time is used in the various material planning methods presented in this chapter.

2 Run-out time planning is a material planning method closely related to re-order point systems. What is in principle the basic difference between the two methods?

3 Why may material requirements planning be a more suitable material planning method than a re-order point method in environments with seasonal variation in demand?

Problems[1]

Problem 11.1

Spade and Rake Ltd manufacture garden tools to stock. Run-out time planning is used to control stock. The following data apply for two different models of spade, SP-1 and SP-4.

Product	Stock on hand	Lead time (days)	Safety lead time (days)	Forecast (pcs/day)
SP-1	37	7	2	5
SP-4	68	12	5	5

a) Should new manufacturing orders be planned to fill stocks for either of the products, and which of them if that is the case?

b) Both SP-1 and SP-4 are manufactured in the same work centre and thus compete for the same capacity. Which of the products SP-1 and SP-4 should be given priority and manufactured first from the material planning aspect if both need to be manufactured to fill up stocks?

Problem 11.2

At Natural Products Health Centre, packets of Vitamineral tablets are kept in stock, among other products. Consumption varies with weekdays and has been forecast in numbers per day as in the table below. Day 1 is Monday, day 2 Tuesday and so on until Saturday, which is day 6. Using material requirements planning, decide when orders must be placed. The optimal order quantity is 10 pcs and stock on hand at the time of ordering in the morning of day 1 is 9 pcs. Lead time from ordering to delivery is one day. There is no safety stock kept due to the short replenishment lead time.

Day	1	2	3	4	5	6	1	2	3	4	5	6	1	2
Forecast	6	6	2	2	2	1	6	6	2	2	2	1	6	6
Stock on hand														
Delivery day														
Order day														

Problem 11.3

Monopoly Ltd uses the purchased item XZ117 continuously in the manufacturing of a large number of products. The item has the following characteristics:

Demand per year	8000 pcs
Purchase price	100 €/piece
Safety stock	$1.6 \times$ lead time demand
Lead time from order to delivery	4 weeks
Purchasing quantity	1000 pcs
Inventory carrying cost per year	20 per cent of the item value

---

1   Solutions to problems are in Appendix A.

A re-order point system that continuously reviews stock on hand is used to generate purchase orders. The product is produced and in demand in all 52 weeks of the year.

a) Determine the re-order point, the average tied-up capital in inbound stock, the inventory turnover rate in the stock, the average run-out time in the stock and the annual inventory carrying cost for the item.

b) Because the company buys several items from the same supplier it is considering changing its review principle from continuous review to periodic review stock on hand of all purchased items every second week. By doing this it is thought that it will be possible to co-ordinate deliveries of several items from this supplier. How will the answers in (a) change if a switch is made to periodic review?

Problem 11.4

The company in the previous problem is considering taking one further step and exchanging the re-order point method for a periodic ordering method for the purchase order planning of the item in question.

Determine the order-up-to level, the average tied-up capital in stock, inventory turnover rate in stock, run-out time in stock and the annual inventory carrying cost using the same data as in problem 11.3. Also determine the order quantity if the stock on hand on review was 1600 pcs.

Problem 11.5

Framer Ltd makes and sells wooden frames. The range of standard frames is made unpainted to stock. The range contains, for example, the model "Antique", with the variants A, B, C and D. Manufacture has two operations: milling and assembly. The time used when planning a manu-facturing order is the milling production time + 0.05 hours per piece for assembly. Only one piece can be made at a time in the milling operation and assembly. The total available capa-city is 80 hours per week in the milling operation and 80 hours per week in the assembly. They work five days a week. The company uses run-out time planning to plan new manufacturing orders. The following table contains current planning data in milling for the four variants:

Variant	Standard time (hrs/piece)	Lot size	Demand per week	Inventory balance	Safety stock
A	0.10	100	35	100	70
B	0.20	150	50	120	100
C	0.25	100	40	130	80
D	0.20	200	40	100	80

a) For what variants should new manufacturing orders be planned?

b) Which of the variants for which manufacturing orders will be planned should be scheduled first, if only stock on hand is considered?

c) Which of the variants for which a manufacturing order will be planned should be scheduled first, if production time and lot size are also considered?

Problem 11.6

A company makes products X and Y. The table below contains the master production schedules for the two products. The weekly figures show the number of products that are planned for production start.

Week	1	2	3	4	5	6	7	8	9	10	11	12
Prod. X		400		400		400		500		300		500
Prod. Y	200		300	300		200	100	100	200		500	

Product X consists of item A (3 pcs) and item B (4 pcs).

Product Y consists of item B (3 pcs) and item C (1 piece).

Item C consists of item A (2 pcs) and item D (1 piece).

Item A consists of item E (2 pcs) and item F (1 piece).

The items have the following planning data:

Item	Lot size	Lead time	Stock on hand
A	3000	2 weeks	2500
B	1500	1 week	1200
C	600	2 weeks	400
D	600	1 week	400
E	300	1 week	200
F	500	1 week	350

Conduct an MRP calculation for a 12-week period for item A. In what weeks should new manufacturing orders be planned for deliveries of item A?

Problem 11.7

The following circumstances exist for an item controlled with material requirements planning:

Week	1	2	3	4	5	6	7	8
Gross requirements	50	75	25	50	70	80	75	70
Scheduled receipts		400						
Projected stock on hand	150	475	450	400	330	250	175	105
Planned order start								

The lot size for the item is 400 pcs, the safety stock is 100 pcs and the lead time is four weeks.

a) Physical inventory counting is conducted at the end of week 1 and the correct stock balance is 30 pcs. What consequence does this have for the MRP plan?

One piece of the item is used in the products X and Y. The master production schedules representing order at start date for X and Y are revised weekly. The assembly lead time is two weeks. In the beginning of week 2 they look like this:

Week	2	3	4	5	6	7	8	9
Product X	25	25	25	30	30	25	40	40
Product Y	50		25	40	50	50	30	50

The following actions take place at the beginning of week 2:

- 150 of the 400 pcs that are scheduled to be finished during week 2 do not pass the quality control and have to be scrapped. Only 250 pcs of the item are delivered to stock.

- The weekly forecast for the recently introduced Product Y is increased by 10 pcs from week 6.

- A large customer order of 110 pcs of product X with delivery in week 7 has been accepted. Due to this large order the marketing department estimates the demand for weeks 6, 7, 8 and 9 will decrease by 20 pcs per week.

- As a result of efficiency improvements in production, lead times have decreased to three weeks and also set-up times have been slightly reduced. With these new conditions the material planning team decides to change future lot sizes to 300 pcs and safety stock to 75 pcs.

b) Conduct an MRP calculation for the item for weeks 2 to 9, taking the above actions into consideration.

Problem 11.8

The company Bicycle Stand has two different products, P1 and P2. A master production schedule has been established for these products as shown in the table below, stating the weeks in which assembly start is planned for each product.

Week	1	2	3	4	5	6	7	8	9	10	11	12
Product P1	50	80	50		80	20	60	80		60	50	70
Product P2	110	200		80	140		50	40	80	120	60	50

To assemble product P1, 2 pcs of item A and 3 of item B are required, and to assemble product P2, 1 pc of item B and 2 pcs of item D. The following data describe the items.

Item	Lot size	Lead time	Stock on hand
A	400	2 weeks	800
B	600	1 week	1000
D	300	1 week	650

No orders have been released for the items previously. Carry out material requirement planning over a 12-week period for item B. In which weeks must new purchase orders be planned to be delivered for item B?

Further reading

Fogarty, D., Blackstone, J. and Hoffmann, T. (1991) *Production and inventory management.* South-Western Publishing, Cincinnati (OH).

Jonsson, P. and Mattsson, S.-A. (2006) "A longitudinal study of material planning applications in manufacturing companies", *International Journal of Operations and Production Management,* Vol. 26, No. 9, pp. 971–995.

Landvater, D. (1993) *World class production and inventory management.* The Oliver Wight Companies, Essex Junction (VT).

Oden, H., Langenwalter, G. and Lucier, R. (1993) *Handbook of material & capacity requirements planning.* McGraw-Hill, New York (NY).

Orlycky, J. (1975) *Material requirements planning.* McGraw-Hill, New York (NY).

Plossl, G. (1985) *Production and inventory control – principles and techniques*. Prentice-Hall, Englewood Cliffs (NJ).

Ptak, C. (2000) *ERP tools, techniques, and applications for integrating the supply chain*. St Lucie Press, Boca Raton (FL).

Rabinovich, E. and Evers, P. (2002) "Enterprise-wise adoption patterns of inventory management practices and information systems", *Transportation Research Part E*, Vol. 38, pp. 389–404.

Silver, E., Pyke, D. and Peterson, R. (1998) *Inventory management and production planning and scheduling*. John Wiley & Sons, New York (NY).

Vollman, T., Berry, W., Whybark, C. and Jacobs, R. (2005) *Manufacturing planning and control for supply chain management*. Irwin/McGraw-Hill, New York (NY).

Wallace, T. (1990) *MRP II: making it happen*. Oliver Wight Publications, Essex Junction (VT).

Wild, T. (2002) *Best practice in inventory management*. Butterworth-Heinemann, Oxford.

CASE STUDY 11.1: MATERIAL REQUIREMENTS PLANNING AT MEDTECH LTD

Medtech Ltd is an internationally operating corporation developing, manufacturing and marketing products for medical care. The products are delivered to hospitals all over the world by a corporate global supply unit. This case study deals with one of the manufacturing plants located in southern Europe, making one model of a heart/lung machine.

Products and production

This particular heart/lung machine model consists of some 450 components supplied from 124 different suppliers, mainly from Europe and China. The lead times for these components are between 3 and 16 weeks. The bill of material for the model has only two levels, the end-product model level and a component level consisting of purchased components common for all variants of the model. Almost 30 per cent of the components used in the machine are also delivered as spare parts. In addition to these spare parts, spare parts intended for past models of heart/lung machines are sold and delivered as well. All spare parts are planned to be delivered from stock the day the customer order arrives, with a service level of 97 per cent.

All delivered products are customer-order specific variants created by adding specific items to the base model, for example electric connecting devices and labels in different languages. Some 60 optional items of this kind exist. These are not included in the bill of material for the model but added to the customer-order specific bills when the customer order arrives.

The production consists mainly of assembly, testing and packing. The assembly operation is carried out at just one assembly line with a planned capacity of 40 machine models on average per week. If needed, it is possible to temporarily increase the capacity to 50 machine models by using overtime and some extra personnel who are available. The lead time for the assembly is about two days. Each assembled machine is tested during three days. Due to low quality of some of the purchased components, the result of this testing quite often is that some components have to be exchanged. These exchanges of erroneous components are made in a separate area in the production department. The delivery lead time to customer is planned to be no more than six days excluding transportation time.

Spare parts are delivered both as individual components and as assemblies of components. The assembly of spare parts takes place in a separate area in the production department. These spare parts are assembled to customer order. Components used for the assembly of machines as well as components used for assembly of spare parts and components sold directly as individual spare parts are stocked together and accounted for as one stock on hand figure. The delivery lead time excluding transportation time for assembled spare parts is planned to be one day.

Forecast and master production schedule

Based on a forecast of future sales of the base model and current customer orders, a master production schedule is made by global supply. This master production schedule is 12 months long and is updated on a monthly basis. In addition to this, new customer orders with specific variants of the base model are sent from global supply to the production department on a daily basis.

The demand for the machines is fairly stable but to maintain the targeted delivery lead times to the customers, some variations in production volumes have to be accepted. To achieve this, the following policy when establishing new master production schedules is applied. The first two weeks in the master production schedule are frozen, i.e. no quantity changes are allowed from one master production schedule to the next. From week 2 up to week 8 a quantity change of 10 per cent is allowed and from week 8 to week 16 a quantity change of 20 per cent is allowed.

Material requirements planning

Based on the established master production schedule and the current customer orders a material requirements planning is carried out once a week, on Monday night. Based on the planned orders received from these MRP runs, new purchase orders are released at the planning department. These orders are sent by fax or email to the suppliers. According to company policy, quantities and due dates for released orders should be updated when order confirmations are received from suppliers. Due to lack of time, however, this is quite often not carried out. All existing planned orders are rescheduled automatically by the ERP system, while released orders only are rescheduled manually based on action messages from the system.

To cover for the spare parts demand, two options are used. For items where spare parts demand corresponds to more than 15 per cent of the total demand, a spare parts forecast is added to the calculated requirements, while for items with less demand no forecasts exist and the spare parts demand is supposed to be covered by the safety stock used.

In the material requirements planning system used by the company, order quantities are calculated as fixed economic order quantities, rounded off to whole containers or pallets. The annual demand used in these calculations is calculated based on the most current master production schedule for the product model. The economic order quantity is revised once a month.

When running the material requirements planning, the stock on hand is reduced by the safety stock before planning to cover gross requirements. The size of the safety stock is calculated as two times the size of the standard picked quantity from the warehouse to the assembly line. The standard picked quantity is manually determined for each item, based on its physical size and its value. This means, for instance, that the standard picked quantities for large and expensive items may correspond to just a couple of days' demand, while for small and inexpensive items the picked quantities may correspond to a month's demand or more.

Discussion questions

1 At Medtech, a re-order point system is used to control the material flow of optional items added to the product model based on customer specifications. Why does the company not use material requirements planning for these items as well as for the common items?

2 A safety stock is used for the common items planned by material requirements planning. What different types of uncertainties is the safety stock supposed to cover in this case? What is your opinion about the method used for safety stock calculations?

3 What would it mean if the material requirements planning was run every night instead of once a week?

CHAPTER 12

Lot Sizing

The balance of supply and demand does not only include a time dimension but also a quantity dimension, and one of the basic issues of **material planning** relates to this quantity dimension. Quantities to be delivered must be determined for every order planned in the flow of materials. In a flow that is ideally synchronised, this order quantity will be equal to the quantity in demand at every point in time, i.e. if there is a demand for three pieces of a certain item on a certain day, then exactly those three pieces will be delivered on that day. Under such conditions there will be no need for stocks.

For different reasons it is often not suitable or possible to only manufacture or procure the quantity required at each point in time. Demand from several consumption occasions must be consolidated into larger quantities. The concepts of **lot sizing** and establishing suitable lot sizes are used. Manufacturing and procuring larger quantities than the immediate demand will cause the build-up of stocks. The type of stock in this case is called turnover stock. It also means that planning material flows has an element of push, even though it is the consuming unit that initiates the flow.

Many lot sizing methods have been developed with the aim of determining lot sizes as efficiently as possible with regard to **tied-up capital**, delivery service to customers and internal utilisation of resources. This chapter sets out the most common methods for determining lot sizes. The characteristic properties of each method are also discussed, as are in which contexts they may primarily be used.

LEARNING OBJECTIVES

After studying this chapter, you should be able to:

- Explain the motives for lot sizing
- Explain how to design and use different types of lot sizing methods
- Categorize and compare the characteristics of different types of lot sizing methods

12.1 Motives for Lot Sizing

The reasons for lot sizing may be roughly divided into financial and non-financial. The non-financial reasons are based on the fact that it is highly appropriate and in certain cases even necessary to manufacture or procure larger quantities than are needed for the moment, even without financial motives. Complete packaged quantities or full pallets when procuring materials are examples of this type of motive. The manufacturing process used may require a certain minimum quantity, as in a vessel for mixing and heating a chemical product, or a foodstuffs product that must utilise its entire volume.

The most common reasons for lot sizes deviating from quantities in direct demand are of a financial nature. Every order and delivery has its associated ordering costs, i.e. all those specific costs that are connected with executing an order process for procurement of items from an external supplier, or from an internal manufacture. The methods for calculating these ordering costs are described in Chapter 6. In general these costs are assumed to be independent of the quantity ordered. The larger the order quantity, the less will be the **ordering costs** per piece. This relationship favours the implementation of large order quantities.

On the other hand, if order quantities are greater than immediate needs, the surplus will need to be stored until it is used or delivered. The costs associated with this are called inventory **carrying costs**, and the elements included in these costs are described in Chapter 6. Storage is also associated with costs for depreciation and obsolescence. In principle, these costs are included in inventory carrying costs, but in certain contexts it may be necessary to give special consideration to the fact that some items, due to limited durability, may not be able to be stored longer than a certain time period. Consequently, they should not be procured or manufactured in quantities greater than the time taken for them to be used or sold within their shelf life. Inventory carrying costs may in general be assumed to be proportional to stock value, i.e. stocked quantity multiplied by standard price or average price for the item in question. In other words, inventory carrying costs rise as the quantity in stock increases.

The relationship between selected lot size and inventory carrying costs is illustrated in Figure 12.1. Since inventory carrying costs increase and ordering costs decrease as lot sizes become greater, the choice of a suitable lot size is obviously a trade-off between these two costs. The sum of inventory carrying costs and ordering costs must be optimised, and the term

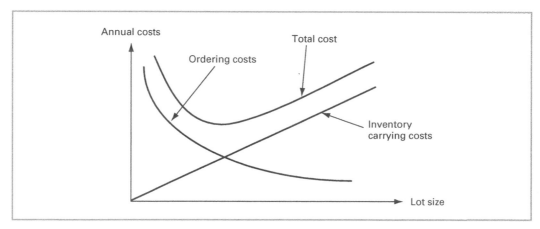

FIGURE 12.1 Relationship between order quantities and ordering costs, inventory carrying costs and total costs for storage

used in this context is **economic order quantity (EOQ)**. Virtually all lot sizing methods for calculating economic order quantities are based on this balance of costs.

As shown in Figure 12.1, the total cost curve is relatively level, meaning that additional costs are not especially large if it is necessary for some reason to deviate from the economically optimal order quantity as calculated. This is very significant in the practical application of lot sizing methods. As a result of this relationship, it is permissible to round off calculated quantities to even numbers, such as rounding up an order quantity to 200 even though the economic order quantity has been calculated as 191. It is also acceptable to round off calculated economic order quantities to multiples of packaging quantities, full load carrying units, etc. Finally, to a certain extent it is possible to adapt order quantities to pricing and discount structures of suppliers. If the order quantity is significant for the price or discount available, it may be more appropriate to use lot sizing methods that can take into consideration existing price and discount structures for calculating economic order quantities.

Since lot sizing involves procurement or manufacturing for future needs, it must be built on an expectation that these needs will arise and any costs that this risk-taking may incur. For companies that make to stock and deliver directly to customers from stock, this risk is greatest since demand is totally based on forecasts. The risk is less if the company has delivery agreements with its customers regarding volumes to be delivered on an annual basis. Likewise, companies that make to delivery schedule will face less risk since delivery schedules from customers will provide a safer indication of possible future quantities than if these are based on internal forecasts.

Lot sizing is not generally relevant for customer-order initiated manufacturing. However, a special type of lot sizing is sometimes practised. Lot sizing may in such cases be completely based on incoming orders. This means that a manufacturing order will not be placed until a sufficient number of orders have come in to cover the smallest acceptable manufacturing order quantity. This approach is used in the steel industry for example, due to very high costs for changing rollers and thus limitations on manufacturing small charges. Ordering costs in this case are not weighed against inventory carrying costs, since the finished products are not stored. Instead they must be balanced with possible loss of income or customers that may take place due to longer delivery lead times caused by the approach.

12.2 Categorisation of Lot Sizing Methods

Lot sizing methods used in industry can be classified according to whether the order quantity is fixed or variable from order to order, and whether the time period that the order is expected to cover is fixed or variable. This classification will produce a four-field matrix as shown in Figure 12.2.

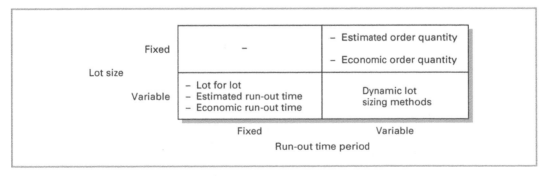

FIGURE 12.2 A structure for lot sizing methods

The combination of a fixed order quantity and a fixed **run-out time** is not a possible alternative, since it would assume that demand was completely constant. In consequence, there are no lot sizing methods that represent this case. The lot sizing methods described below belong to the other three categories.

Using a fixed order quantity means that on every ordering occasion the same fixed quantity is ordered. Since the quantity is fixed, the time between ordering occasions must vary. A fixed order quantity, however, does not mean that it never will be changed, only that it is not adapted to the short-term demand situation. Due to changes in demand over time, or due to other changes in the planning environment, it must be revised from time to time; for example, a couple of times per year. The same is true for fixed run-out time.

12.3 Lot for Lot

Lot for lot is the simplest lot sizing method used. The order quantity corresponds to a unique demand in this method. In other words, this method means that there is no lot sizing at all in the sense of consolidating demand. Thus, it does not involve any storage or tied-up capital. The purest form of this method, lot for lot, is quantity based, time invariant and discrete.

Lot for lot is most often used in a somewhat limited sense in conjunction with **material requirements planning**. It then stands for the total net requirement in an individual period, such as the net requirement during one day. This net requirement may be a sum of more than one demand though, or a consolidation of demand. Since this consolidation of demand only covers a certain time period, it is not usually considered as lot sizing. In this particular form, lot for lot is time based, time invariant and discrete, since the unchanged quantity corresponds to the run-out time of one period.

Using the lot sizing method lot for lot may mean that the order quantity in certain cases is very small, and there are many ordering occasions. The lot for lot method is mainly used in order-based material flows, for expensive products and other items, and in planning environments with short set-up times and thus low ordering costs, i.e. environments similar to a typical just-in-time environment. When planning items such as semi-finished items and purchased components, it is normally more often used at the high levels in the product structure, i.e. closer to the end product, than at lower levels. The application of the lot for lot method is illustrated in Figure 12.3.

12.4 Estimated Order Quantity

The method of estimated order quantity is built on a fixed order quantity being intuitively estimated with the aid of judgements based on experience. The estimations are based on

Week	1	2	3	4	5	6	7	8	9
Net requirement	10	6	19	8	13	27	17	12	6
Lot size	10	6	19	8	13	27	17	12	6
Stock on hand	0	0	0	0	0	0	0	0	0

FIGURE 12.3 Illustration of projected stock on hand when lot for lot is applied

estimated annual consumption, price, resource needed for executing the order process, the risk of obsolescence, etc. The method is principally quantity based and time invariant. However, in some cases it is also used as a time-variant method. This means that the order quantity is re-estimated on every ordering occasion, and consequently may vary from time to time.

Theoretically at least, an estimated order quantity is always inferior to an economic order quantity as a method, since it is virtually impossible to balance ordering costs and inventory carrying costs to achieve a reasonably optimised order quantity only on the grounds of experience. The method may be used, though, in simpler contexts when there is no system support for making calculations such as that provided by an ERP system. It may also be appropriate in situations where the information required to make systematic financial calculations is not available. For example, this may be the case when a new product is to be introduced and there is no historic demand data available. In addition, it may be more suitable to establish fixed order quantities on the basis of experience in situations where there are large restrictions on quantity. In the case of procurement, the order quantity may need to be adapted to a large unit load (package, pallet, container, etc.). There may also be weight, shelf life and volume restrictions on internal storage that will dictate any order quantities established.

One disadvantage of determining order quantities on the basis of experience is that it is difficult in practice as well as time-consuming to update order quantities as conditions and demand change. This can lead to delays in updating and increasing deviation from reasonably optimised order quantities over time. The situation is completely different when order quantities are calculated using some form of financial optimisation. Updating may then be carried out at regular intervals almost automatically.

12.5 Economic Order Quantity

The most commonly used lot sizing method for determining a fixed order quantity with the aid of calculations is the **square root formula**, or **Wilson's formula**. Calculations are based on minimisation of total costs for keeping the stock that an order gives rise to and for executing the ordering process. With the help of these calculations, an economic order quantity is obtained. In the same way as estimated order quantity, economic order quantity is quantity based, time invariant and non-discrete. That the order quantity is time invariant, i.e. fixed over time, does not mean that it never changes, but only that it is not adapted to the short-term demand situation. As a result of changes in demand, or due to other changes in the planning environment, it must be revised from time to time.

The formula for calculating an economic order quantity is based on the following assumptions:

■ Demand per time unit (D) is constant and known.

■ Lead time for stock replenishment is constant and known.

■ Stock is replenished momentarily with the entire order quantity on delivery.

■ Ordering costs per order occasion (S) are constant and known, and are independent of order quantity.

■ Inventory carrying costs per piece and time unit are constant and known, and are independent of order quantity ($I \cdot C$).

■ The price or cost of the item ordered is constant and known, and independent of order quantity and purchase occasion/manufacturing occasion.

As a result of the rather simplified assumptions for the EOQ formula, the term optimisation cannot be used in a strict sense with respect to prevailing conditions in normal planning situations.

Week	1	2	3	4	5	6	7	8	9
Net requirement	10	6	19	8	13	27	17	12	6
Lot size	30		30			30	30		
Stock on hand	20	14	25	17	4	7	20	8	2

FIGURE 12.4 Illustration of stock development when using fixed estimated order quantity or economic order quantity

Considering the flat total cost curve, at least the formula may be said to provide a satisfactory basis for decisions when determining suitable order quantities from a cost perspective. The formula is widely used in industry. In addition to taking into consideration the costs associated with material flows, the lot sizing method compared with estimated order quantity also has the advantage of being able to be calculated and stored in an ERP system. It is therefore easier and more rational to carry out updates of what are suitable order quantities when conditions change in the planning environment, for example when demand increases or decreases.

The use of a fixed order quantity can be illustrated with the aid of Figure 12.4. This figure reflects the situation for an item that is part of a product. Material requirements planning is applied, and consequently there is information about expected future demand. The order quantity of this item is 30 pcs. The conditions are the same in principle even if other material planning methods are used, or if it concerns products for sale. Conditions are the same whether the order quantity has been estimated manually or calculated with the aid of the square root formula.

In stocks with constant demand and instantaneous replenishment of the entire order quantity on recurrent occasions, stock on hand development will have the appearance of a saw tooth pattern, as illustrated in Figure 12.5.

Since demand, lead times and order quantities are known and constant, no safety stock will be required. The maximum stock level will correspond to the order quantity Q and average stocks will be equal to half the order quantity $Q/2$. Since demand per time unit is D, one order quantity will be consumed during the time period Q/D, which corresponds to the interval between replenishments, or the stock replenishment cycle.

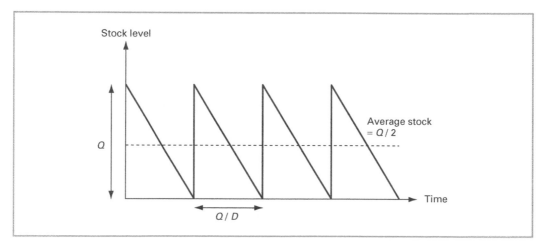

FIGURE 12.5 Stock level development under the assumptions that apply for Wilson's formula

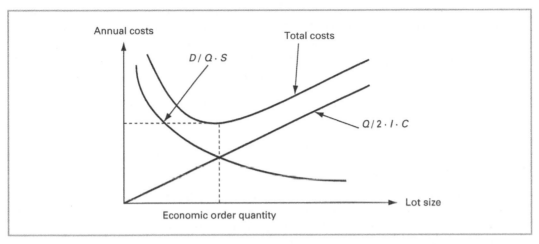

FIGURE 12.6 Relationship between order quantity and ordering costs, inventory carrying costs and total derived costs

With the above assumptions, the annual inventory carrying costs will be equal to the average stock $(Q/2)$ multiplied by the inventory carrying costs per piece and year $(C \cdot I)$. The variation of the annual inventory carrying costs with order quantity is illustrated in Figure 12.6. As shown in this figure, the cost is proportional to the order quantity.

In a similar fashion, the annual ordering costs will be equal to the number of orders per year (D/Q) multiplied by the ordering costs per ordering occasion. Consequently, fewer ordering occasions will mean that annual ordering costs decrease, i.e. the larger the quantities ordered, the less ordering costs per year. This relationship is also illustrated in Figure 12.6.

The annual total costs for storage and ordering can then be expressed with the aid of the formula below. The order quantity at the local minimum point constitutes the optimal quantity in financial terms, i.e. the economic order quantity EOQ.

$$Total\ cost = TC = \frac{Q}{2} \cdot I \cdot C + \frac{D}{Q} \cdot S$$

A formula for calculating the optimum order quantity is obtained by differentiating the total cost function with respect to the order quantity.

$$\frac{dTC}{dQ} = \frac{I \cdot C}{2} - \frac{D}{Q^2} \cdot S = 0$$

$$\frac{I \cdot C}{2} = \frac{D}{Q^2} \cdot S$$

$$Q = \pm \sqrt{\frac{2 \cdot D \cdot S}{I \cdot C}}$$

$$EOQ = \sqrt{\frac{2 \cdot D \cdot S}{I \cdot C}}$$

where D = demand per time unit
           Q = order quantity
           EOQ = economic order quantity
           S = ordering costs per ordering occasion
           I = inventory carrying costs per time unit
           C = item value per stock unit

As can be seen in Figure 12.6 and as was pointed out above, the total cost curve is very flat, meaning that the sum of derived costs is relatively insensitive to order quantities that deviate from the optimal value. The calculation of economic order quantities is accordingly relatively insensitive to faults in parameters included in the formula. If, for example, demand increases by 10 per cent then EOQ will increase by $\sqrt{1.10} = 4.88$ per cent. If the original value of EOQ had been used despite the increase in demand, total derived costs would increase, but only by 0.11 per cent. If inventory carrying costs decrease by 20 per cent then EOQ will decrease by 11.8 per cent. If, despite this, the original value of EOQ is maintained, derived costs will only increase by 0.62 per cent.

Example 12.1

A purchased item has an annual demand of 10 000 pcs that is evenly distributed over the entire year. Inventory carrying costs are 20 per cent per year. The item is valued at its average purchase price of 100 €/piece and ordering costs are estimated to be 30 € per ordering occasion. The economically optimal order quantity according to Wilson's formula will then be as follows:

$$EOQ = \sqrt{\frac{2 \cdot 10\,000 \cdot 30}{0.20 \cdot 100}} = 173 \text{ pieces}$$

Considering that the total cost curve is relatively flat and additional costs will not be especially large if, for different reasons, it is necessary to deviate from the economically optimal calculated order quantity, it is possible to round off calculated quantities to even figures, such as rounding up the calculated order quantity to 200.

Example 12.2

A stocked item has a purchase price of 100 €/piece. Demand for the item has been forecast at 50 000 pcs/year. The inventory carrying costs and ordering costs have been calculated as 1.6 €/piece/year and €100 per order respectively.

The optimal order quantity according to Wilson's formula and average stocks for the item will be as follows:

$$EOQ = \sqrt{\frac{2 \cdot 50\,000 \cdot 100}{1.6}} = 2500 \text{ pcs}$$

$$Average\ stock = \frac{2500}{2} = 1250 \text{ pcs}$$

Every piece of the item has a volume of 20 dm³ and the maximum storage volume is 45 000 dm³. The product has a limited shelf life and may only be stored for a maximum of 30 days. Due to cash flow problems the company has decided to limit the average stock value to a maximum of €100 000. Taking these restrictions into account, the following order quantities are the maximum permissible.

Volume restriction → Max stock = 45 000 / 20 = 2250 pcs

Shelf life restriction → Max stock = 50 000 · 30 / 360 = 4166 pcs

Cash flow restriction → Max stock = 100 000 · 2 / 100 = 2000 pcs

Both the volume restriction and cash flow restriction mean that the optimal order quantity cannot be used in this case. Since the cash flow restriction is the greatest, this will determine what order quantity can be used, i.e. the quantity that should be ordered is 2000 pcs.

12.6 Economic Order Quantity with Quantity Discounts

A large number of variants of the traditional formula for calculating economic order quantities have been developed and are published in textbooks and scientific papers. The majority of these variants have been developed by eliminating or modifying one or more of the assumptions that lie behind the formula. Here, the variant considering **quantity discounts** is presented.

In cases where suppliers offers quantity discounts if order quantities exceed a certain figure, it may be financially optimal to deviate from the optimal order quantity as calculated by the EOQ formula and instead accept the larger quantity at which a discount will be given off the purchase price. A corresponding condition applies if a supplier has a price structure that favours large quantities, i.e. a lower price is given if a larger quantity is purchased. The reason for suppliers wishing to influence customers to purchase certain minimum quantities may be because they wish to minimise their ordering costs, that they do not wish to break packaged quantities, or that they wish to fill a whole loading unit during transportation.

If the quantity required for a discount off the purchase price is higher than the calculated economic order quantity without giving consideration to discounts, total inventory carrying costs and ordering costs will increase if the higher quantity is accepted. For the total costs to be lower with quantity discounts, annual savings of purchase costs through quantity discounts must exceed the increase in inventory carrying costs and ordering costs, i.e. the difference between TC_{opt} and $TC_{discount}$ in Figure 12.7 is negative. Costs per year in the figure include purchasing costs, inventory carrying costs and ordering costs. The relationships are illustrated with the aid of an example.

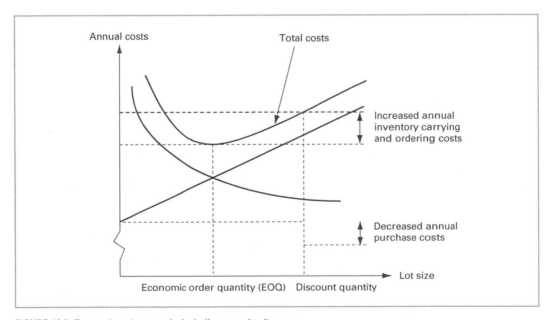

FIGURE 12.7 Economic order quantity including quantity discount

Example 12.3

Determine the economic order quantity and the annual costs for inventory carrying and ordering for a purchased item if the annual demand is 10 000 pcs. Inventory carrying costs are 20 per cent per year, the purchase price of the item is 100 €/piece without discounts and the ordering cost is €30 for each order. There are no order quantity restrictions, but the supplier allows a 1 per cent quantity discount if the order quantity corresponds to one packaging unit of 1000 pcs.

$Q = EOQ$:

$$Q_{opt} = \sqrt{\frac{2 \cdot 10000 \cdot 30}{0.20 \cdot 100}} = 173\,pcs$$

$$TC_{opt} = \sqrt{2 \cdot 10000 \cdot 30 \cdot 0.20 \cdot 100} = 3464\,€/year$$

$$Purchase\ cost = 10\,000 \cdot 100 = 1\,000\,000\,€/year$$

$$Total\ cost_{548} = 3464 + 1\,000\,000 = 1\,003\,464\,€/year$$

$Q = 1000$ pcs:

$$TC = \frac{10000}{1000} \cdot 30 + \frac{1000}{2} \cdot 0.2 \cdot 100 \cdot 0.99 = 10\,200\,€/year$$

$$Puchase\ cost = 10\,000 \cdot 100 \cdot 0.99 = 990\,000\,€/year$$

$$Total\ cost_{1000} = 10\,200 + 990\,000 = 1\,000\,200\,€/year$$

$$Total\ cost_{548} - Total\ cost_{1000} = 1\,003\,464 - 1\,000\,200 = 3264\,€/year$$

If the quantity discount is chosen, the total ordering and inventory carrying costs will increase by 6736 €/year (10 200 – 3464 €/year) while the discounts on the purchase price will be 10 000 €/year. Since the discount for one year exceeds the annual increase in derived costs, the total cost will be lower if a whole package of 1000 pcs is chosen instead of the optimal order quantity according to Wilson's formula, i.e. 173 pcs.

12.7 Estimated Run-out Time

When using the lot sizing method of estimated run-out time, an order quantity is selected so that it covers a number of planning periods, such as weeks or days. In the same way as for estimated order quantity, the selection of the number of periods is based on a manual assessment through experience of what may be a suitable time interval to cover needs each time. The estimations here, too, are based on estimated annual demand, price, how expensive the ordering process is, risk of obsolescence, etc.

The use of estimated run-out time as a lot sizing method means that the order quantity is specified as a time length expressed in a number of periods in the ERP system. The order quantity is then calculated at every ordering occasion based on this number of periods and the current demand during the periods. There are two different alternatives. In the material requirements planning environment, the order quantity is calculated by adding forecast demand, allocations and exploded material requirements over a number of periods. If usage-based methods are used instead, such as the **re-order point system**, the order quantity is calculated as the number of periods multiplied by the average demand per period. Estimated run-out time is a time-based, time-invariant and discrete lot sizing method. Time invariant means that the run-out time does not change from order to order. However, the order quantity changes since the demand to be covered during the run-out time varies. This is particularly the case when the method is used together with material requirements planning.

Example 12.4

Annual demand for an internally manufactured item is estimated at 260 pcs. For lot sizing in conjunction with material planning, a run-out time of two weeks is used. If usage takes place during 52 weeks of the year, the order quantity may be calculated in the following way:

$$Q = 260 \cdot \frac{2}{52} = 10 \text{ pcs}$$

12.8 Economic Run-out Time

The only real difference between estimated run-out time and **economic run-out time** is the way in which the run-out time is established. Economic run-out time can be calculated as a trade-off between ordering costs and inventory carrying costs that arise so that the total costs associated with storage are minimised. This is achieved most simply by first calculating the economic order quantity and subsequently calculating the economic run-out time in periods, ERT, with the aid of the following formula:

$$ERT = \frac{EOQ}{D}$$

where D = average demand per period

By calculating the run-out time in this way, order quantities on average will be equivalent to the economic order quantity. The use of run-out time as a lot sizing method is illustrated in Figure 12.8 with the same demand data as for lot sizing methods illustrated above. In this example the fixed run-out time is three weeks. Note that order quantities in this planning environment are adapted to whole weeks' requirements.

In Example 12.5, a comparison is made between the ordering costs and the inventory carrying costs that arise when the economic order quantity is used instead of the economic run-out time.

The example illustrates, in all its simplicity, one of the strengths of run-out time compared with economic order quantity, since it allows order sizes to automatically adapt to changes in demand to a certain extent. Consequently, run-out times will not need to be updated as often as order quantities. If, for example, demand increases, the quantity required during the run-out time will increase and, as a result, the order quantity will also increase – which is only reasonable considering the increase in demand.

Week	1	2	3	4	5	6	7	8	9
Net requirement	10	6	19	8	13	27	17	12	6
Lot size	35			48			35		
Stock on hand	25	19	0	40	27	0	18	6	0

FIGURE 12.8 Illustration of stock development when using run-out time as a lot sizing method

Example 12.5

An internally manufactured item has an annual demand of 4800 pcs distributed over the next 12 weeks according to the table. Due to long set-up times, ordering costs are €40 and inventory carrying costs 2.4 €/piece/year, i.e. 0.05 €/piece/week. Opening stock on hand is 0.

Week	1	2	3	4	5	6	7	8	9	10	11	12
	90	50	50	100	120	220	150	150	120	70	50	30

If lot sizing is carried out based on the economic order quantity (EOQ) formula, the economic order quantity will be as follows:

$$EOQ = \sqrt{\frac{2 \cdot 4800 \cdot 40}{2.4}} = 400 \text{ pcs}$$

If economic run-out time (ERT) is used for lot sizing instead and is calculated on the basis of economic order quantities, the following is obtained:

$$ERT = \frac{400}{100} = 4 \text{ weeks}$$

If for the sake of calculation it is assumed that demand for one whole period is withdrawn from stock at the beginning of the week, at the same time as a stock replenishment takes place, the consequences of using the economic order quantity will be as shown in Table 12.1. The table shows which orders have been generated, the size of stock, inventory carrying costs and ordering costs per period. The average stock will be 167 pcs, the annual inventory carrying costs €401 and the annual ordering costs €480. Total derived costs will be €881 per year.

	Outcome when using economic order quantities											
Week	1	2	3	4	5	6	7	8	9	10	11	12
Demand	90	50	50	100	120	220	150	150	120	70	50	30
Order	400				400			400				
Stock on hand	310	260	210	110	390	170	20	270	150	80	30	0
Inventory carrying costs	15.5	13	10.5	5.5	19.5	8.5	1	15	7.5	4	1.5	0
Ordering costs	40				40			40				

TABLE 12.1 Inventory carrying costs and ordering costs when lot sizing using economic order quantities

When using economic run-out time, orders are generated three times during the planning period covered by the calculations, each for order quantities corresponding to four weeks' requirements. Using the same assumptions as for economic order quantities, the outcome will be as shown in Table 12.2. The ordering costs will be equal to those with economic order quantities, but the average stock will decrease to 140 pcs (16 per cent reduction), which results in €65 lower inventory carrying costs per year. Total derived costs will be 816 €/year.

	Outcome when using economic run-out time											
Week	1	2	3	4	5	6	7	8	9	10	11	12
Demand	90	50	50	100	120	220	150	150	120	70	50	30
Order	290				640				270			
Stock on hand	200	150	100	0	520	300	150	0	150	80	30	0
Inventory carrying costs	10	7.5	5	0	26	15	7.5	0	7.5	4	1.5	0
Ordering costs	40				40				40			

TABLE 12.2 Inventory carrying costs and ordering costs when lot sizing using economic run-out time

The example illustrates, in all its simplicity, one of the strengths of run-out time compared with economic order quantity, since it allows order sizes to automatically adapt to changes in demand to a certain extent. Consequently, run-out times will not need to be updated as often as order quantities. If, for example, demand increases, the quantity required during the run-out time will increase and as a result the order quantity will also increase – which is only reasonable considering the increase in demand.

12.9 Dynamic Lot Sizing Methods

The third category of methods in the matrix in Figure 12.2 is called dynamic lot sizing methods. Using these methods, quantity and run-out time both vary between consecutive orders. The methods demand more advanced calculations but provide also, at least in theory, more optimal order quantities. **Least unit cost method**, least period cost method and the **Silver-Meals method** are examples of dynamic lot sizing.

The theory behind the methods is built on inventory carrying costs being as large as ordering costs for the optimal order quantity. Every time a new order is created, successive calculations period for period are carried out to check if the quantity for one further period will be covered by the order. It is above all in material requirements planning environments with reasonably certain, known requirements that the advantages of dynamic methods are most apparent in comparison to other methods. Using dynamic lot sizing methods, order quantities are calculated so that they are the equivalent of a full number of periods' requirements, in a similar way to run-out time.

Silver-Meals method

The Silver-Meals method, also known as the least period cost method, is one of the most widely used dynamic lot sizing methods. Just as the other dynamic methods (e.g. the **least total cost**

method), it is characterised by being time variant, with respect to both quantity and time, and it is discrete.

Using the Silver-Meals method, the average sum of inventory carrying costs and ordering costs per period determines whether the order quantity will correspond to one period's requirements, two periods' requirements, three periods' requirements and so on. The order quantity is equal to the number of periods' requirements which result in the lowest average derived cost per period. It is a sequential method, since it successively considers demand in future periods. When demand in period 2 is considered, it determines whether it is economic to plan a new order with delivery during that period. If this is not the case, it checks instead whether the new delivery should take place in period 3, and so on. With the same conditions as in the example using economic run-out time, the Silver-Meals method will result in the following steps in calculations:

1 Calculate how many weeks' demand the order with delivery in week 1 will cover.

   Average total of derived costs per week will be as follows:

   If one week is included: €40
   If two weeks are included: $(40 + 50 \cdot 0.05 \cdot 1) / 2 = 21.30$ €/week
   If three weeks are included: $(40 + 50 \cdot 0.05 \cdot 2 + 50 \cdot 0.05 \cdot 1) / 3 = 15.8$ €/week
   If four weeks are included: $(40 + 100 \cdot 0.05 \cdot 3 + 50 \cdot 0.05 \cdot 2 + 50 \cdot 0.05 \cdot 1) / 4 = 15.60$ €/week
   If five weeks are included: $(40 + 120 \cdot 0.05 \cdot 4 + 100 \cdot 0.05 \cdot 3 + 50 \cdot 0.05 \cdot 2 + 50 \cdot 0.05 \cdot 1)/5 = 17.30$ €/week.

   The lowest average derived cost per week is obtained if the order quantity is equivalent to four weeks' demand, i.e. 290 pcs. A new order must therefore be planned for delivery in week 5.

2 Calculate how many weeks' demand the order to be delivered in week 5 must cover.

   Average total of derived costs per week will be as follows:

   If one week is included: €40
   If two weeks are included: $(40 + 220 \cdot 0.05) / 2 = 22.50$ €/week
   If three weeks are included: $(40 + 150 \cdot 0.05 \cdot 2 + 220 \cdot 0.05) / 3 = 22.00$ €/week
   If four weeks are included: $(40 + 150 \cdot 0.05 \cdot 3 + 150 \cdot 0.05 \cdot 2 + 220 \cdot 0.05 \cdot 1)/4 = 22.10$ €/week.

   The lowest average derived cost per week is obtained if the order quantity is equivalent to three weeks' demand, i.e. 490 pcs. A new order must therefore be planned for delivery in week 8.

3 Calculate how many weeks' demand the order to be delivered in week 8 must cover.

   Average total of derived costs per week will be as follows:

   If one week is included: €40
   If two weeks are included: $(40 + 120 \cdot 0.05) / 2 = 23.00$ €/week
   If three weeks are included: $(40 + 70 \cdot 0.05 \cdot 2 + 120 \cdot 0.05) / 3 = 17.70$ €/week
   If four weeks are included: $(40 + 50 \cdot 0.05 \cdot 3 + 70 \cdot 0.05 \cdot 2 + 120 \cdot 0.05) / 4 = 15.10$ €/week
   If five weeks are included: $(40 + 300 \cdot 0.05 \cdot 4 + 50 \cdot 0.05 \cdot 3 + 70 \cdot 0.05 \cdot 2 + 120 \cdot 0.05 \cdot 1)/5 = 13.30$ €/week.

	Outcome when using the Silver-Meals method											
Week	**1**	**2**	**3**	**4**	**5**	**6**	**7**	**8**	**9**	**10**	**11**	**12**
Demand	90	50	50	100	120	220	150	150	120	70	50	30
Order	290				490			420				
Stock on hand	200	150	100	0	370	150	0	270	150	80	30	0
Inventory carrying costs	10	7.5	5	0	18.5	7.5	0	13.5	7.5	4	1.5	0
Ordering costs	40				40			40				

TABLE 12.3 Inventory carrying costs and ordering costs when lot sizing using the Silver-Meals method

The lowest average derived cost per week is obtained if the order quantity is equivalent to five weeks' demand, i.e. 420 pcs.

When using the Silver-Meals method, orders will be generated three times during the period covered by the calculations. Refer to Table 12.3. Ordering costs will be €480 and inventory carrying costs €300. Total derived costs will thus be 780 € for the planning period covered.

Using the Silver-Meals method of lot sizing with the above conditions will result in lower inventory carrying costs compared with economic run-out time. The total derived costs will be €36 lower per year, equivalent to almost 4.4 per cent.

12.10 Method Comparison

A number of different lot sizing methods have been described above. Their properties differ in a number of ways and thus they are more or less applicable, depending on the context. A summary of their properties is shown in Table 12.4.

All of the lot sizing methods included in Table 12.4, except for the Silver-Meals method, are represented in virtually all ERP systems on the market. Many ERP systems also include some form of dynamic lot sizing method. However, in practice these methods are seldom used. When they are used it is primarily in combination with material requirements planning.

Property variables	Lot for lot	Estimated order quantity	Economic order quantity	Estimated run-out time	Economic run-out time	Silver-Meals method
Quantity versus time based	–	Quantity	Quantity	Time	Time	–
Variable versus fixed quantity	Variable	Fixed	Fixed	Variable	Variable	Variable
Variable versus fixed time	Fixed	Variable	Variable	Fixed	Fixed	Variable
Discrete versus non-discrete	Discrete	Non-discrete	Non-discrete	Discrete	Discrete	Discrete

TABLE 12.4 A comparison of methods with respect to different properties

12.11 Summary

This chapter has set out motives and considerations for lot sizing. Many lot sizing methods have been developed with the aim of determining order quantities as efficiently as possible with respect to tied-up capital and internal utilisation of resources. The most common methods for determining order quantities were described: lot for lot, estimated order quantity, economic order quantity, estimated run-out time, economic run-out time and one of the more common dynamic lot sizing methods. The characteristic properties of methods were outlined and the contexts in which they are normally used.

🔒 Key concepts

Carrying cost 242	Ordering cost 242
Economic order quantity (EOQ) 243	Quantity discount 249
Economic run-out time 251	Re-order point system 250
Least total cost method 253	Run-out time 244
Least unit cost method 253	Silver-Meals method 253
Lot for lot 244	Square root formula 245
Lot sizing 241	Tied-up capital 241
Material planning 241	Wilson's formula 245
Material requirements planning 244	

Discussion tasks

1 A number of lot sizing methods are available. Among these lot for lot, economic order quantity and estimated run-out time are frequently used in industry. Which of these methods are the most appropriate to apply when using material requirements planning in situations characterized by (a) negligible set-up and other ordering costs in manufacturing, (b) medium-sized ordering costs and seasonal demand for purchased items and (c) medium-sized ordering costs and stable and frequent demand for manufacturing items?

2 What are the practical difficulties of properly determining the inventory carrying cost in a company?

3 How will different costs of a supermarket change if they increase lot sizes of purchased food stuff? Discuss what it must do in order to not get an increased total cost after the change.

Problems[1]

Problem 12.1

Johnson & Sons Ltd purchase the following items, among others, from the same supplier. The following conditions apply for both items:

	Item A	Item B
Demand per year	3200 pcs	4500 pcs
Ordering costs per order	€25	€35
Purchase price	€30	€25

The fact that ordering costs for item B are larger than for item A is a result of the more exhaustive quality control that must be carried out for each lot delivered. An inventory carrying cost of 15 per cent applies at the company.

a) The company now receives items once a month. How will the sum of inventory carrying costs and ordering costs per year be influenced if item A is ordered 20 times a year and item B 30 times a year?

b) How large is the economic order quantity for each item? How large will the sum of inventory carrying costs and ordering costs be per year if the company chooses to purchase lot sizes equivalent to the economic order quantity? How many times a year must each item be ordered if these order quantities are used?

c) The supplier offers a discount if full pallets are purchased each time. The number of items A per pallet is 1000 pcs. The number of items B per pallet is 500 pcs. Should the company purchase full pallets of each item if the supplier offers a 1 per cent quantity discount off the purchase price?

d) In order to decrease annual ordering costs, the company is considering always ordering both items at the same time. Joint ordering costs will be €50 for simultaneous ordering. How much change will there be in the sum of inventory carrying costs and ordering costs per year if both items are ordered the same number of times per year as for item A in question (b) above? What is the optimal economic number of purchases per year if the items are ordered equally often? What will be the size of any changes in costs?

Problem 12.2

The following data apply for an item purchased by a manufacturing company:

Demand per year	10 000 pcs
Ordering costs per order	€50
Purchase price	€50
Inventory carrying costs per year	15 per cent of goods' value

---

1   Solutions to problems are in Appendix A.

a) Determine the economic order quantity for the item.

b) Every piece of the item has a volume of 30 dm^3 and the maximum storage volume is 5500 dm^3. The item has a limited shelf life and may be kept in stock for a maximum of 20 days. Due to demands of good solvency, it has been decided not to allow a stock value that exceeds €7500. With due consideration to the above restrictions, determine how large lot sizes should be purchased each time.

Problem 12.3

Svenssons' Mechanical Workshop is evaluating what lot sizing method should be used with its MRP system. The choice is between economic order quantity (EOQ), economic run-out time (ERT) and Silver-Meals method. In order to analyse the suitability of the different methods, it has been decided to test how well they would operate for a typical item over eight historic periods. The following data apply to the item:

Week	1	2	3	4	5	6	7	8	9
Demand	10	45	55	80	60	100	50	20	30

Ordering costs per order	€50
Inventory carrying costs	€10 per piece and year
Stock on hand at start of period 1	0 pcs

Which lot sizing method will result in the lowest sum of inventory carrying costs and ordering costs? Assume that one whole week's demand is withdrawn from stocks at the beginning of the week and that inbound deliveries are put in stock at the beginning of the week.

Problem 12.4

The Plastic Nicolas Company manufactures a plastic product for finished goods stocks. The product is delivered to stocks as it is manufactured. Stock withdrawals take place continuously. Production is in batches and the production rate is 30 pcs per hour. The start-up material is granules, which are also used for a large number of other products. The product is delivered to many different customers relatively evenly over the year. The following conditions apply:

Demand per year	90 000 pcs
Manufacturing costs per piece	€100
Inventory carrying costs	15 per cent of the goods' value
Raw materials costs	€50 per kg
Raw materials usage	1 kg per manufactured product
Cost of man-hours	€30 per hour
Cost of machine-hours	€50 per hour

Installing a tool in the process machine takes 3 hours, and removing the tool and cleaning the machine takes 0.5 hours. In both cases two men work simultaneously. The first production hour is needed for "fine tuning". The products manufactured during the tuning process are scrapped without being recycled. After installing the tool the machine is then served by one man. The workshop has a continuous three-shift system for a full workload. Every manufacturing batch produces a number of full pallets, which are transported one by one to stocks. For each pallet shipped there is a truck cost estimated at €100. Quality control is carried out for each manufacturing batch and takes 30 minutes per manufactured batch plus 10 minutes per 100 pcs. The hourly cost of the quality control is €30.

a) How large is the ordering cost per manufacturing batch and the economic order quantity?

b) Assume that the time taken for setting up a tool has been reduced to 0.5 hours, but that it is still necessary for two men to work simultaneously. What will be the ordering cost per manufacturing batch, and the economic order quantity?

c) Assume that the workload in the workshop has decreased and that there is only a 60 per cent utilisation rate of the machines. What will be the ordering costs per manufacturing batch, and the economic order quantity? Assume that the time for installing the tool is 3 hours, as in question (a).

So far the company has used a general inventory carrying factor of 15 per cent to determine inventory carrying costs. The company now wishes to calculate the real inventory carrying costs, based on the following data:

Calculated interest	10 per cent per year
Inventory carrying costs	€50 per pallet and month
Insurance costs	€1 per stocked product and year
Estimated obsolescence	3 per cent of the manufacturing batch with average run-out time of finished goods stock > 1 month
	0 per cent of the manufacturing batch with average run-out time of finished goods stock < 1 month

One pallet holds 100 of the products. Products that have been phased out are valued at €0 and all are put in "blocked stocks", to be remanufactured for second grade granules or to be scrapped. As stated above, it is estimated that all products can be sold if the average run-out time of finished goods stocks is less than one month. There are no safety stocks.

d) How large is the economic order quantity if inventory carrying costs are calculated as above and other conditions are the same as in question (a)?

Problem 12.5

Clockmaster Ltd calculates economic order quantities using the EOQ formula when ordering stocked spare parts for wall clocks. Ordering costs are estimated at €35 and inventory carrying costs 25 per cent per year. For one particular spare part, the purchase price is €100 and its demand estimated to be 1200 pcs per year. The packaging size provided by the supplier contains 50 pcs.

a) What quantities should be ordered at a time?

b) What consequences does the rounding off to an even packaging quantity have for total inventory carrying costs and ordering costs?

Problem 12.6

At John's Health Food Store, run-out time is used to determine how much must be ordered each time since there is great variation in demand during the six sales days of the week. The product Glentamin requires three days' run-out time. How large will the first three order quantities be if the demand of the product is estimated from the table below and the first order is for delivery on day 1?

Day	1	2	3	4	5	6	7	8	9	10	11	12	13	14	15
Forecast	4	6	10	12	15	20	4	6	10	12	15	20	4	6	10

Problem 12.7

At the Gift Shop, sales vary a great deal during the year and run-out time is used to determine the size of order quantities from the supplier on each ordering occasion. The economic order quantity has been calculated for two of the company's products as in the table below, which also shows forecast sales for the 48 sales weeks of the year.

Item	Economic order quantity (pcs)	Forecast per year (pcs)
Glass vase Antique 2	12	96
Glass vase Nature experience	13	150

a) How long will the economic run-out time be for the two items?

b) Assume that both items are purchased from the same supplier and that orders are co-ordinated to reduce transportation costs. What run-out times should then be applied?

Further reading

Ballou, R. (2004) *Business logistics/supply chain management.* Prentice-Hall, Upper Saddle River (NJ).

Bernard, P. (1999) *Integrated inventory management.* Oliver Wight Publications, Essex Junction (VT).

Bowersox, D., Closs, D. and Cooper, B. (2002) *Supply chain logistics management.* McGraw-Hill, New York (NY).

Fogarty, D., Blackstone, J. and Hoffmann, T. (1991) *Production and inventory management.* South-Western Publishing, Cincinnati (OH).

Jonsson, P. (2008) *Logistics and supply chain management.* McGraw-Hill, New York.

Lambert, D., Stock, J., Ellram, L. and Grant, D. (2005) *Fundamentals of logistics management.* McGraw-Hill, New York (NY).

Muller, M. (2003) *Essentials of inventory management.* AMACOM, New York (NY).

Silver, E., Pyke, D. and Peterson, R. (1998) *Inventory management and production planning and scheduling.* John Wiley & Sons, New York (NY).

Vollmann, T., Berry, W., Whybark, C. and Jacobs, R. (2005) *Manufacturing planning and control for supply chain management.* Irwin/McGraw-Hill, New York (NY).

Car Accessories is a supplier of car accessories with manufacturing facilities in a couple of European countries. The company develops, manufactures and markets products such as roof racks and roof top boxes. The case study is limited to the manufacturing facility making roof racks and parts necessary for rack installations on cars. The number of manufactured products is close to 1000. The manufacturing strategy used at this production unit is make to stock. The manufactured products are stocked in a warehouse close to the production site. This warehouse is used as a central warehouse for further distribution to a number of warehouses in Europe.

Demand characteristics and forecasting

The demand for most of the rack products is highly seasonal with a major peak during the autumn and a minor peak during Easter. The total demand for all products during three autumn months is close to 30 per cent higher than the average demand per month over a whole year, and for some products the seasonal variation is even bigger. This demand pattern has been stable for many years. The peak season during the autumn is associated with large wholesalers building up their stocks in good time before the ski season.

Forecasting plays a big role at the company and is carried out once a month. For the 200 biggest and most important products the first step in the forecasting process is to scrutinise previous years' sales. One designated person within the sales organisation analyses the figures and tries to identify trends and possible changes in the seasonal pattern. A preliminary forecast regarding sales per month for each product during the coming 12 months is prepared. In the next step sales managers and product managers look at sales irregularities and consider various aspects of future sales, for example as a consequence of planned sales promotions. In the final step, sales history and future expectations are put together and a final forecast established.

For the remaining products a more simple forecasting process is applied. These products are simply forecasted by multiplying last year's sales per month by a percentage to reflect future trends on the market. The percentages are estimated per product group by the sales manager.

Inventory control system

The currently used inventory control system for managing end products at the central warehouse is a time-phased order point system. The system is run once a week and generates planned inventory replenishment orders for the coming six months. The established forecasts, safety stocks calculated from the forecast and current stock on hand constitute the basis for this planned order generation. The normal replenishment lead time for most of the products is about 15 days.

Seasonally varying demand means varying manufacturing capacity requirements. The opportunities to be flexible enough and adapt the capacity to the seasonal variation in demand is, however, quite low at Car Accessories. Accordingly, the company has to apply a strategy with a production rate that is rather even and instead build up a seasonal inventory during low seasons. To accomplish this, replenishment orders planned by the inventory control system are rescheduled to be delivered at earlier dates than needed by planners at the planning department.

2   Based on Hansson Rahnboy, H. and Högberg, A. (2007) "How to boost inventory performance", master thesis TMTP-07/5621, Lund University, Lund, Sweden.

Establishing order quantities

The order quantities for replenishment orders are set by the manufacturing department product by product for each planned order. It is based on a manually estimated so-called order multiple. Estimated sales volumes and manufacturing conditions are considered when making these estimates. The order multiple is set when a product is introduced on the market and is fixed during the entire product life cycle. The order multiple quantity is entered into the ERP system used at the company for the inventory control.

The current order quantities when planned orders are released are set equal to a number of the determined order multiple. The number used may change depending on the situation at hand, for instance depending on the current season and the load situation. It is the person responsible for each production line who decides what number to use in each situation. According to a policy applied, the number of order multiples must not lead to order quantities exceeding one day of production.

Considering alternative ways to determine lot sizes

The current way to determining lot sizes has been questioned by the planning department for some time. The lot sizing methods Economic order quantity and Economic run-out time have been discussed as alternatives. To apply these methods the ordering cost has been calculated to be 125 € for all products since the set-up time is roughly the same for all of them. The inventory carrying cost has been set to 25% per year of the product cost.

The planning manager has also been thinking about using more efficient policies when rescheduling planned orders to be carried out earlier than needed to make the capacity requirements more even before and during the high demand seasons. He thinks that it might for instance be possible to reduce the seasonal inventory by selecting products to manufacture in advance by considering differences in costs, capacity requirements and future demand for the various products. Today no specific rules are applied when selecting products to be manufactured in advance. It is up to judgments by the individual planners.

Discussion questions:

1 What is your opinion about the lot sizing method currently used at Car Accessories?

2 Assume that the company can adjust the manufacturing capacity to correspond to the capacity requirements from the seasonal variation in demand. What will then the consequences for the annual ordering cost and carrying cost be when using Economic order quantity as opposed to Economic run-out time.

3 The set-up times when manufacturing the rack products are quite high. How might that influence the choice between Economic order quantity and Economic run-out time?

4 Which policies should be applied when selecting products to manufacture in advance instead of the current ones to be able to reduce the seasonal inventory?

Determining Safety Stocks

I t is almost impossible to perfectly synchronise the demand and supply of material flows. This is especially true for the time dimension, i.e. to exactly co-ordinate the times for inbound and outbound deliveries of material due to demand uncertainties. Disturbances in the material flows from supplying to consuming units also make the supply side uncertain. Material flow uncertainties, consequently, both concern supply and demand and refer to quantity and time. In order to hedge against these uncertainties, different kinds of buffering mechanisms, mainly **safety stocks** and **safety lead times**, are applied.

Uncertainties in future demand mainly concern quantity uncertainty, i.e. uncertainty regarding what quantities will be demanded and sold in the future. On product level, these uncertainties can concern volumes of unique products or product families, but also the distribution of demand on different variants of a specific product model.

The supply uncertainty mainly concerns stock on hand inaccuracies, the extent to which suppliers will deliver ordered quantities, and the scrap levels of supplied items because they do not meet quality requirements. Compared to the uncertainties on the demand side, these uncertainties are normally comparatively marginal. The largest supply uncertainties instead concern the time uncertainty, i.e. delivery precision uncertainty.

This chapter describes different approaches for uncertainty **hedging** in **material planning**. Five different types of methods for estimating safety stocks are defined, illustrated and compared.

LEARNING OBJECTIVES

After studying this chapter, you should be able to:

- Explain the principles of uncertainty hedging
- Explain the financial motives for safety stock and safety lead times
- Explain how to design and use different safety stock methods
- Describe and compare the characteristics of different safety stock methods

13.1 Principles of Uncertainty Hedging

There are two principles of uncertainty hedging in material planning. Quantity hedging is one of these. It means having larger quantities available than what on average is expected to be needed. This is called safety stocks. The other principle is **time hedging**, where a delivery is made earlier than the expected time of need. This is called safety lead time. Both principles lead to extra stock, so-called safety stocks.

Figure 13.1 illustrates the difference between quantity and time hedging. In the upper half of the figure, quantity hedging is used with a safety stock. The stock on hand of 60 pieces is decreased with the safety stock amount before the projected available stock on hand is calculated each week. The calculations result in a new order released being planned for delivery in week 5, so that the 10 pieces of safety stock can be used without shortage. Time hedging is used in the lower half of the figure. Here no safety stock subtractions are conducted. Instead, deliveries are planned one week before the stock is expected to be empty, for example, in order to hedge against delayed deliveries. In both cases, the planned order gets the same delivery time. This is because the safety stock of 10 pieces in the upper half of the figure is equal to one week of demand, i.e. one week earlier delivery which is used as uncertainty hedging mechanism in the lower half of the figure.

Quantity hedging is normally the preferable principle for quantity uncertainties and time hedging for time uncertainties. Quantity hedging can be a good alternative also when time uncertainty occurs if the material requirements are small in relationship to the annual demand. The safety stock can in such situations cover several periods of demand to hedge against delayed deliveries. The larger the demand per individual period in relation to the annual demand, the less appropriate is it to hedge against time uncertainties with quantity-based safety stocks. Very large safety stocks would be required to cover individual period demand.

Safety lead time is the only available alternative in a make-to-order environment, because the order quantities are equal to the actual demand. The same applies to situations where the lot sizing method lot for lot is used. Sometimes it is necessary to use some quantity hedging also in these cases, for example, if some of the delivered quantity is scrapped. A larger quantity than needed is ordered, to cover for such quality defects of delivered items. The uncertainty in this

Week		1	2	3	4	5	6	7
Demand		10	10	10	10	10	10	10
Stock on hand	60 – 10	40	30	20	10	0	–10	
Planned delivery							40	

Week		1	2	3	4	5	6	7
Demand		10	10	10	10	10	10	10
Stock on hand	60	50	40	30	20	10	0	–10
Planned delivery							40	

FIGURE 13.1 Differences in quantity and time hedging

case is because one cannot predict the exact number of scrapped items. The extra quantity, therefore, has to cover situations with higher scrap levels than the estimated average. This safety quantity is, however, not put in stock. If the scrapped items are fewer than the number of extra pieces ordered, these extra pieces are also scrapped. This is because the item is a customer-specific item that does not have any other demand or because the item is not planned to be stored.

Safety stocks can be expressed in terms of quantity but also in terms of run-out time. The time in periods, for example days or weeks, is then determined by dividing the estimated safety stock by the average demand per period. The calculated time is registered in the ERP system. During material planning, when a safety stock figure is needed, the time is multiplied by the then actual demand per time period. This way of expressing safety stock has the benefit that the size of safety stock is automatically adjusted to demand changes, for example due to seasonal variations. The safety stock is, however, not automatically adjusted to changes in the demand uncertainty. It is important to notice that the calculated run-out time is not the same as a safety lead time. It is just another way of expressing a safety stock quantity.

13.2 Financial Motives for Safety Stocks and Safety Lead Times

Economic order quantities are determined by balancing ordering and inventory **carrying costs**. In the same way, safety stocks can be determined by balancing shortage and inventory carrying costs. The larger the safety stock is, the higher the inventory carrying costs and the lower the **shortage costs** are. The economically optimum safety stock could consequently be determined by minimising the sum of these two costs.

Shortage costs are all costs associated with an item not being available in stock when needed. Shortage costs could, for example, be lost revenue due to lower sales than predicted, fees due to delayed delivery or costs for backlog and express transport. Shortage costs could also be due to production disturbances and lower work centre utilisation because input material is not available when needed. It is normally very difficult to make correct estimates of shortage costs. They are often not predictable and their sizes are often dependent on the actual situation when the shortage occurs. Therefore, shortage costs are seldom used as the basis for safety stock estimation. Instead, policy determined service levels, as described in Chapter 6, are used for balancing shortage costs and inventory carrying costs.

Because safety stocks and safety lead times result in extra cost and resource usage, there are reasons for differentiating the size of the safety stocks and safety lead times in order to get optimum value of a given investment, i.e. in terms of **tied-up capital**. This is accomplished in some of the safety stock methods described in Section 13.3, by considering demand variation, forecast error, lead time length, etc. This consideration concerns individual items and stock-keeping units (SKUs). In several situations there may also be reasons for differentiating the safety stock and the safety lead time sizes from a more holistic perspective. For products sold to customers it may, for example, be motivated to have relatively larger safety stocks of products sold in large volumes or with high contribution margins. It may also be motivated to have larger safety stocks of spare parts of critical components in products delivered to customers or of parts where competitors make substitute items. In the same way, longer safety lead times may be motivated for suppliers which have not delivered on time in the past.

Differentiated safety stocks and safety lead times are normally also motivated for items used in production. Here, uncertainty hedging is a way of dealing with problems related to synchronising material flows of individual items in convergence points where several items are used in an operation. For items for which material flows meet in such convergence points, shortage

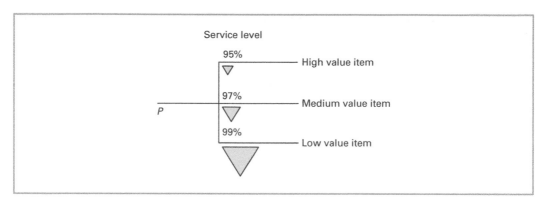

FIGURE 13.2 Differentiating service levels based on item value

costs arise if not all involved items are available when needed in the operation. The size of the shortage cost is normally not dependent on the cost of a missing item. A production disturbance may occur no matter if a cheap or expensive item is missing. The shortage cost is related more to the type of production disturbance than to what item is causing it. The most economic way of allocating safety stocks and safety lead times to hedge against production disturbances is to have relatively larger quantities and longer lead times for low value items compared to high value items. This can be achieved by differentiating service level according to value. The principle of differentiating safety stocks for items used in production is illustrated in Figure 13.2. The example shows a product P, made from three items, a high value item with service level 95 per cent, a medium value item with service level 97 per cent and a low value item with service level 99 per cent.

13.3 Determining Safety Stocks

There are two main types of methods for determining buffer sizes. The first is based on manual estimates and simple calculations. The safety stocks/safety lead times are not results of specific target measures, for example, minimum service level or minimum inventory carrying cost and shortage cost. Manually determined safety stocks and safety lead times and safety stocks calculated as the percentage of the demand during the lead time are examples of such simple methods.

The second type of method is based on more advanced calculations and information about the level of uncertainty. The aim is to optimise the safety stocks in relation to a desired service level, or to minimise the inventory carrying costs and shortage costs that are related to the size of the safety stock.

The following safety stock methods are described:

- Manually estimated safety stocks
- Safety stock as percentage of lead time demand
- Safety stock based on **cycle service**, i.e. probability of no shortage during an inventory cycle (between two successive inventory replenishments)
- Safety stock based on demand fill rate, i.e. proportion of demand that can be delivered directly from stock
- Safety stock based on optimising shortage and inventory carrying cost

Item A:
Demand per week: 18 – 21 – 19 – 20 – 20 – 18 – 22 – 20 – 19 – 23
Average demand per week: 20 pcs

Item B:
Demand per week: 44 – 0 – 4 – 8 – 12 – 0 – 48 – 20 – 0 – 64
Average demand per week: 20 pcs

FIGURE 13.3 Example of items with different demand variations but the same average demand

Manually estimated safety stock or safety lead time

The simplest way of determining safety stock or safety lead time is to use experience and manual estimates. During such estimates, one tries to consider the tied-up capital and cost effects of inventories, and the consequences of inventory shortages and delayed deliveries. The manually estimated safety stocks and safety lead times have to be manually registered in the item files of the ERP system. This makes it quite resource-consuming to revise safety stocks and safety lead times in order to make them reflect changing planning environments, for example with changes in demand, scrap levels or supplier behaviour and performance.

Safety stock as percentage of lead time demand

A simple way of determining the safety stock level is to calculate it as a percentage of the demand during the lead time. The size of the safety stock will be related to the size of the demand and the length of the lead time. The safety stock will then easily be updated when the demand or lead times change. The method also allows systematic differentiation of safety stock quantities, by using various percentages for different types of item groups. The drawback with the method is that it does not take the demand variation or forecast error into consideration. The result of this is illustrated in Figure 13.3. The time series for the two items contain historical demand per week during 10 weeks. Because the items have the same average demand per week they will get the same safety stock if they have the same lead time. The demand variation of item B is, however, considerably larger than that of item A. The safety stock for item B should therefore be larger than for item A, in order to represent the same buffer for future demand variations.

Demand distributions

In order to apply the more advanced safety stock methods that base calculations on an expected service level, alternatively minimising the inventory carrying and shortage costs, information is needed about the demand variation during the lead time, i.e. from order to delivery. Such demand variations are specified by various standard distribution functions. The most common distribution used for safety stock calculations is the **normal distribution**. A normal distribution is defined by a mean value (x) and **standard deviation** (σ) and is expressed as $N(x, \sigma)$. It is a symmetric distribution where 68.27 per cent of all demand values are within ± 1 standard deviation from the mean value, 95.45 per cent within ± 2 standard deviations from the mean value and 99.73 per cent within ± 3 standard deviations. For a normal distribution, there is an equal probability for higher and lower values than the mean value. The shape of the normal distribution is illustrated in Figure 13.4.

Because a normal distribution is defined by its mean value and standard deviation, it can be used for determining safety stocks, based on the demand statistics, by calculating the average

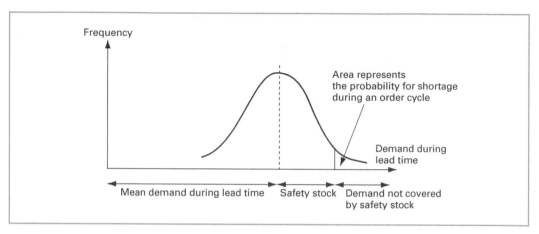

FIGURE 13.4 Illustration of demand variation and safety stock requirement

demand per time period and standard deviation during the lead time. A simplified and more practical way of estimating the standard deviation during the lead time is to calculate the **mean absolute deviation (MAD)** per period and to use a standard deviation approximation of $1.25 \cdot \text{MAD}$.

Normally, it is not only the demand per period that varies; the lead time can also vary from time to time. During such circumstances, the standard deviation calculation has to be adjusted in the following way. If the demand has a normal distribution with the average demand D and standard deviation σ_D, and the lead time has a normal distribution with the average lead time LT and standard deviation σ_{LT}, the demand during the lead time will also have a normal distribution with a mean value of $D \cdot LT$ and the standard deviation can be calculated thus:

$$\sigma_{DDLT} = \sqrt{LT \cdot \sigma_D^2 + \sigma_{LT}^2 \cdot D^2}$$

where LT = average lead time in periods from order to delivery
        D = average demand per period
        σ_D = standard deviation of demand per period
        σ_{LT} = standard deviation of lead time

In situations with small lead time variations, the total standard deviation of the demand during the lead time can be approximated as follows:

$$\sigma_{DDLT} = \sigma_D \cdot \sqrt{LT}$$

Whether the demand is normally distributed can be tested with statistical methods. A simple rule of thumb is to consider the normal distribution to be appropriate if the average demand during the lead time is larger than two standard deviations.

When the demand is low, for example for spare parts, a Poisson distribution may be a more appropriate model of the demand than the normal distribution. The Poisson distribution is defined by its mean value, and the standard deviation equals the square root of the mean value. This distribution is, in contrast to the normal distribution, not symmetric around its mean value, but has more values to the "right" of the mean value than to the "left", i.e. more values that are larger than the mean value. It is consequently skewed to the right. A simple way of testing whether the demand has a Poisson distribution is to compare the standard deviation with the square root of the mean value. It is considered appropriate to use the Poisson distribution if the difference between these two figures is within a margin of error of +/–20 per cent.

Safety stock based on cycle service

The service level measure cycle service is defined as the probability of being able to deliver directly from stock during one inventory cycle, i.e. between two successive inventory replenishments.

$$Service\ level\ cycle\ service\ in\ \% = \left(1 - \frac{Number\ of\ inventory\ cycles\ with\ shortage}{Total\ number\ of\ inventory\ cycles}\right) \cdot 100$$

In order to achieve this specific service level, the stock level when a new customer order is placed must be the average demand during the lead time plus a quantity representing a specified number of standard deviations. The number of standard deviations is decided so that the probability of shortage during the inventory cycle is equal to the specified service level. The relationship is illustrated in Figure 13.4.

The safety stock equals the standard deviation of the demand during the lead time multiplied by the **safety factor** that represents the acceptable probability of stock-out during the inventory cycle. The formula for calculating safety stocks, based on the service level defined as cycle service, is thus:

$$SS = k \cdot \sigma_{DDLT}$$

where SS = safety stock
        k = safety factor
        σ_{DDLT} = standard deviation of demand during lead time

The background to the formula is illustrated in Figure 13.5. Graph A shows the maximum possible demand during the lead time, graph B the most probable demand and graph C the minimum possible demand. Graph D shows the probability of demand to be between the minimum and maximum values.

Table 13.1 shows the safety factor for some different service levels when the demand during the lead time has a normal distribution. Appendix B includes more safety factor values. It can be observed that the safety factor is zero if not using safety stocks, i.e. using a service level of 50 per cent. The table also shows that an infinite safety stock is needed to achieve a service level of 100 per cent.

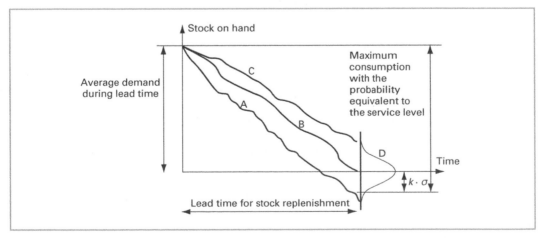

FIGURE 13.5 Demand variation and safety stock estimation

Service level in %	Safety factor
50.00	0.00
75.00	0.67
80.00	0.84
85.00	1.04
90.00	1.28
95.00	1.65
98.00	2.05
99.00	2.33
99.50	2.57
99.99	4.00

TABLE 13.1 Example of service levels and corresponding safety factors with normally distributed demand during lead time

Example 13.1

The demand for an item is assumed to have a normal distribution with a mean value of 2000 pcs per week and a standard deviation of 1000 pcs per week. The lead time has a normal distribution with a mean value of two weeks and a standard deviation of half a week. Determine appropriate safety stocks if the goal is to have a service level (cycle service) of 95 per cent, 98 per cent and 99.8 per cent, respectively.

Average demand during the lead time is:

$$D \cdot LT = 2000 \cdot 2 = 4000 \text{ pcs}$$

Standard deviation of demand during lead time:

$$\sigma_{DDLT} = \sqrt{LT \cdot \sigma_D^2 + \sigma_{LT}^2 \cdot D^2} = \sqrt{2 \cdot 1000^2 + 0.5^2 \cdot 2000^2} = 1732 \text{ units/week}$$

Necessary safety stock for achieving 95 per cent service level is calculated as:

$$k \cdot \sigma_{DDLT} = 1.65 \cdot 1732 = 2858 \text{ pcs}$$

Necessary safety stock for achieving 98 per cent service level is calculated as:

$$k \cdot \sigma_{DDLT} = 2.06 \cdot 1732 = 3568 \text{ pcs}$$

Necessary safety stock for achieving 99.99 per cent service level is calculated as:

$$k \cdot \sigma_{DDLT} = 2.86 \cdot 1732 = 4954 \text{ pcs}$$

The example shows that the size of the safety stock increases progressively with increased service level, and decreased probability of stock-out during the inventory cycle. This means that a small increase in an already high service level will result in much higher tied-up capital in the safety stock.

When using a Poisson distribution, the safety stock determination is conducted somewhat differently. The size of the safety stock is achieved through Poisson tables of the cumulative probability that the demand does not exceed a quantity equal to the average inventory plus the safety stock. The sizes of the necessary safety stocks for achieving various service levels at various demand mean values are shown in Table 13.2. Because the Poisson distribution is a

Average demand during lead time	80%	85%	90%	95%	97%	98%	99%
1	1	1	1	2	2	2	3
2	1	1	2	2	3	3	4
3	1	2	2	3	4	4	5
4	2	2	3	3	4	5	5
5	2	2	3	4	5	5	6
6	2	2	3	4	5	5	6
7	2	3	3	5	5	6	7
8	2	3	4	5	6	6	7
9	2	3	4	5	6	7	8
10	3	3	4	5	6	7	8

TABLE 13.2 Example of service levels and corresponding safety stock with Poisson distributed demand

discrete distribution (compared to the normal distribution which is continuous), for which the demand can only be represented by integer numbers, the safety stocks can also only be expressed by integer numbers. Therefore, it is not possible to determine a safety stock that corresponds to an exact service level, and it is almost always necessary to round off.

The safety stock values in Table 13.2 are calculated so that the minimum service level is fulfilled with the actual safety stock. For example, a safety stock of three pieces will result in a service level of at least 95 per cent if the demand during the lead time is four pieces.

Safety stock based on demand fill rate

The safety stock can also be calculated on the basis of another service level measure, demand **fill rate service**, defined as the proportion of demand that can be delivered directly from stock.

$$\frac{\text{Service level demand}}{\text{fill rate in \%}} = \left(1 - \frac{\text{Demand not directly fulfilled from stock}}{\text{Total demand}}\right) \cdot 100$$

This service level definition, in contrast to the cycle service definition, takes into consideration that the number of inventory replenishments per year is different for different items, and hence that the number of stock-out occasions differs. This service level definition, consequently, gives a more correct measure of the actual service level over time and not only during a single inventory cycle. The method is, however, more difficult to calculate compared to the previous one.

The expected number of shortage pcs during an inventory cycle can be expressed as $\sigma_D \cdot E(z)$, where σ_D is the standard deviation of the demand during lead time and $E(z)$ is the so-called service loss function. Table 13.3 (and Appendix B) shows different service loss function values for different safety factors when the demand is normally distributed. Demand fill rate can, by using the service loss function, be expressed as:

$$\text{Demand fill rate} = 1 - \frac{\dfrac{D}{Q} \cdot \sigma_{DDLT} \cdot E(z)}{D} = 1 - \frac{\sigma_{DDLT} \cdot E(z)}{Q}$$

where D = demand per year
  σ_{DDLT} = standard deviation of demand during lead time
  Q = average order quantity

Service loss values	Safety factors
0.40	0
0.30	0.22
0.25	0.35
0.20	0.49
0.15	0.67
0.10	0.90
0.05	1.26
0.01	1.94
0.005	2.19
0.0015	2.59

TABLE 13.3 Some service loss function values at different safety factors

The above function implies that $E(z)$ equals:

$$E(z) = \frac{(1 - Fill\ rate) \cdot Q}{\sigma_{DDLT}}$$

The service loss function for a specific service level can thus be determined if one knows the standard deviation of the demand during the lead time and the order quantity. Safety factors can then be obtained from the service loss function in Table 13.3 (and Appendix B).

The safety stock is then calculated in the same way as the previous method, with the following formula:

$$SS = \sigma_{DDLT} \cdot Z$$

Example 13.2

The demand for an item is normally distributed with a mean value of 2000 pcs/week and standard deviation of 1000 pcs/week. The order quantity is 10 000 pcs. Determine appropriate safety stocks if the goal is to have a service level (fill rate service) of 95 per cent, 98 per cent and 99 per cent, respectively.

$$Fill\ rate = 0.95 \rightarrow E(z) = \frac{(1 - SERV2) \cdot Q}{\sigma_{DDLT}} = \frac{(1 - 0.95) \cdot 10\ 000}{1000} = 0.50 \rightarrow Z = 0$$

$$SS = Z \cdot \sigma_D = 0$$

$$Fill\ rate = 0.98 \rightarrow E(z) = \frac{(1 - 0.98) \cdot 10\ 000}{1000} = 0.20 \rightarrow Z = 0.50$$

$$SS = 0.50 \cdot 1000 = 500\ units$$

$$Fill\ rate = 0.99 \rightarrow E(z) = \frac{(1 - 0.99) \cdot 10\ 000}{1000} = 0.10 \rightarrow Z = 0.90$$

$$SS = 0.90 \cdot 1000 = 900\ units$$

Safety stock based on cost optimisation

To use safety stocks in order to minimise shortages in inventories means that the shortage costs can be reduced while the inventory carrying costs are increased. The safety stock can consequently be determined by comparing these inventory shortage costs and carrying costs. An economically optimal safety stock can then be achieved by minimising the sum of these two inventory cost types. This sum has its minimum when the expected cost for carrying inventory equals the expected cost for inventory shortages. Based on this relationship, the following formula can be used for calculating the probability that stock-out does not occur during an inventory cycle $\Phi(k)$, assuming that a stock-out results in lost sales:

$$\Phi(k) = \frac{SC}{SC + IC \cdot \dfrac{Q}{D}}$$

where k = safety factor
Q = order quantity
IC = inventory carrying cost per unit and time period
SC = shortage cost per unit
D = demand per time period

If a stock-out results in a back order and delivery on a later occasion, i.e. the stock-out does not lead to lost sales, the following formula should be used:

$$\Phi(k) = 1 - \frac{IC \cdot Q}{D \cdot SC}$$

These relationships are generally applicable, regardless of demand distribution. However, if assuming normal distribution, a normal distribution table can be used to obtain the corresponding safety factors. The safety stock can then be determined with the following formula, in the same way as for the previous methods:

$$SS = k \cdot \sigma_{DDLT}$$

Example 13.3

An item has an annual demand of 200 pcs. The order quantity when replenishing inventory is 80 pcs and the item value is €50 per unit. The standard deviation of the demand during the lead time is 11 pcs and the inventory carrying cost is 25 per cent of the item value per year. The shortage cost is estimated as €20 per unit. A shortage normally results in lost sales. This results in:

$$\Phi(k) = \frac{20}{20 + \dfrac{0.25 \cdot 50 \cdot 80}{200}} = 0.8$$

The table in Appendix B gives $k = 0.84$, which results in a safety stock of:

$$SS = 0.84 \cdot 11 = 9 \text{ units}$$

13.4 Comparing Service Levels and Shortage Costs

Basing the safety stock calculations on the service level definitions cycle service and fill rate service, respectively, can result in very different safety stock sizes. It is therefore very important to know which definition is used and to choose the correct formula and calculation procedure for that definition. The following example shows the effects of mixing service level definitions and calculation procedures.

Example 13.4

An item has an annual demand of 500 pcs, a lead time of two months and a standard deviation of the demand during the lead time of 25 pcs. The order quantity is 100 pcs and the service level requirement is 95 per cent.

For service level cycle service a 95 per cent service level corresponds to a safety factor of 1.65 (Table 13.1). This gives a safety stock of 1.65 · 25 = 41 pcs. For service level fill rate the service loss function equals (1 – 0.95) · 100 / 25 = 0.2 which gives a safety factor of 0.49. This gives a safety stock of 0.49 · 25 = 12 pcs.

There is also a relationship between service levels and shortage costs, i.e. by choosing a specific service level the actual shortage cost is indirectly determined. For service level cycle service, defined as the probability of no stock-out during an inventory cycle, the theoretical shortage cost can be calculated with the following formula:

$$SC = \frac{IC \cdot Q}{D \cdot \left(1 - \dfrac{Cycle\ service}{100}\right)}$$

where SC = shortage cost per unit
IC = inventory carrying cost per unit and time period
D = demand per unit and time period

Example 13.5

An item with an annual demand of 500 pcs, a lead time of two months, a standard deviation of the demand during lead time of 25 pcs and a price of €100 has a service level (cycle service) objective of 95 per cent. The order quantity on every order occasion is 100 pcs and the inventory carrying cost is 25 per cent of the item value per unit and year. The equivalent shortage cost is:

$$SC = \frac{0.25 \cdot 100 \cdot 100}{500 \cdot \left(1 - \dfrac{95}{100}\right)} = €100 \text{ per unit}$$

If using the other service level definition (fill rate service) the shortage cost can be calculated as:

$$SC = \frac{IC \cdot Q}{D \cdot \left(1 - \Phi(k)\right)}$$

where SC = shortage cost per unit
IC = inventory carrying cost per unit and time period
D = demand per time period and unit
$\Phi(k)$ = probability that the demand during an inventory cycle is less than the average demand plus safety stock

Example 13.6

For the same item as in example 13.5, but with fill rate service level of 95 per cent, the safety factor is 0.49. This corresponds to a probability of no shortage during an inventory cycle, i.e. $\Phi(k)$, of about 69 per cent. The equivalent shortage cost is:

$$SC = \frac{0.25 \cdot 100 \cdot 100}{500 \cdot \left(1 - \dfrac{69}{100}\right)} = €16 \text{ per unit}$$

13.5 Method Comparison

Two "simple" methods and three more "advanced" methods for safety stock determination have been presented. Table 13.4 compares some characteristics of the methods. In addition to these characteristics, it should be mentioned that only the first two methods can be used without software support. These are also the only methods that do not require some basic knowledge about theory of probability and statistics.

Safety stock method	Demand variation consideration	Shortage occasions consideration	Service level consideration	Incremental cost consideration
Manual estimation	Intuitively	No	No	Intuitively
Percentage of lead time demand	No	No	No	No
P (no shortage during order cycle)	Yes	No	Yes	No
Proportion demand from stock	Yes	Yes	Yes	No
Cost optimisation	Yes	Indirectly	No	Yes

TABLE 13.4 Comparison of safety stock method characteristics

All five safety stock methods presented here express the safety stock as a quantity, and not as a run-out time. It is, as explained in the beginning of this chapter, easy to translate quantities into run-out times, by dividing the safety stock quantity by the demand per time period. The calculated safety lead time will then be stored in the ERP database and is re-calculated during material planning as a quantity by using the actual demand. The benefit of expressing the safety stock as a lead time is consequently that the safety stock quantity automatically adjusts to systematic variations in demand. The safety stock will, for example, vary with seasonal variations without new service level or cost optimisation calculations. This is, therefore, an alternative way to adjust safety stock quantities to some types of demand variations.

13.6 Summary

This chapter has described different approaches for uncertainty hedging in material planning. Principles for lead time and quantity hedging and financial motives for safety stocks were discussed. Further, the characteristics and use of five different types of methods for estimating safety stocks were presented. Special focus has been put on the methods for estimating safety stocks based on cycle service and demand fill rate, respectively.

 Key concepts

Carrying cost 265	Safety factor 269
Cycle service 266	Safety lead time 263
Fill rate service 271	Safety stock 263
Hedging 263	Shortage costs 265
Material planning 263	Standard deviation 267
Mean absolute deviation (MAD) 268	Tied-up capital 265
Normal distribution 267	Time hedging 264

Discussion tasks

1 Safety stocks can be calculated based on cycle service and demand fill rate service. How will the safety stock be affected when using each of these two service level definitions and corresponding calculations if the number of replenishment orders per year increases and decreases respectively?

2 What impact on the safety stocks will a change from storing a product at several regional distribution centres to storing it at one central distribution centre and conducting direct distribution to end customers have?

3 In practice, rather few companies calculate the safety stock levels based on service levels or cost. Instead, they use more simple rules when setting safety stock parameters, e.g. manual estimates, a percentage of the lead time demand, etc. What is the effect of using such simple rules?

 Problems[1]

Problem 13.1

Pump & Co. makes and sells drainage pumps. Distribution of spare parts is also an important part of the operations.

a) The demand for a spare part has a Poisson distribution with a mean demand of 8 pcs during the two weeks' lead time. Determine the safety stock in order to have a probability for stock-out during the inventory cycle (cycle service) of less than 10 per cent.

---

1 Solutions to problems are in Appendix A and safety stock tables are in Appendix B.

b) By changing to a cheaper supplier lead time will increase to four weeks. At the same time the lead time variation increases and is estimated to have a normal distribution with standard deviation of one week. Determine necessary safety stock in order to keep the same service level (cycle service) as in (a). Calculate with a normal distributed demand with a mean of 4 pcs per week and a standard deviation of 2 pcs per week.[2]

c) What service level, defined as the proportion of demand that directly can be delivered from stock (fill rate service), gives the safety stock calculated in (b)? The average order quantity is 20 pcs.

Problem 13.2

A stocked product has a forecasted demand of 2500 pcs per week with a standard deviation of 500 pcs per week. The lead time from order to delivery into the stock has during the last 10 order cycles been 2, 3, 1, 5, 3, 5, 4, 2, 8 and 3 weeks, respectively. The demand and lead time are expected to have normal distributions.

a) Determine necessary safety stock if the probability for stock-out during the order cycle (cycle service) should not exceed 5 per cent.

b) Determine necessary safety stock if the probability of not being able to deliver a unit from stock (1-fill rate) should be no higher than 1, 5 and 15 per cent, respectively. The ordering cost is €80 and inventory carrying cost is €10 per unit and year. The company uses the EOQ formula for determining lot sizes.

c) What service level (fill rate service) would the safety stock in (a) have resulted in?

Problem 13.3

A spare part delivered from a warehouse has the following planning data:

Forecasted annual demand: 120 pcs
Standard deviation of demand per 4 weeks: 25 pcs
Delivery time from supplier: 2 weeks
Standard deviation of the lead time: 1 week
Order quantity from supplier: 20 pcs
Inventory value per unit: €50

A re-order point system is used for inventory control in the spare parts warehouse. Customer deliveries are carried out 48 weeks of the year and is considered to be normal distributed. For the actual spare part the service level, defined as the proportion of demand that can be directly fulfilled from stock (fill rate), should be 95 per cent.

a) Determine the appropriate re-order point in order to reach the service level goal.

b) How much should the safety stock be increased if changing to a supplier with a lead time of four weeks and a standard deviation of two weeks?

c) How would the safety stock, re-order point, average tied-up capital in stock and inventory turnover rate be affected if this new supplier cannot deliver order quantities smaller than 30 pcs?

---

2 The standard deviation of a Poisson distribution equals the square root of the mean value which is 8 pcs during a two-week period. When approximating a Poisson distribution with a normal distribution the mean value consequently becomes 4 pcs per week and the standard deviation 2 pcs per week.

Problem 13.4

A periodic ordering system is used to plan purchase orders for two products from the same supplier. Both products are purchased at the same review occasion and the common ordering cost is €100. Planning data for the products:

	Product A	Product B
Demand per week	200 pcs	50 pcs
Standard deviation of demand	100 pcs/week	70 pcs/week
Lead time	1.5 weeks	1.5 weeks
Purchase price per unit	€25	€20
P (stock-out during order cycle)	95%	80%
Inventory carrying cost	15%	15%

a) Design a periodic ordering system for the products.

b) Determine the average inventory level for the respective product.

Problem 13.5

The company Store & Distribute AB distributes a product through four regional warehouses. The demand in the respective region is on average 30, 50, 25 and 100 pcs per week, respectively. The demand can be estimated to have a normal distribution. The standard deviation of the demand is 20, 50, 20 and 40 pcs per week for the respective region. The warehouses are controlled with re-order point systems. The safety stocks are determined in order to receive a service level (fill rate) of 98 per cent and the replenishment quantities are calculated based on EOQ and are 250, 322, 227 and 455 pcs for each respective warehouse. All warehouses have the same inventory carrying cost per unit and the same ordering cost per order. The lead time from order to received delivery is four weeks for all warehouses. The lead time is more or less constant.

a) Determine the necessary safety stock and the average tied-up capital for a product with a value of €1000 per unit.

The company considers shutting down three warehouses and will instead distribute all products through a central warehouse. The central warehouse will also be controlled with a re-order point system with the same service level and lead time (i.e. 98 per cent and four weeks). The order quantities will be calculated with EOQ. Ordering and carrying costs are not changed.

b) Determine necessary safety stock and the average tied-up capital for a product with a value of €1000 per unit.

c) Determine necessary safety stock and average tied-up capital for the central warehouse solution if the lead time can be decreased from four to two weeks (and the item value still is €1000).

Problem 13.6

The re-order point system is used to control replenishment of stocked products at the company Building Materials Ltd. Average usage per day of one of its products is 5 pcs and lead time is two days. Variations in sales are normally distributed with a standard deviation of 4.46 pcs during the lead time. Cycle service is used to calculate safety stocks.

a) How large will the safety stock be for the product with service levels of 90, 95 and 99 per cent respectively?

b) What will the re-order point be if the various service levels are used?

Further reading

Ballou, R. (2004) *Business logistics/supply chain management*. Prentice-Hall, Upper Saddle River (NJ).

Bernard, P. (1999) *Integrated inventory management*. Oliver Wight Publications, Essex Junction (VT).

Fogarty, D., Blackstone, J. and Hoffmann, T. (1991) *Production and inventory management*. South-Western Publishing, Cincinnati (OH).

Lambert, D., Stock, J., Ellram, L. and Grant, D. (2005) *Fundamentals of logistics management*. McGraw-Hill, New York (NY).

Silver, E., Pyke, D. and Peterson, R. (1998) *Inventory management and production planning and scheduling*. John Wiley & Sons, New York (NY).

Vollmann, T., Berry, W., Whybark, C. and Jacobs, R. (2005) *Manufacturing planning and control for supply chain management*. Irwin/McGraw-Hill, New York (NY).

CASE STUDY 13.1: CALCULATING AND USING SAFETY STOCKS AT ALFA LAVAL AB[3]

Alfa Laval is a Swedish company developing and manufacturing products sold in more than 100 countries around the world, in more than half of these countries through own sales organisations. The product range includes heat exchangers, valve, pump and tank-products and separation products. This case study concerns the after sales operation and specifically the spare parts operation at one of the distribution centres. All spare parts that are sold to end customers are delivered from distribution centres. After sales represents some 25 per cent of the company's total turnover.

Spare parts classification

All spare parts at Alfa Laval are classified into three categories: stocked items, non-stocked items and request items. Stocked items are items kept in and delivered from stock. Non-stocked items are items not kept in stock but delivered to customer order with a standard lead time and at a list price. Request items are also delivered to customer order but at a quoted delivery lead time and a quoted price established for each individual customer request. The three classes are defined as follows:

- Stocked items   ≥ 4 customer orders per year
- Non-stocked items   2–3 customer orders per year
- Request items   ≤ 1 customer order per year

Reclassification is carried out once a year. The stocked items are in turn grouped into three classes, A, B and C. This classification is based on the number of customer orders per item and is updated every month automatically in the ERP system. The classification is made so that

---

3 Based on Frödå, E. and Magnusson, O. (2006) "Item classification of spare parts at Alfa Laval", master thesis TMIO-06/5247, Lund University, Lund, Sweden.

A-items stand for 80 per cent of all orders, B-items for 15 per cent of all orders and C-items for 5 per cent of all customer orders.

Managing the spare parts

To control the stock of stocked items Alfa Laval uses a re-order point system of periodic review type with fixed order quantities and variable ordering frequencies. For most of the items the order quantity is calculated as an economic order quantity by the EOQ formula in the ERP system. The order quantities for some other items are for various reasons estimated and manually entered into the system.

To cover for various kinds of uncertainties Alfa Laval uses a safety stock calculated in the following way.

$$SS = k \cdot \sigma \cdot \sqrt{LT + RT}$$

where k = safety factor
σ = standard deviation in month
LT = lead time in months
RP = review period in months

The review period is one week. The safety factor is based on decided cycle service levels. The cycle service is currently set to 97 per cent for A-items, 89 per cent for B-items and 81 per cent for C-items.

For some items with very lumpy demand, the standard deviation tends to become very high, generating unacceptable big safety stocks. To avoid this, a limit to the size of the standard deviation has been implemented. This limit is set to 10 per cent of the forecasted demand per year.

Monitoring the service level

After sales is an important part of the business for Alfa Laval. It is also important to provide excellent after sales service to successfully sell the end products. Delivery service is accordingly a big concern for the company and delivery performance measured as availability is continuously monitored. Availability at Alfa Laval is based on order lines and defined as the probability that a customer can receive an order line directly from the distribution centre without any delay and expressed in per cent. This means, for example, that if a customer orders 20 pieces of an item and only 15 pieces are available to deliver off the shelf, the availability is 0 per cent. If, on the other hand, a customer orders 10 different items and nine of these can be delivered without delay as requested, the availability is 90 per cent. The availability is measured on a one-week basis and a 10-week basis. The measure is only used for stocked items.

Discussion questions

1 What is the reason to differentiate the service level for A-, B- and C-items?

2 What is your conclusion regarding adding one review period to the lead time when calculating the safety stock?

3 The service level used for calculating the safety stock and the service level expressed as availability are not defined in the same way. What are the major differences and what are the major implications of using cycle service?

Capacity Planning

I n all production operations, resources are required to bring about added value – in other words, to manufacture products for sale from raw materials, purchased components and semi-finished items. The extent in volume to which a company's resources can achieve such added value is called capacity. The mere fact of having production resources available in a company for adding value is associated with costs, whether the resources are used or not. There will also be costs in the form of lost income if, due to demand from customers, there is a need to produce more than existing resources can achieve. It is therefore vital to balance supply of capacity with requirement for capacity. The function in a company that deals with activities for balancing the supply of capacity and the demand for capacity is called **capacity planning**.

This chapter will treat capacity planning at the various planning levels that have been previously defined. What is meant by the supply of capacity and the demand for capacity are defined, and ways of expressing and calculating both of these concepts are considered. Different strategies for changing capacity and for adapting the utilisation of capacity are discussed. Alternative action variables for achieving a balance between supply and demand of capacity in the medium- and short-term perspective are examined. Finally, an overall review is made of a number of commonly used methods for capacity planning.

14.1 Points of Departure for Capacity Planning

Capacity planning in this context means that the need for capacity is calculated and compared with currently available capacity. An adjustment of available capacity or demand for capacity is then made until an acceptable balance between the two is obtained. Decisions concerning the supply of capacity relate to both the size of volumes to be produced and the time at which these volumes can be produced. Capacity requirements also have a volume dimension and a time dimension, relating to how much capacity is needed and when it is needed.

The concepts of capacity and capacity requirements

For any one unit of resource, i.e. a single machine, a **work centre**, a manufacturing cell or an entire manufacturing department or foreman area, we can talk about two different types of capacity: volume capacity and **throughput capacity**. Volume capacity refers to how many hours or other capacity units a given resource unit can produce during one time period; for example, how many working hours can be performed per week in a work centre. Volume capacity is crucial for capacity planning.

Throughput capacity is a measurement of how many hours per time period can be set aside to perform a certain manufacturing operation within one resource unit. This measurement is an expression of how long it takes, in days and weeks, to perform a value-adding activity rather than how many hours in total can be performed during that time period. For resource units with only one machine or person, the throughput capacity is the same as the volume capacity. If there are many producing units in one resource unit, the throughput capacity will be equal to the volume capacity per manufacturing unit multiplied by the number of manufacturing units which can be used simultaneously and in parallel for manufacturing a certain product. If, for example, a work centre consists of two machines which are each able to be used 40 hours per week, the volume capacity for this work centre will be 80 hours. If, on the other hand, a certain order can only be planned for one of the two machines due to a single tool being available, the average throughput capacity of the work centre will be 40 hours. The concept of throughput capacity is primarily of interest when planning throughput, scheduling of operations for priority control and for delivery time monitoring.

As terms for expressing the demand for manufacturing capacity, the concepts of current workload and capacity requirements both exist. Current workload means the capacity to perform released manufacturing orders. This means that a management decision lies behind the required capacity when it is called current workload. The concept of capacity requirements is of a more general nature and covers also the need for capacity to execute **production plans** in general. The term capacity requirement is mainly used here since it better represents both decided and planned production operations. Figure 14.1 shows an example of capacity and capacity requirements for a work centre.

Capacity planning at different planning levels

Capacity planning in the sense of creating a balance between demand for capacity and supply of capacity is relevant at all planning levels, and it is carried out at every level parallel with the planning of future production. The approach that is generally used involves the drawing up of production plans and planning of manufacturing orders without consideration to current capacity limitations. The aim here is to make an initial calculation of capacity requirements in order

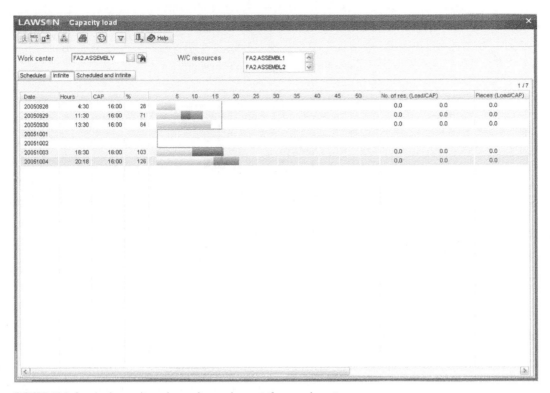

FIGURE 14.1 Graph of capacity and capacity requirements for a work centre

The line in the picture shows the daily capacity in work centre FA2.Assembly. The bars show the total capacity requirements. In two of the periods the requirement exceeds the capacity. The illustration shows one function of Lawson M3 Multi Site Planner

to gain information on what the ideal capacity supply would be, i.e. what capacity would be required to completely match the existing demand for capacity. This is an important general starting point, especially at the upper planning levels where the time perspective is so long that it is possible to adjust available capacity if it does not correspond to requirements. If production planning in this situation is to finite capacity, decisions on increases or cutbacks in capacity cannot be based on the real demand for capacity as related to current orders and delivery plans. Neither would conditions be such that measures could be balanced in an optimal way with regard to costs for increasing or decreasing capacity, or increasing stocks or manufacturing less than demand from the market.

Even though capacity planning affects all planning levels, the detailed approach and the methods which can be used for capacity planning vary somewhat at the different levels. The options for adapting capacity to existing requirements also vary a great deal since the planning time perspective ranges between single weeks to several years, which will affect the possibilities of making changes to available capacity.

At the level of sales and operations planning, capacity requirements are based on long-term planned production, which must be achieved so that planned deliveries can be made. Planned deliveries are normally based on forecasts. Due to the long time perspectives involved at this planning level, often one year or more, the possibilities in terms of time for making adjustments

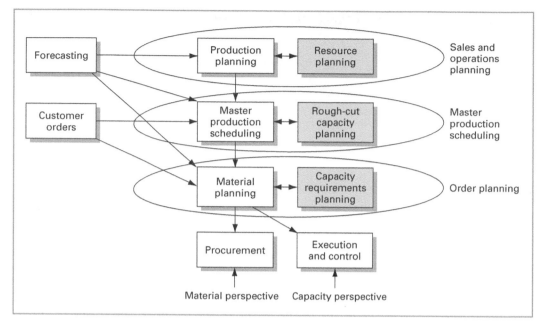

FIGURE 14.2 Capacity planning at different planning levels

to available capacity are almost unlimited. Capacity decisions taken at this level will also affect investments in new production plants or the phasing out of existing plants. Decisions will also cover changes in manning levels and the distribution of competence among staff.

At the master production scheduling level, too, capacity requirements are based on planned production volumes, but for certain types of company it is necessary to calculate capacity requirements on the basis of planned and released manufacturing orders. Planned deliveries, such as on the sales and operations planning level, may be based on forecasts, but in many companies real orders play a large role at the master production scheduling level. The planning horizon is shorter at this level, normally in the magnitude of some months up to one year. The possibilities of adapting available capacity are thus more limited. Capacity decisions are also limited to individual work centres that are known from past experience to be bottlenecks rather than the entire production plant.

At the order planning level, capacity requirements are based entirely on released and planned manufacturing orders. Orders originate from **master production schedules (MPSs)** drawn up at the master production scheduling level or they are a direct consequence of orders from customers, depending on the type of company in question. With a typical planning horizon of a number of months, the possibilities of adjusting available capacity are even more limited at this planning level. It is largely a question of marginal changes to existing capacity, such as using **overtime** or re-allocating labour between different manufacturing departments.

The relationships between capacity planning at these three different planning levels and other planning processes are illustrated in Figure 14.2. It is this aspect of capacity planning that is examined in this chapter. Capacity planning at the workshop planning level is largely an issue of adapting capacity requirements since the planning horizon here is so short, generally one or a few weeks, that there is little scope for adjusting available capacity at all. Capacity planning at this planning level is described in Chapter 16.

14.2 Calculation of Capacity and Capacity Requirements

Capacity is calculated or estimated for each work centre or other manufacturing unit in the company and constitutes a measurement of how much each centre can produce. The most commonly used units for capacity are man-hours per time period or machine-hours per time period. These units state how many hours of production can be expected to be performed during a certain time period. The choice of man-hours or machine-hours depends on how machine-intensive or labour-intensive the manufacturing process is. Other units, such as quantities, kilograms and euros per time period, are also used. The most important aspect is that the unit chosen is representative for operations and operation times in the work centre or other manu-facturing unit, and that capacity requirements are expressed in the same unit.

Calculation of capacity

The theoretical maximum capacity of a work centre is the capacity that could be obtained from the work centre if manufacturing continued around the clock every day of the year. This is seldom the case in practice so this information on capacity is not generally of interest. Instead, a calculation is made of available capacity from which a company normally counts on using. This so-called nominal capacity is most often stated in the form of four variables: the number of machines or other manufacturing units in the work centre, the number of shifts per day, the number of hours per shift and the number of working days per time period. The first three vari-ables are stated in the work centre file. The number of days per time period is obtained from the **shop calendar**. This covers all the days of the year and has information on which days are working days, Sundays, public holidays, vacation days or other non-working days. The principles of a shop calendar are illustrated in Figure 14.3. The days in bold type are non-working days and the numbers in the lower right-hand corner of each square are consecutively numbered workdays. For example, 15 April is a working day with the serial number 119.

Nominal capacity can be calculated with the aid of this data. If there are four machines in a work centre, work is carried out in two shifts, each shift consists of 7.5 hours and a certain week contains five working days, the nominal capacity of the work centre will be $4 \cdot 2 \cdot 7.5 \cdot 5 = 300$ hours in that week.

Month	Week	Mon	Tue	Wed	Thu	Fri	Sat	Sun
April	14	**1**	2 /110	3 /111	4 /112	5 /113	**6**	**7**
	15	8 /114	9 /115	10 /116	11 /117	12 /118	**13**	**14**
	16	15 /119	16 /120	17 /121	18 /122	19 /123	**20**	**21**
	17	22 /124	23 /125	24 /126	25 /127	26 /128	**27**	**28**
	18	29 /129	**30**	**1**	2 /130	3 /131	**4**	**5**

FIGURE 14.3 Illustration of a shop calendar

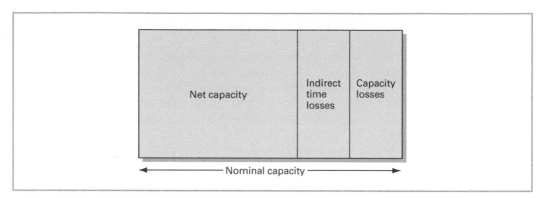

FIGURE 14.4 Different capacity levels in a work centre

The nominal capacity is generally not fully available and for the calculation of available capacity it is necessary to take consideration to different forms of reductions (see Figure 14.4). It may be a question of various capacity losses due to machine breakdowns, short-term absence, maintenance activities and so on. After taking into consideration such capacity reductions, we can speak of remaining gross capacity. However, allowance must also be made for different types of indirect time losses that almost always occur, such as waiting time for materials, time taken to go through work with supervisors and suchlike. Capacity may also be required for non-planned operations such as extra manufacturing to cover defective goods or for urgent orders. After having adjusted the nominal capacity in this respect too, we arrive at something which may be called **net capacity**, which represents the capacity calculated as available for performing planned manufacturing activities.

Enterprise resource planning systems usually take these different types of capacity reductions into account by using **utilisation rates**. These are a measurement of the proportion of nominal capacity which is expected to be available for use in a work centre. The utilisation rate is stored in the work centre file, and can be used to calculate net capacity as in the following formula:

$$Net\ capacity = Utilisation\ rate \cdot Nominal\ capacity$$

With a nominal capacity of 300 hours, as calculated in the example above, and a utilisation rate of 85 per cent, net capacity will be equal to 0.85 · 300 = 255 hours.

Calculation of capacity requirements

There are three basic problems involved in the calculation of capacity requirements. Two of these are illustrated in Figure 14.5. The first basic problem is how the requirements of capacity should be expressed relative to the quantities in production plans and order quantities on manufacturing orders. To be able to add capacity requirements from plans and manufacturing orders for different products, these quantities must be converted into uniform and comparable measurements of capacity requirements. Manufacturing hours in the form of man-hours or machine-hours are examples of such measurements.

The second basic problem involves the planning periods in which capacity requirements, due to current throughput times, must be placed when the equivalent planned quantities or order quantities are to be completed. The different methods for capacity planning are presented later in this chapter and all have their own methods of handling these basic problems.

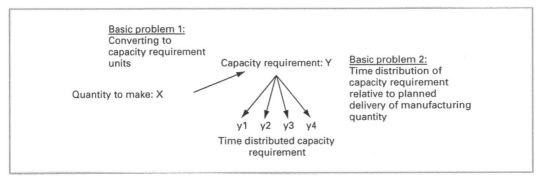

FIGURE 14.5 Two of the basic problems involved in capacity requirements planning

For calculations of capacity requirements, information is generally obtained about the number of manufacturing hours required to manufacture a product or internally manufactured item or semi-finished item from the operation times that are stored in the operation file of the ERP system. The third basic problem is what sort of time will be used to express these manufacturing hours. When calculating and estimating operation times, these are normally expressed in **standard times** or calculated times, i.e. the time it would take if the work was performed under normal conditions. If manufacturing operations are not completely automated, there is often an addition to these norm times, so that actual times will be shorter than planned norm times in operation structuring. In the area of capacity planning it is the assessed actual operation times that must be compared with current capacity in work centres. This is achieved by converting planned operation times to estimated actual times with the use of **efficiency factors**. These factors are normally estimated and stated for each work centre and express the normal relationship between planned time and actual time for operations in the work centre. Operation times for capacity requirements are calculated with the aid of the following formula:

Estimated operation time = Standard time for operation / Efficiency factor

If the standard time for an operation in a work centre is 0.5 hours per piece, the **set-up time** is 2 hours, the order quantity is 20 pieces and the efficiency factor in the work centre in which the operation will be performed is 1.12, the estimated operation time will be equal to (2 + 20 · 0.5) / 1.12 = 10.92 hours.

14.3 Strategies for Capacity Adjustments

All companies experience rises and falls in demand, resulting in imbalances between the capacity available to produce and the delivered volumes equivalent to current demand. These imbalances can be handled to a certain extent by increasing or decreasing the stock of finished products in companies that make to stock, or by increasing or decreasing the backlog of orders through changes in delivery times in companies that make to order. If this is not possible, not appropriate or cannot be performed to the extent necessary, capacity must be increased or the utilisation of capacity must be decreased, depending on whether demand is too large or too small in comparison with the manufacturing capacity available. Different strategies can be used, some when changes in the capacity level are necessary and some when the existing capacity level must be utilised.

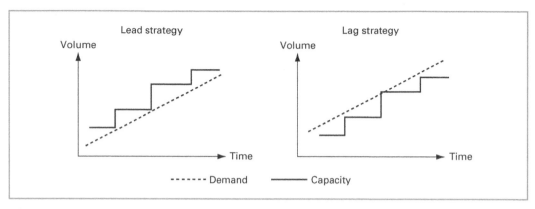

FIGURE 14.6 Illustration of different strategies for adjusting capacity

Strategies for capacity changes

There are two fundamental strategies for making changes in capacity: **lead strategy** and **lag strategy**. Both alternatives are illustrated in general terms in Figure 14.6 for the case when demand increases over time. Lead strategy means that capacity is increased or decreased before demand rises or falls. It is thus a proactive strategy, meaning that access to capacity is adapted before the need has arisen or disappeared. With rises in demand, this strategy provides volume flexibility which enables a company to gain market share and thereby expand its operations. However, this will also entail taking great risks in the utilisation of capacity invested in. If demand is falling, the strategy may reinforce the decline as the company may find it difficult to retain its market share due to capacity shortage. The benefit is that risks associated with costs for unutilised capacity are decreased.

Lag strategy is a reactive strategy and means that investments in new capacity are not made until a change in the size of demand is stated and real. This strategy limits volume flexibility with rising demand and puts greater demands on being able to utilise changes in stock or changes in delivery times to avoid losing sales and market shares. Thus, the strategy results in higher average stock levels and longer order backlogs in comparison with the previous strategy. One advantage of the strategy is to decrease the risk of landing in a situation with costly over-capacity if there is uncertainty about how large and durable the rise in demand will be. In the case of falling demand, the relationship will be the opposite.

Both of these strategies represent two extremes. Both are characterised by either more or less having constant overcapacity or constant undercapacity. In practice it is more common to use some form of compromise, switching between working with overcapacity and undercapacity to a greater or lesser extent.

Strategies for adjustment of capacity utilisation

Within the framework of the manufacturing capacity available when planning, different strategies may also be used for adjusting the utilisation of capacity. In this case, too, there are two main strategies: **level strategy** and **chase strategy**.

Level strategy means that either the stock or the delivery time is fully utilised to gain a completely level utilisation of capacity, i.e. manufacturing capacity is not adapted to variations

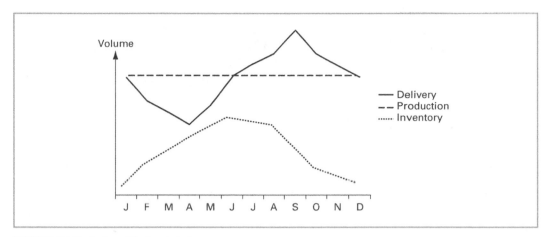

FIGURE 14.7 Illustration of the level strategy

in demand at all. The production volume per period that is required to achieve this may be calculated in a simplified fashion using the following formula for companies that make to stock:

$$Production\ volume\ per\ period = \frac{Closing\ stock - Opening\ stock + Delivery\ volume}{No.\ of\ planning\ periods}$$

The number of planning periods is equal to the number of periods in the planning horizon. The delivery volume is the planned total delivered volume during this number of periods. To ensure that the calculated production volume per period will be correct, stocks must not be allowed to become negative in any single period. If this is the case, the calculation result must be corrected. A corresponding formula may be derived for make-to-order companies. The level strategy is illustrated in general terms in Figure 14.7.

The advantage of level strategy is that costly capacity adjustments such as overtime, sub-contracting and underemployment can be bypassed. Productivity downturns that are frequent with volume changes can also be avoided. Its largest downsides include stocks and tied-up capital that must be built up during periods with low demand and planned for use in periods of high demand.

The second main strategy for capacity adjustments, chase strategy, means that capacity utilisation is completely adapted to current demand. In other words, when using this strategy no stocks of finished products are necessary other than in the form of cycle stock if a company makes to stock. When making to order, delivery times to customers do not need to be varied as a result of workloads being too great or too small. The chase strategy is illustrated in general terms in Figure 14.8.

The advantages of chase strategy correspond to the disadvantages of level strategy, and vice versa. These two strategies represent extremes, and as in the case of strategies for capacity changes, it is often necessary to use some form of compromise or combination strategy. This will involve using stocks or delivery times to a greater or lesser degree in order to decrease the need to fully adapt capacity utilisation.

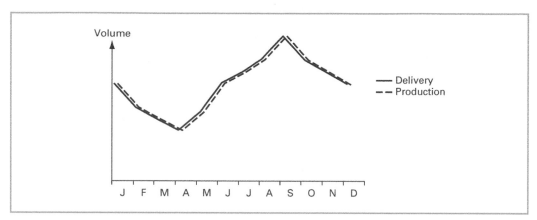

FIGURE 14.8 Illustration of the chase strategy

14.4 Alternative Action Plans when Planning Capacity

In the long-term perspective, it is generally possible to adapt available capacity to the capacity requirements that are expected to occur. If there are no financial limitations or other special restrictions, capacity can always be increased by investing in new plants or individual production resources, or it can be decreased if necessary by closing down existing plants and manufacturing resources. The same applies to the number of employees. Strategies described in the previous section may be applied, and such options are generally at the level of sales and operations planning. The time perspective is tighter at other planning levels and the scope for action is thus more limited. As a result, companies are forced to adapt capacity at the margins of existing capacity, or influence capacity requirements.

Capacity planning involves allocating capacity requirements over time to different planning periods and comparing these allocated capacity requirements with available capacity at the work centres in question. This comparison may indicate that the demand for capacity does not correspond with the supply of capacity, either during individual periods or during the whole planning horizon. Corrective measures must then be taken, either with regard to capacity supply or capacity demand, or both of these. Four different types of planning situations can be identified by differentiating between accumulated correspondence and periodic balance, as illustrated in Figure 14.9. Accumulated correspondence describes the situation in which the accumulated available capacity corresponds with the accumulated capacity demand over the entire planning horizon. In other words, an accumulated correspondence means that the total capacity demanded during all planning periods is equal to the total supply of capacity. Periodic correspondence means that there is a balance between capacity demand and the supply of capacity in every planning period. In ideal capacity planning operations, there should be correspondence within each period as well as during the whole planning horizon. If this is not the case, different measures must be used to achieve as much correspondence as possible. Such general action plans are shown in the figure and described in more detail below. Only action plans that are relevant in the short and medium terms are included, i.e. those that are applicable at the master production scheduling level and below. Planning situations equivalent to the two right-hand boxes in Figure 14.9 are most likely to occur at the master production scheduling level, while planning situations equivalent to the bottom left box are more likely to occur at the order planning and workshop planning levels, especially if master production scheduling has been performed efficiently and a long-term balance has been achieved.

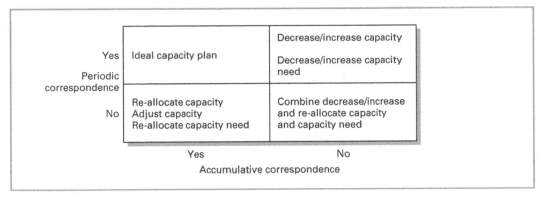

FIGURE 14.9 Types of planning situations with respect to correspondence between supply of capacity and capacity demand

To increase capacity, new personnel may be hired and temporary employment may be used if the shortage in capacity is not expected to last. To a certain extent it may be possible to gain time before investing in new machines or other manufacturing equipment. Increasing the number of shifts is another option to increase capacity. This option has the effect of increasing the utilisation of machines while the utilisation of labour remains constant. Increased shiftwork may take place by transferring personnel from manufacturing areas with lower workloads or by hiring new personnel. Finally, capacity can be increased by **subcontracting**, i.e. arranging for an external supplier to take over certain parts of manufacturing for shorter or longer periods of time. Capacity may be decreased by taking opposite measures such as decreasing the number of shifts, introducing a short week, laying off personnel or taking back subcontracted work.

Decreasing or increasing the need for capacity in cases where there is not an accumulated correspondence between capacity supply and capacity requirements is primarily a question of adjusting production plans at a company. Decreasing capacity requirements can be achieved by deliberately making production plans for smaller volumes than demanded by the market. In a corresponding fashion, capacity requirements in manufacturing may be increased by raising demand with the aid of campaigns or other marketing activities.

Re-allocation of capacity is one possible measure which may be used in the short term to adjust capacity to current capacity requirements, and thereby balance the utilisation of capacity between different planning periods. This may be achieved by transferring manufacturing personnel from departments with low workloads to departments with high workloads. For this to be possible in the short term, either the type of work must be relatively unskilled or the personnel involved must be versatile and trained in advance for different types of work.

Short-term adjustment of capacity is also one alternative for solving the problem with periodic non-correspondence between capacity supply and capacity demand. This alternative involves temporarily, for one week or two, increasing or decreasing available capacity. Overtime is one such option. Compared with using more than one shift, this means that the rate of utilisation will increase both for labour and for machines. This option of using overtime is limited by industrial law, however. Other short-term measures include delaying service and maintenance on machines and carrying out those activities in periods with low capacity requirements. Subcontracting may also be used in certain contexts for short-term capacity adjustments.

The option of re-allocating capacity requirements has a wide range of different possibilities. The principles are illustrated in Figure 14.10 in the case of accumulated but not periodic correspondence between capacity supply and capacity demand. The problem with overload in

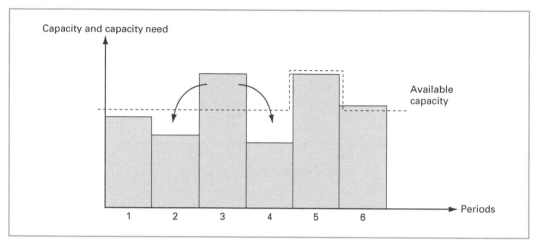

FIGURE 14.10 Principles for adjusting capacity and capacity requirements through short-term capacity boosting and re-allocation of capacity requirements

period 3 can be solved by bringing forward or postponing some manufacturing to periods 2 and 4. In period 5 the problem has been solved through temporary capacity adjustment.

Postponing or bringing forward manufacturing relative to its ideal timing is one main alternative for re-allocation of capacity requirements. This is achieved by changing the start time and finish time for scheduled **manufacturing orders**. To be able to bring forward manufacturing, start-up materials must be available earlier than originally planned and this alternative leads to extra tied-up capital since the items manufactured are finished earlier than they are needed. Postponement can bring about shortages of materials in final manufacturing/assembly. When manufacturing end products, the option of bringing forward manufacturing mainly comes up in the context of anticipation stock being built up during periods of low demand, and thus low capacity requirements, to be used in periods of high demand. This approach will thus include postponement of capacity requirements. Correspondingly, postponement of capacity requirements may be achieved in companies that make to order by increasing delivery lead times to customers.

Using alternative work centres or alternative routings, and thus relieving manufacturing resources that are overloaded and utilising instead those that have low workloads, is another way of re-allocating capacity requirements. This option means that production resources aimed to be used for manufacturing of certain operations are temporarily replaced by resources which, in normal conditions, are less optimal from the manufacturing economy point of view. Increasing or decreasing order quantities is another way of re-allocating capacity requirements. For example, by cutting back on order quantities fewer products are manufactured during the period in question and more must be manufactured in a later period in order to satisfy total demand.

Overlapping and **order splitting** are also possible measures for temporarily re-allocating capacity requirements. Overlapping operations means that one already manufactured part of the order quantity is moved to the next operation before the entire order quantity is complete. Operation splitting, or running operations in parallel, means that the execution of one step of refinement is split up between two or more places of manufacturing that are all able to perform the same type of refinement. In both cases there is a reduction of throughput time, which means that manufacturing order operations are moved between planning periods. Order splitting also involves re-allocation in another sense; manufacturing is divided between several machines.

Type of alternative	Medium term	Short term
Increase/decrease capacity	Hire/fire employees New machines Number of shifts/short week Subcontracting	Subcontracting Extra shift
Increase/decrease capacity need	Change master production Schedule increase/decrease stock	
Re-allocate capacity	Re-allocation between work centres	Re-allocation between work centres
Adjust capacity		Over time Postponed maintenance
Re-allocate capacity requirements	Make to anticipation stock Change delivery lead times	Earlier/postponed order start Changed order quantities Alternative work centres Overlapping/order splitting

FIGURE 14.11 Summary of options for creating correspondence between capacity supply and capacity requirements

A summary of some of the most important options for creating correspondence between capacity supply and capacity requirement in the short and medium term is illustrated in Figure 14.11.

14.5 Methods for Capacity Planning

As described above, capacity planning means attempting to achieve correspondence between requirements for capacity and supply of capacity. For this to be possible a company must be able to calculate and express capacity supply and capacity demand with the aid of suitable methods. Calculating the supply of capacity is comparatively straightforward and has been described in Section 14.2. Methods for capacity planning generally refer only to methods that are intended for the calculation of capacity requirements. These methods differ mainly with respect to what the calculations are based on and at what planning levels they are suitable for use.

This section provides an overview of commonly used methods for capacity planning. Variants of each method are also described. The character of the various methods, calculation methodology, their application and primary areas of use are also discussed.

Capacity planning using overall factors

The simplest way of expressing capacity requirements from a production plan or a number of manufacturing orders is to use the same units as production volumes are expressed in. A production plan for an item will then be a direct measurement of the demanded capacity as obtained by adding together the production plans for all products to be manufactured in the same production plant. Capacity planning is carried out for each end product since the manufacturing unit expresses the aggregate capacity need for end-product manufacturing as well as manufacturing all incorporated components and semi-finished items.

One requirement for being able to calculate capacity requirements expressed in overall factors is that the products are of the same type from the manufacturing point of view with regard to the unit that is used, otherwise capacity requirements from different products cannot be added together to give a total requirement. Since capacity requirements are expressed in overall factors, detailed information is not necessary on the structure of each product or how it is manufactured, i.e. the type of information that is commonly stored in item, bills of material and routing files. This method of capacity planning is primarily used at planning levels with low detail and long planning horizon, i.e. the planning levels of sales and operations planning and master production scheduling. At this planning level it is common to convert monthly or quarterly production plans into capacity requirements. The aim is to create data to balance, in the long term, investments in machines, facilities available, man-hours and so on with expected demand. The calculation of capacity requirements with the aid of overall factors is illustrated by the following example.

Example 14.1

A company manufactures two products with a production plan as in Table 14.1.

	Planning period – month											
	1	**2**	**3**	**4**	**5**	**6**	**7**	**8**	**9**	**10**	**11**	**12**
Product A	170	160	150	150	150	180	200	200	200	200	200	200
Product B	70	100	100	100	120	120	120	120	120	120	120	120
Total	240	260	250	250	270	300	320	320	320	320	320	320

TABLE 14.1 Production plan for products A and B

The only resource category that is critical for the company is manning, and the resource requirements for the two products are approximately the same, so pieces are used in production planning. Accumulated real production volumes during the previous 12-month period were 2000 pcs of product A and 1000 pcs of product B. This was the equivalent of 167 and 83 pcs per month respectively (in total 250 pcs per month).

As shown in Table 14.2, the company will have a shortage of capacity from month 5 onwards. From month 7 there will be approximately 70 pcs of the products produced per month if it is assumed that capacity during the following year is the same as the current year. The accumulated shortage of capacity on an annual basis is 490 pcs, or two months' manufacturing capacity, or 17 per cent of the historic capacity. If the shortage of capacity is judged to be continuous, a reasonable strategy would be to employ more operators. If it is judged to be temporary and not continuing into the following year, other strategies such as utilising overtime may be possible instead.

	Planning period – month											
	1	**2**	**3**	**4**	**5**	**6**	**7**	**8**	**9**	**10**	**11**	**12**
Total capacity requirement	240	260	250	250	270	300	320	320	320	320	320	320
Available capacity	250	250	250	250	250	250	250	250	250	250	250	250
Difference	+10	−10	0	0	−20	−50	−70	−70	−70	−70	−70	−70

TABLE 14.2 Capacity requirements and available capacity

Method variants

Method variants that occur within the framework of this capacity planning method are related to the unit that is used to express capacity requirements. Examples of common capacity requirements units are overall factors such as number of pieces, area, volume, etc. Overall factors in the form of production value are also common, such as refinement value, manufacturing value or invoicing value.

Characteristic properties

The method using overall factors as a basis for calculating and expressing capacity requirements may be characterised in the following ways:

- The method is simple and easy to understand.
- The method allows simulation of different production plans and the effects of more customer orders either manually or with the aid of simple tools such as Excel.
- The calculation of capacity requirements is very approximate since it relates to the entire company or factory/workshop.
- Since the capacity that capacity requirements are compared with is based on historic outcomes, the method does not take into consideration productivity improvements achieved through rationalisation or changes in technology.
- Unless all the products covered by the calculation of capacity requirements have similar resource requirements, the method has low reliability if changes are made to product mix.
- It may be difficult to establish comparable capacity levels. Capacity measurements must often be based on experience values, i.e. an assessment of what can normally be produced through experience. This difficulty is compounded by the fact that the method does not use a capacity-neutral unit, i.e. a quantity of a product can represent another capacity requirement than a quantity of another product. There are particular difficulties when production values are used, since these are subject to variations due to inflation, which increase the level of approximation with regard to available capacity.

Primary application environments

Capacity planning with the aid of overall factors is mainly appropriate in the following planning environments:

- *No detailed production preparations.* The method is primarily intended for planning environments without detailed production preparations since it does not demand any capacity requirements data.
- *Simulation capabilities desired.* It is also particularly suitable in conditions where simulation with simple tools is desirable. The method provides a rough picture of future capacity requirements very quickly and easily.
- *Short accumulated lead times.* The planning method is most attractive in environments with short accumulated lead times or long planning periods so that its shortcomings with respect to lack of lead time offsetting do not affect results. Its strength lies in uniform manufacturing environments and in cases of low refinement levels at order placement, since it is not dependent on production preparations.

Item number	Description	Unit
12 789	UPL 150AC	Piece

Department/work centre		Time per piece
01	Machining	27.0
02	NC work	18.3
05	Finishing	5.7
06	Component assembly	8.1
07	Final assembly	6.5

FIGURE 14.12 Example of a capacity bill

■ *Sales and operations planning and master production scheduling.* The method is mainly used for capacity planning at the sales and operations planning level, and in master production scheduling.

Capacity planning using capacity bills

If a more detailed calculation of capacity requirements is required and it is not possible to assume that the products utilise manufacturing resources in proportion to their respective volumes, **capacity bills** may be used as an alternative. A capacity bill expresses the total capacity requirement per item of a product in terms of the resources required for its manufacture. Capacity requirement is often stated in man-hours or machine-hours per manufacturing department or equivalent planning unit in a company. Capacity requirements can then be calculated at the departmental level and not only as a total for the entire company. Units other than time may be used. An example of a capacity bill is shown in Figure 14.12. The bill shows that 27.0 hours of machine work is required for each piece of product UPL 150AC to produce all the internally manufactured parts in the product, 18.3 hours of NC work, etc., and finally 6.5 hours to assemble the product.

Capacity planning using capacity bills is often based on production plans expressed as quantities per time period, but capacity bills can also be created for customer orders or manufacturing orders, depending on what context they will be used in. The level of detail of the method is only limited by the number of manufacturing departments that it is possible to estimate operation times for.

Capacity planning using capacity bills can also be carried out for individual products, and the method is most often used for capacity planning at the levels of sales and operations planning and master production scheduling. Since capacity requirements in the profile are expressed in hours per item of each product, the total capacity need is obtained by multiplying this number of hours by the quantity to be produced. The method of using capacity bills for capacity planning is illustrated in general terms in Figure 14.13 for the two products, RVC 100 and UPL 150. A capacity bill is drawn up for each of these products and a production plan is established. For example, the capacity requirement in April will be $60 \cdot 4.2 + 120 \cdot 3.9$ hours, or 720 hours.

In contrast to the previous method, this method links capacity requirements for individual products directly with individual departments. Data on total capacity usage per work centre will be required for this task. Example 14.2 illustrates the calculations used in the method.

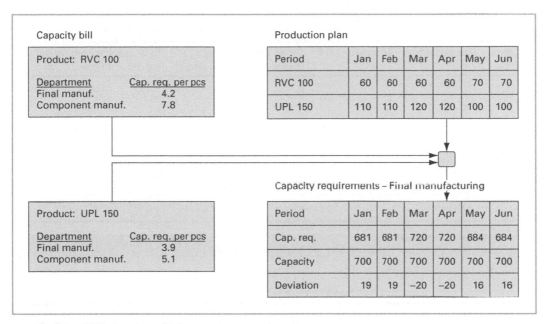

FIGURE 14.13 Illustration of a capacity requirements calculation with the aid of capacity bills

Example 14.2

The production plan for a company's two products (A and B) is expressed in weekly planning buckets in Table 14.3.

	Planning period – week									
	1	2	3	4	5	6	7	8	9	10
Product A	40	40	40	50	50	40	40	30	30	40
Product B	20	20	15	15	20	20	30	30	30	20

TABLE 14.3 Production plan for products A and B

Product A consists of one C item and two D items, whereas product B consists of one C item and one E item. Item E consists in turn of three F items. If there is access to detailed data on set-up times and operation times in the routing and work centre files, this may be used as the basis for calculating capacity bills which state the capacity requirements per work centre for each product. The data required for such a calculation is shown in Table 14.4.

In certain cases access is not available to such detailed data as in Table 14.4 from the routing and work centre files. The method may still be used, but estimates of capacity requirements per work centre must instead be based on assessments from experience.

With the aid of the capacity bills shown in Table 14.5, capacity requirements to fulfil the production plan for each work centre and planning period are calculated. For week 1, for example, the capacity requirement for WC 100 will be 128.6 hours (40 · 1.88 + 20 · 2.67), for WC 200 60.6 hours (40 · 0.78 + 20 · 1.47) and for WC 300 30.8 hours (40 · 0.34 + 20 · 0.86). The capacity requirements are then compared with available capacity as in Table 14.6.

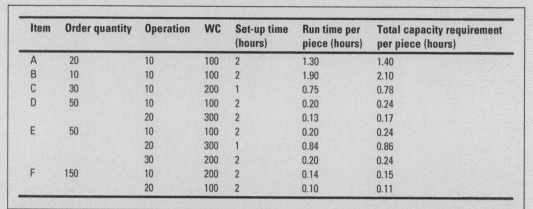

Item	Order quantity	Operation	WC	Set-up time (hours)	Run time per piece (hours)	Total capacity requirement per piece (hours)
A	20	10	100	2	1.30	1.40
B	10	10	100	2	1.90	2.10
C	30	10	200	1	0.75	0.78
D	50	10	100	2	0.20	0.24
		20	300	2	0.13	0.17
E	50	10	100	2	0.20	0.24
		20	300	1	0.84	0.86
		30	200	2	0.20	0.24
F	150	10	200	2	0.14	0.15
		20	100	2	0.10	0.11

TABLE 14.4 Data required to generate capacity bills

Work centre/department	Capacity requirement	
	Product A	**Product B**
100	$1.40 + (2 \cdot 0.24) = 1.88$	$2.10 + 0.24 + (3 \cdot 0.11) = 2.67$
200	0.78	$0.78 + 0.24 + (3 \cdot 0.15) = 1.47$
300	$2 \cdot 0.17 = 0.34$	0.86
Total capacity requirement/piece	3.00 hours	5.00 hours

TABLE 14.5 Capacity bills for products A and B

	Planning period – week									
	1	**2**	**3**	**4**	**5**	**6**	**7**	**8**	**9**	**10**
WC 100:										
Capacity requirement	128.6	128.6	115.2	134.0	147.4	128.6	155.3	136.5	136.5	128.6
Available capacity	128	128	128	128	128	128	128	128	128	128
Deviation	−0.6	−0.6	+12.8	−6.0	−19.4	−0.6	−27.3	−8.5	−8.5	−0.6
WC 200:										
Capacity requirement	60.6	60.6	53.2	61.0	68.4	60.6	75.3	67.5	67.5	60.6
Available capacity	40	40	40	40	40	40	40	40	40	40
Deviation	−20.6	−20.6	−13.2	−21.0	−28.4	−20.6	−35.3	−27.5	−27.5	−20.6
WC 300:										
Capacity requirement	30.8	30.8	26.5	29.9	34.2	30.8	39.4	36.0	36.0	30.8
Available capacity	36	36	36	36	36	36	36	36	36	36
Deviation	+5.2	+5.2	+9.5	+6.1	+1.8	+5.2	−3.4	0	0	+5.2

TABLE 14.6 Capacity requirements and available capacity

The result from capacity planning using capacity bills is somewhat different compared with the result based on overall planning factors. The reason is that the capacity requirements for the two products do not have the same proportional distribution between the three departments. When consideration is given to the real capacity requirement in the different departments, it becomes apparent that WC 200 is constantly over-utilised by an average of 24 hours per week. The workload is slightly unbalanced for WC 100 even though the department as a whole has an undercapacity of 6 hours per week. WC 300, on the other hand, has an average overcapacity of 3.5 hours per week.

Method variants
The difference between the method variants that exist lies in how capacity requirements data are created. One variant produces capacity requirements data on the basis of assessments of capacity requirements through experience for each department/resource unit. Capacity requirements are registered and maintained manually. In general this is the only possibility when working in an engineer-to-order environment.

To further increase the level of detail, a variant of capacity bills called load profiles may be used. Using this variant, it is possible to offset the lead time for capacity requirements in relation to the period by which the product must be completed. For example, it is possible to take into consideration the fact that machine processing for the product UPL 150AC in Figure 14.12 must be carried out some weeks before final manufacture for reasons of throughput time.

If there is access to an ERP system with basic data files, capacity bills can also be created and maintained automatically for standardised products with the aid of calculation programs in the system. The principle for such calculations is illustrated in Figure 14.14.

Characteristic properties
Capacity planning based on the use of capacity bills may be characterised in the following ways:

- In the same way as the previous method, it is simple and easy to understand. It allows simulation and analysis of workload effects from incoming orders using simple tools such as Excel.

- When calculating or estimating capacity requirements per resource unit in terms of capacity bills, set-up times are treated in rather general terms. In principle, calculations are based on manufacturing always taking place using the same order quantity for every item so that an average set-up time for units manufactured can be determined and added on to direct manufacturing times in question.

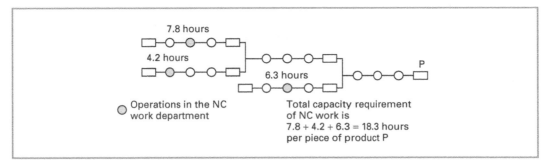

FIGURE 14.14 Generation of capacity bills from bill of material and operation data

- The method allows the use of completely capacity-neutral capacity requirements measurements that are not influenced by changes to the product mix.

- For the variant based on automatic generation of capacity bills from the basic data files, consideration can be easily given to productivity improvements brought about by rationalisation or changes in manufacturing technology.

- When the method is used to plan capacity requirements on the basis of manufacturing orders, it is more difficult to achieve precision in the short term compared with the following method, **capacity requirements planning**, since it is more difficult to deduct capacity requirements at the same rate as manufacturing takes place through **labour reporting**. This is particularly decisive in the case of long operation times.

Primary application environments

The primary environments in which the method may be applied are described below:

- *No detailed production preparations.* Capacity planning with the aid of capacity bills is mainly useful in companies without detailed production preparations since it places small demands on such preparations. The demand for uniform manufacture is considerably less than for the previous method, since capacity bills take consideration to the variation in capacity requirements for different products.

- *Master production scheduling.* The method is a main alternative for master production scheduling in companies with order-based products and manufacturing.

- *Simulation capabilities desired.* It is also particularly suitable in conditions where simulation with simple tools is desirable. The method provides a rough picture of future capacity requirements very quickly and easily.

- *Accumulated throughput time shorter than a couple of planning periods.* One important demand for the method to give satisfactory accuracy at a more detailed level is that the accumulated throughput times of products are not much longer than the period length used in capacity planning. Fluctuations in capacity requirements will otherwise be given a misleading allocation of time. Since capacity requirements from set-up times are calculated in relatively standard terms, the method has advantages in planning environments with short set-up times and thus comparatively small order quantities.

Capacity requirements planning

All the methods presented above for the calculation of capacity requirements are mainly intended for planning situations in conjunction with sales and operations planning and master production scheduling, although they are also useful to a certain extent at the levels of order planning. The planning objects are primarily customer orders or production plans, although some of the methods may be used for manufacturing orders. The circumstances are the opposite for capacity requirements planning (CRP). This method works with manufacturing orders and is intended above all for the levels of order planning and production activity control. It allows considerably more detailed calculations to be made, per day or per hour, and for individual work centres and workstations. Another difference compared with the two previous methods is that capacity requirements planning is not carried out for complete products. Instead, capacity requirements are calculated individually from the manufacture of each of the incorporated items separately. The rather exhaustive and complex calculations involved make

Work centre 245–7		Lathe			
Week	Delayed	1	2	3	4
Released order	25	35	10	5	0
Planned order		0	24	32	38
Capacity requirement	25	35	34	37	38
Capacity		45	45	45	45
Utilisation (%)		78	76	82	84
Cum. available capacity		–15	–4	4	11

FIGURE 14.15 Presentation of calculated capacity requirements from released and planned manufacturing orders

it difficult to use the method with traditional ERP systems to simulate different planning situations. For this reason the method is not so widely used at higher planning levels.

Capacity requirements planning means that the requirement of capacity is calculated from the starting point of operations associated with released manufacturing orders as well as planned manufacturing orders. Material requirements planning is normally used to generate information on start times and delivery times for manufacturing orders. To calculate in which planning period each order will utilise a specific work centre, information is used regarding the routings, set-up times and operation times for each operation, **queue times** in each work centre, transportation times between work centres and the number of working days included in each planning period. This information is fetched from files in the ERP system and from the workshop calendar. The calculated capacity requirement from each of these manufacturing orders is totalled for each work centre and may be presented as in Figure 14.15.

There are two different methods of distributing capacity requirements for a certain operation between planning periods. In the first method, all capacity requirements are placed in the planning period which contains the planned finish time for the operation. The second is more exact and the capacity requirement is allocated proportionally to the planning periods in which the operation is scheduled to be ongoing, as shown in Figure 14.16. With short operation times the method selected is of minor importance.

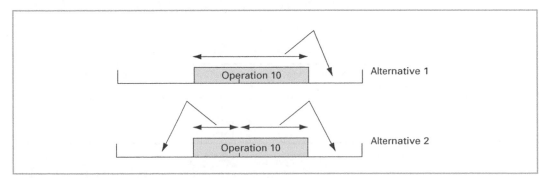

FIGURE 14.16 Allocation of capacity requirements between periods

Capacity planning using capacity requirements planning

To illustrate the principles of calculation in capacity requirements planning, a simplified example of the use of the method is presented below.

Example 14.3

A company manufactures the products A and B and the components C, D, E and F that are part of the A and B products. Capacity requirements planning is used to calculate the required capacity in work centres WC 100, WC 200 and WC 300. It is now Monday morning in week 18. The example shows how the capacity requirements for one released and one planned order are calculated and scheduled with the aid of capacity requirements planning. In order to carry out the planning, information is required on routings, queue times, set-up times, **run times**, transportation times and start times and finish times for the orders. Information is also required on available capacity in the work centres used.

Table 14.7 states routings, set-up times and run times for manufacturing the items in question.

Table 14.8 shows available capacity and queue times in work centres. In order to schedule operations it is necessary to know which calendar days are workdays and how many work hours are available in one workday. This information is shown on the workshop calendar in Figure 14.17.

Item	Order quantity	Operation	Work centre	Set-up time (hours)	Set-up time (hours)	Run time/ piece (hours)	Total capacity requirement/ piece (hours)	Total capacity requirement per order
A	20	10	100	2	0.10	1.30	1.40	28.0
B	10	10	100	2	0.20	1.90	2.10	21.0
C	30	10	200	1	0.03	0.75	0.78	23.4
D	50	10	100	2	0.04	0.20	0.24	12.0
		20	300	2	0.04	0.13	0.17	8.5
E	50	10	100	2	0.04	0.20	0.24	12.0
		20	300	1	0.02	0.84	0.86	43.0
		30	200	2	0.04	0.20	0.24	12.0
F	150	10	200	2	0.01	0.14	0.15	22.5
		20	100	2	0.01	0.10	0.11	16.5

TABLE 14.7 Information on routings, set-up times and run times from the basic data file

Work centre	Number of machines	Nominal capacity per day	Workload	Capacity per day	Planned queue time
WC 100	4	32 h	100%	32 h	8 h
WC 200	2	16 h	100%	16 h	12 h
WC 300	1	8 h	100%	8 h	11 h

TABLE 14.8 Work centre data

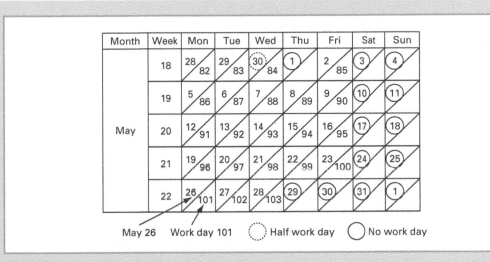

FIGURE 14.17 Workshop calendar for the month of May

Information on transportation times between work centres and warehouse is also necessary for scheduling operations. This information is shown in Table 14.9.

In capacity planning, released orders are normally distinguished from planned orders. Table 14.10 shows the order status for the released order 131. Operation 1 was completed at the end of week 17 and the order is now waiting to be moved from WC 100 to WC 300.

		To work centre/stores			
		WC 100	**WC 200**	**WC 300**	**Warehouse**
From work centre	WC 100	0 h	1 h	2 h	2 h
	WC 200	1 h	0 h	2 h	2 h
	WC 300	2 h	2 h	0 h	2 h

Order	Item	Order quantity	Due date	Operation	Work centre	Total capacity requirement per piece (hours)	Total capacity requirement (hours)	Order status
131	E	50	14 May	10	100	0.24	12	Ready
				20	300	0.86	43	
				30	200	0.24	12	

					Planning period – week						
	18	**19**	**20**	**21**	**22**	**23**	**24**	**25**	**26**	**27**	
Planned delivery			150		150			150			
Planned order start	150		150			150					

TABLE 14.11 Information from material requirements planning for item F

Table 14.11 shows an extract from material requirements planning. The start of one manufacturing order with an order quantity of 150 pcs of item F is planned for weeks 18, 20 and 23. The lead time for the item is two weeks and inbound deliveries of finished items are scheduled for two weeks after order start.

The planned manufacturing order for item F with a planned delivery in week 20 must go through two steps of operations. Scheduling of operations for manufacture of the order is shown in Figure 14.18.

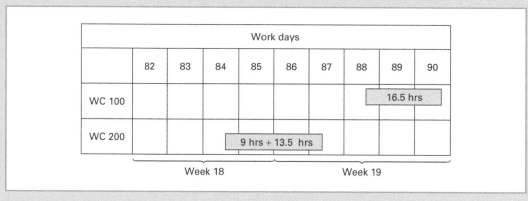

FIGURE 14.18 Scheduling of operations for manufacture of the planned order of 150 pieces of item F with delivery in week 20

For the manufacturing order to be completed by the beginning of week 20, it must be delivered to the warehouse at the end of Friday in week 19, i.e. day 90. If **backward scheduling** is used to schedule the operations and if a manufacturing order can only utilise one machine at a time, and 8 hours a day maximum, the last operation in WC 100 will be completed 2 hours before the end of day 90, since the transportation time from WC 100 to the warehouse is two hours. The 16.5 hours that correspond to set-up and run time in WC 100 will utilise the work centre for 6 hours on day 90, 8 hours on day 89 and 2.5 hours on day 88. There is 8 hours' queue time before the operation and one hour's transportation time. This slack time is scheduled for days 87 and 88 and means that the previous operation step in WC 200 is scheduled for completion when there are 3.5 hours remaining in day 87. The 22.5 hours for set-up time and run time in WC 200 are then planned for 5.5 hours in day 87, 8 hours in day 86, 8 hours in day 85 and 1 hour in day 84. Since the stated queue time for WC 200 is 12 hours and day 84 only has 4 working hours, the order is planned to be started in day 82. Planned capacity requirements for the manufacturing order of item F with planned delivery in week 20 will then be 9 hours in week 18 for WC 200, 13.5 hours in week 19 for WC 200 and 16.5 hours in week 19 for WC 100.

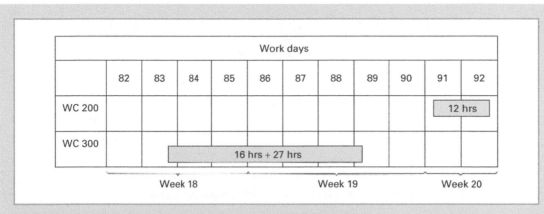

FIGURE 14.19 Scheduling of operations for the manufacture of released order 131

Capacity requirements and scheduling of operations for the manufacture of order 131 are shown in Figure 14.19. The calculations are carried out in the same way as was described in detail for the planned order above. For WC 200, 16 hours are scheduled for week 18 and 27 hours for week 19.

The capacity requirements for all orders are added together per work centre and planning period and are reconciled with available capacity. As an example, capacity requirements and available capacity for WC 200 are shown in Figure 14.20.

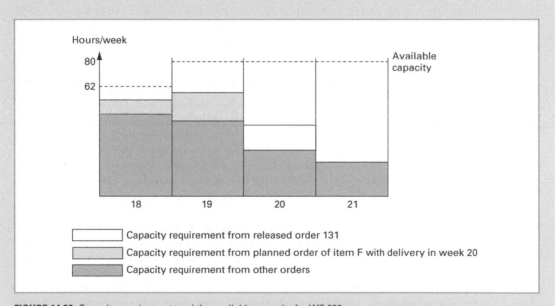

FIGURE 14.20 Capacity requirements and the available capacity for WC 200

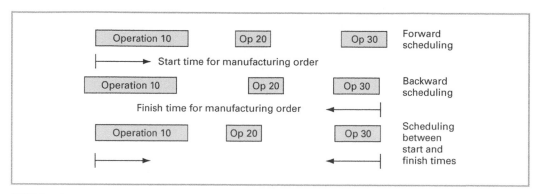

FIGURE 14.21 Different methods for scheduling operations in capacity requirements planning

Method variants

Existing method variants differ with respect to how operations associated with a manufacturing order are scheduled as a basis for calculating capacity requirements per planning period. There are three different variants: **forward scheduling**, backward scheduling, and scheduling between material-planned start times and finish times. These variants are illustrated in Figure 14.21.

Forward scheduling of operations means that operations belonging to a manufacturing order are successively scheduled forwards in time from the start time of the order, e.g. as it was planned by material planning. Subsequent operations are scheduled to start successively at the scheduled finish time of the previous operation plus a general increment to cover transportation time between consecutive work centres and the queue time that may normally be expected to arise before the next work centre. Information on queue times and transportation times are available in most ERP systems in the work centre file or transportation time file. This scheduling alternative may mean that the scheduled finish time of the last operation may not coincide with the planned finish time of the manufacturing order due to lead times used in material planning not being correctly updated or because the order quantity has been changed, for example.

Operations are successively scheduled backwards in the backward scheduling variant, i.e. the finish time of the last operation coincides with the planned finish time for the manufacturing order. Scheduling of earlier operations is generally carried out in the same way as with forward scheduling, that is, the scheduled start time for an operation minus estimated times for queues and transportation will be the finish time for the previous operation. There is a risk with this scheduling alternative that the start time of the first operation may deviate from the start time of the manufacturing order.

These two alternatives are the most commonly occurring variants. However, there is a third alternative that in principle is more exact and does not entail the same risk of start and finish operations not being scheduled in compliance with the planned start and finish times of the equivalent manufacturing order. Operations are spread out between the planned start time of the manufacturing order and its finish time. The spread of operations takes place by distributing slack time between operations. This slack time is equal to the time difference between the start and finish time of the order minus the total operation time for all operations. It thus represents the sum of queue times and transportation times for all operations involved. There are two main variants for distributing slack time. The first variant distributes this time equally between operations while the second distributes slack time between operations in proportion to the queue and transportation times for each as stored in the basic database.

Characteristic properties

Capacity requirements planning as a tool for planning the use of capacity may be characterised in the following ways:

- The method requires extensive processing in an ERP system. In general the ERP system must also include a function that can generate planned manufacturing orders, such as material requirements planning.

- Due to the extensive and time-consuming calculations required, the method is less suitable for simulating and comparing capacity requirements from different master productions schedules, at least if an advanced planning and scheduling (APS) systems is not available.

- The level of detail in capacity requirements calculations may be considerably higher than in the previous methods described, and in principle down to the level of single machines and workstations.

- The calculated capacity requirement is more exact due to more accurate lead time off-setting since it is based on the scheduling of individual orders, and because capacity requirements include set-up times per order and not general estimates.

- The method allows a distinction to be made between capacity requirements from released orders on the one hand, and capacity requirements from planned orders on the other hand.

- Since capacity requirements are calculated from manufacturing orders, it is possible to deduct capacity requirements as manufacturing progresses with the aid of labour reporting. Capacity requirements in the short term will be considerably more correct than those achieved by other capacity planning methods.

Primary application environments

Capacity requirements planning is useful in the following environments:

- *High capacity planning accuracy desired.* Capacity requirements planning is above all suitable in planning situations where there is a need for greater precision and level of detail. It is the only capacity planning method that correctly takes into consideration available stock on hand and stock changes for items at structural levels under end products. The method also has large relative advantages in environments with long set-up times and with large and varying order quantities.

- *Detailed production preparations.* The method requires detailed production preparations and high basic data quality for its benefits to be apparent.

14.6 Properties and Characteristics

In the same way as for material planning methods, capacity planning methods are more or less applicable in different planning environments. Characteristics unique to each different method are summarised in Table 14.12.

Property variables	Overall factors	Capacity bills	Capacity requirements planning
Primary planning objects	Production plan	Production plan Customer orders	Manufacturing orders
Scope of objects	Individual items Group items	Individual items Group items	Individual items
Consideration to stock on hand	No	No	Yes
Consideration to lead time	No	No	Both allocation and lead time offsetting
Primary planning level	Sales and operations planning Master production scheduling	Sales and operations planning Master production scheduling	Order planning Production activity control
Capacity grouping	Whole company	Departments Foreman areas	Individual work centres

TABLE 14.12 Comparison of methods with respect to properties

14.7 Summary

This chapter dealt with capacity planning at the different planning levels. Capacity supply and capacity requirements were defined and different ways of expressing and calculating the two concepts were described. Different strategies for changing capacity and adapting the utilisation of capacity were discussed. A number of action variables used to achieve a balance between supply of capacity and capacity requirements in the short and medium term were accounted for. The chapter described in detail the most commonly used capacity planning methods: capacity planning using overall factors, capacity bills and capacity requirements planning. The method of calculating capacity requirements using each method was explained. Finally, the characteristic properties of each method were compared, together with their utility in different planning situations.

🔐 Key concepts

Backward scheduling 304	Forward scheduling 306
Capacity bill 296	Labour reporting 300
Capacity planning 281	Lag strategy 288
Capacity requirements planning (CRP) 300	Lead strategy 288
Chase strategy 288	Level strategy 288
Efficiency factor 287	Manufacturing order 292

Discussion tasks

1 Capacity planning using capacity bills and using capacity requirements planning are two different methods to calculate future capacity requirements. In what ways are these two methods different in the following respects: (a) product versus manufacturing order oriented; (b) consideration to set-up times; and (c) lead time offsetting of capacity requirements?

2 In what industries and situations would you think a chase strategy is the most suitable and in what situations would you think the levelled strategy is suitable?

3 Why is it problematic to lose an hour of capacity in a bottleneck resource?

Problems[1]

Problem 14.1

Sludge Removal Ltd manufactures three different types of sludge evacuators: A, B and C. The master production schedule in pieces per month for the next year is shown in the table.

Month	1	2	3	4	5	6	7	8	9	10	11	12
Product A	130	130	130	130	120	110	110	120	130	130	130	130
Product B	80	80	85	90	90	80	80	80	80	75	75	70
Product C	90	90	90	90	90	90	90	100	110	120	120	130

The only resource category that is critical for the company is personnel, since manufacturing consists mainly of assembly and testing. Resource requirements per unit for the three products are approximately the same. During the year the company employed 42 operators and the average rate of production was 122 pieces per month of product A, 84 pieces per month of product B and 96 pieces per month of product C.

What will the capacity requirements be for next year, and what consequences will there be if capacity is the same as during the current year?

---

1 Solutions to problems are in Appendix A.

Problem 14.2

Product Precision Ltd has gathered the following data regarding man-hours and machine-hours for the manufacturing of the item Precision-WZ, which is manufactured in two models, X and Y. The information is shown in the table.

	Year 1	Year 2
Production volume, model X (pcs)	1200	1150
Production volume, model Y (pcs)	580	520
Man-hours, model X (h)	3800	3700
Man-hours, model Y (h)	750	700
Machine-hours, model X (h)	1300	1100
Machine-hours, model Y (h)	1350	1250

Manufacturing takes place in two work centres, WC 100 and WC 200. All machine processing and 30 per cent of the man-hours are carried out in WC 100. The remaining man-hours are carried out in WC 200.

a) Calculate the appropriate capacity planning factors for year 3.

b) How large will capacity requirements be in the two work centres next year for the production plan below showing planned manufacturing in pieces per quarter?

Quarter	1	2	3	4	Total
Model X	400	500	400	450	1500
Model Y	200	150	200	200	700

Problem 14.3

The production plan for Swan Island Ltd in pieces per week for the product Seagull in weeks 11–20 is shown in the table.

Week	11	12	13	14	15	16	17	18	19	20
Product Seagull	400	500	450	550	450	400	400	450	500	450

Seagull is manufactured in several work centres, but the two work centres WC 980 Welding and WC 990 Assembly are critical with respect to available capacity. Every piece of the product utilises 0.3 hours in WC 980 and 0.4 hours in WC 990. There are no other products manufactured in these work centres. Net capacity in WC 980 is 140 hours per week and in WC 990 is 200 hours per week. Analyse the critical resources by carrying out capacity planning.

Problem 14.4

Deltatool Ltd manufactures one end product 111, which consists of the two internally manufactured items 888 (1 piece) and 999 (1 piece). The bill of materials for product 111 is shown in the figure.

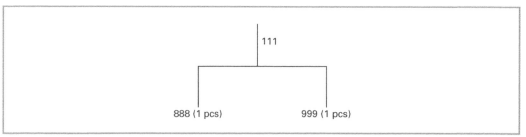

Manufacturing takes place in two work centres, M 1 and M 2. Items 111 and 999 are manufactured in one operation and item 888 in two operations, as shown in the table.

Item	Lot size	Operation	Work centre	Set-up time (hours)	Run time per piece (hours)
111	5	10	M 1	1.00	0.20
888	10	10	M 2	1.00	0.15
		20	M 1	1.00	0.10
999	20	10	M 1	2.00	0.25

Processing in the work centres takes place in two shifts, each shift consisting of 7.5 hours. There are four machines in work centre M 1 and one machine in M 2.

Utilisation rate is 85 per cent and the efficiency factor is 100 per cent for both work centres. All weeks have five working days.

a) Draw up a capacity bill for product 111 with the aid of the above information.

b) Calculate the nominal capacity and net capacity for the work centres M 1 and M 2.

c) What will be the capacity requirements and available capacity in the work centres M 1 and M 2 in periods 21–26? The production plan in pieces per week for products 111 is shown in the table.

Week	21	22	23	24	25	26
Product 111	220	250	250	300	220	250

Problem 14.5

Green Valley Ltd manufactures an end product C, which consists of the manufactured items X, Y and Z as in the figure.

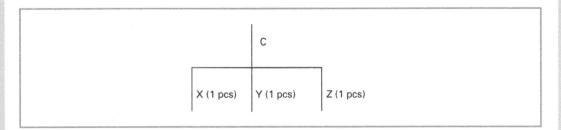

All manufacturing takes place in two work centres, WC 1 and WC 2. Item C is manufactured in one operation, item X in two operations, item Y in one operation and item Z in two operations, as shown in the table below.

Item	Lot size	Operation	Work centre	Set-up time (hours)	Run time per piece (hours)
C	5	10	WC 1	1.5	2.0
X	30	10	WC 2	1.8	0.5
		20	WC 1	0	0.5
Y	60	10	WC 2	2.4	0.2
Z	30	10	WC 1	0.9	0.4
		20	WC 2	0.6	0.5

The scheduling of operations is shown in the figure.

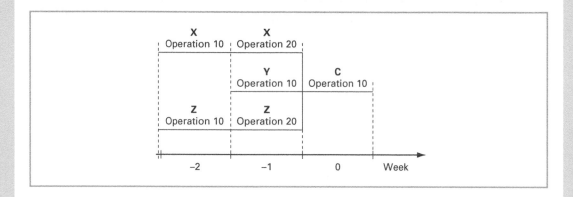

a) Draw up a resource profile (i.e. a capacity bill with lead time offsetting) for the end product C.

b) Calculate the capacity requirements for WC 1 and WC 2 during the periods 14–18, using the derived resource profile. The production plan in pieces per week for end product C is shown in the table.

Week	14	15	16	17	18	19	20
End product C	25	20	30	25	30	30	30

Problem 14.6

Big Products Ltd manufactures an end product K, with a bill of materials as shown in the figure.

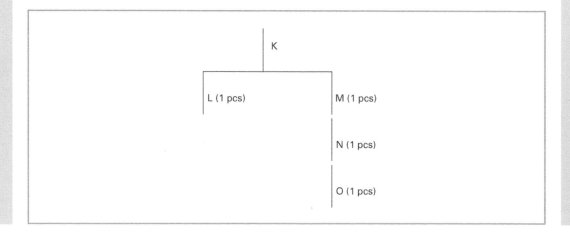

All manufacturing takes place in two work centres, WC 200 and WC 300. All items are manufactured in only one operation and data for the operations are shown in the table.

Item	Lot size	Work centre	Set-up time (hours)	Run time per piece (hours)	Lead time (weeks)
K	10	WC 200	4.0	1.5	1
L	10	WC 300	3.0	1.0	1
M	10	WC 200	2.0	1.0	1
N	10	WC 300	1.0	0.5	1
O	10	WC 300	1.0	2.0	1

Develop a resource profile (capacity bill with lead time offsetting) and determine the capacity requirements in the work centres WC 200 and WC 300 that will result from the following production plan in pieces per week.

Week	11	12	13	14	15
End product K	10	20	10	0	20

Problem 14.7

Placard Ltd manufactures a number of products. One important product is Q, which consists of two internally manufactured items R and S as in the figure.

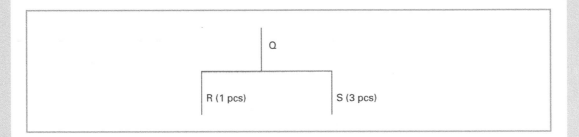

Item Q is assembled in an assembly station, A 10; item R is manufactured in work centre WC 30, and item S is manufactured in work centre WC 70. Operation data for product Q and items R and S are shown in the table.

Item	Lot size	Work centre	Set-up time (hours)	Run time per piece (hours)	Lead time (weeks)
Q	As needed	A 10	0	1.7	1
R	50	WC 30	3.0	1.0	1
S	150	WC 70	2.0	1.0	1

The following table shows material requirements planning stated for product Q and its components.

Item	Week		1	2	3	4	5	6
Q	Gross requirements		40	40	40	30	40	30
	Scheduled receipts							
	Stock on hand	50	10	0	0	0	0	0
	Net requirements			30	40	30	40	30
	Planned delivery in			30	40	30	40	30
	Planned order start		30	40	30	40	30	
R	Gross requirements							
	Scheduled receipts		50					
	Stock on hand	10						
	Net requirements							
	Planned delivery in							
	Planned order start							
S	Gross requirements							
	Scheduled receipts		130					
	Stock on hand	0						
	Net requirements							
	Planned delivery in							
	Planned order start							

a) Determine capacity requirements for product Q in the work centres A 10, WC 30 and WC 70 for the next five weeks on the basis of material requirements planning carried out.

b) The work centre A 10 is only used for product Q. The net capacity in this work centre is 60 hours per week. Make a proposal for measures to solve any shortages in capacity that may arise.

Problem 14.8

Epsilon Ltd uses capacity requirements planning to calculate capacity requirements. Two of their manufactured items are item 880 and item 990. These items are only manufactured in the work centres WC 730, WC 740 and WC 750. Several of the company's other items are also manufactured in these work centres.

Carry out capacity requirements planning for the month of October (weeks 40–43). Monday to Friday are whole working days with 8 working hours per day. Transportation time between the two work centres and from one work centre to stock is estimated at 1 hour. There is one released order for item 990. The following data are taken from the company's ERP system.

Operation data for items 880 and 990:

Item	Order quantity	Operation	Work centre	Set-up time (hours)	Run time per piece (hours)
880	15	10 cutting	WC 730	1	0.3
		20 turning	WC 740	2	0.9
990	90	10 turning	WC 740	2	0.2
		20 grinding	WC 750	1.5	0.3

Work centre data:

Work centre	No. of machines	Net capacity (hours/week)	Planned queue time (hours)
WC 730	2	70	6
WC 740	2	70	10
WC 750	1	35	15

Extract from the file of manufacturing orders:

Manufact. order	Item	Order quantity	Due date	Operation	Work centre	Operation status
3456	990	90	Monday, week 41	10 20	WC 740 WC 750	Ready

Information from material requirements planning (deliveries are assumed to take place at the beginning of the week):

Item		Week					
		40	41	42	43	44	45
880	Planned delivery in			15		15	
	Planned order start		15		15		
990	Planned delivery in				90		90
	Planned order start		90		90		

The final table shows capacity requirements in hours per week from capacity requirements planning for the other items manufactured in the work centres WC 730, WC 740 and WC 750.

Week	40	41	42	43
WC 730	65	60	68	64
WC 740	58	34	67	50
WC 750	6	34	12	33

Further reading

Blackstone, J. (1989) *Capacity management.* South-Western Publishing, Cincinnati (OH).

Burcher, P. (1992) "Effective capacity planning", *Management Services*, Vol. 36, No. 10, pp. 22–27.

Fogarty, D., Blackstone, J. and Hoffmann, T. (1991) *Production and inventory management.* South-Western Publishing, Cincinnati (OH).

Hyer, N. and Wemmerlöv, U. (2002) *Reorganizing the factory.* Productivity Press, Portland (OR).

Jonsson, P. and Mattsson, S. A. (2002) "Use and applicability of capacity planning methods", *Production and Inventory Management Journal*, Vol. 43, No. 3/4, pp. 89–95.

Landvater, D. (1993) *World class production and inventory management.* The Oliver Wight Companies, Essex Junction (VT).

Oden, H., Langenwalter, G. and Lucier, R. (1993) *Handbook of material & capacity requirements planning.* McGraw-Hill, New York (NY).

Plossl, G. (1985) *Production and inventory control – principles and techniques.* Prentice-Hall, Englewood Cliffs (NJ).

Vollmann, T., Berry, W., Whybark, C. and Jacobs, R. (2005) *Manufacturing planning and control for supply chain management.* Irwin/McGraw-Hill, New York (NY).

CASE STUDY 14.1: CAPACITY PLANNING AT ELROD SYSTEMS LTD

Elrod Systems Ltd is a manufacturer of telecommunication systems. In its plant, about 150 product models (each related to one of three brands), with some 3000 possible product configurations, are made and distributed to customers all over the world.

The average product life cycle is one to two years and there are a couple of engineering changes per product per year. The demand is some 30 to 40 per cent higher during the winter compared to the summer months. November and December are the peak months but there is also extensive random demand variation within individual weeks. Three assembly lines run during two day shifts and can build any product but each line is primarily used for making one of three brands. Most suppliers deliver components every day, but some deliver twice a day and some once a week.

The capacity strategy

Elrod wants to keep a fixed order-to-delivery cycle of five days, at the same time as it faces high demand variation. It therefore has developed a volume flexibility strategy, based on a combination of Elrod employees and temporary contract workers. During peak demand periods more than half of the 500–600 workers could be temporary contractors. The volume flexibility strategy is based on the following principles:

- Production equipment should always have 15–20 per cent overcapacity.
- Own personnel is used for production of the "base volume", i.e. non-peak volumes.
- Temporary contract workers are used for peak and seasonal production volumes.
- All temporary works have to finish internal education and training. New arrivals undergo two to three days' education and training before performing the simple operations. Elrod has a contract with Workpower, a temporary staffing company, which supplies Elrod with most temporary workers. Some of the extra workers are friends to employees. Workpower has a very high personnel turnover. Several of the workers hired from Workpower have been working as temporary employees once before but most are first-time workers at Elrod.
- Two-shift is used for "normal" production volumes. Night shifts and weekends are only used as extra capacity and production in high volume periods.
- Overtime is only used following internal or supply process disruptions, and cannot be used as planned extra capacity.
- If capacity requirements exceed the base capacity over a three-month period, employment of more own personnel is considered.
- If capacity exceeds the requirements over a six-month period, laying off personnel is considered.
- If capacity requirement exceeds the base capacity over a six-month period, investment in new equipment is evaluated.

Example of applying the strategy for manufacturing of a specific product:

* Increase production volume from 110 to 150 products per day is possible within two weeks with 25 extra temporary workers.
* Increase production volume from 150 to 195 products per day is possible within two weeks with 30 extra temporary workers and the use of two expansion areas in the factory.
* Increase production volume from 195 to 275 products per day is possible within two months with 70 extra temporary workers, for night-shift work.
* The capacity can be decreased to the base level within one week.

The capacity planning

Capacity is defined as products produced per day in the respective assembly line. Long-term capacity planning is part of the global supply chain manager's responsibility, and is conducted as part of the SOP process. It is a 12-month rolling planning process where an understanding of the capacity required for the daily production at the different lines is developed and a balance of what products to be made and delivered to the markets and the available capacity is established. Monthly forecasts per product are used for developing delivery plans. The respective line contains four different production departments. The capacity planning is based on capacity bills showing the standard hours per product in each production department. Balance is achieved by adjusting the production plan or by increasing or decreasing the workforce.

Short-term capacity planning in the respective production department is based on capacity and material requirements planning. The capacity requirements planning reports are reviewed by the production manager in the respective line, to identify needed capacity adjustments. The plans have 20-week horizon and daily planning buckets during the two most recent weeks and weekly buckets during weeks 3 to 20. They show the amount of work hours required on each machine and assembly cell.

Discussion questions

1 Which of the presented capacity adjustment strategies could be considered in the SOP process and which could be considered in the short-term capacity planning process?

2 Elrod has had some problems with temporary employees not producing as high quality output as their own workers. Another problem is that Workpower during the last six months has not been able to supply Elrod with as many temporary workers as it has asked for. In order to reduce the negative effects of using temporary workers, Elrod therefore considers combining temporary employees with subcontracting work to an external manufacturer. What issues should Elrod include in a cost–benefit analysis of switching to a combined temporary employees and subcontracting strategy?

3 The three production managers have brought up the idea that the company should produce some of the products to stock, using a levelled production plan. This would reduce Elrod's need for temporary employees, the production cost would decrease and the product quality would improve, they argue. The supply chain manager, sales manager and marketing manager who also participate in the SOP meetings where these issues have been discussed, however, do not like the idea and argue that they have to continue working with a chase strategy. What should be the key motives for sticking to the actual chase strategy?

15

Execution and Control in Pull Environments

In a pull-based execution and control system, the planning, execution and control of production and materials movement are conducted by a type of material planning method that is characterised by that material requirements arise in a consuming unit and more or less directly initiates production and/or delivery from an upstream material supply unit. In this way a continuous and smooth material flow, generated by a production schedule at the most downstream operation, is generated through the process.

Material planning, execution and control are consequently not separate functions in a **pull system**, but the functions are conducted in an integrated manner. Several planning and control activities can thereby be eliminated and the resulting planning and control function simplified. For example, no traditional production or procurement order is necessary for initiating production or materials movement. Sequencing and priorities on the shop floor are also kept to a minimum since the order sequence at a single operation equals the order in which replenishment signals are received from the direct downstream operation. The degree to which the material planning method directly initiates production or material movement in a pull system depends on the size of the inventory buffers which must be used for various reasons. These reasons include lead times being far too long to be able to order with immediate delivery, or that set-up times and ordering costs are far too high to justify production and/or transportation of a direct demand. Restrictions of this type force companies to a certain extent to produce and/or deliver quantities that cover several direct needs. Figure 15.1 illustrates the integration of material planning and execution and control in a pull system. It also shows that a pull system has the same relationship to master production scheduling and capacity planning as other planning approaches.

This chapter describes what a pull system is and the necessary preconditions for setting up and using a pull system. Pull-based execution and control can be very effective if used in an appropriate way and in an appropriate planning environment. If not using a pull system in an appropriate way it can however be very defective. Different types of methods can be used for generating pull signals and authorising production and materials movement. **Kanban** is the most common authorisation method. Different types of kanban systems, ways of designing and using kanban, as well as alternative authorisation methods, are described in the chapter.

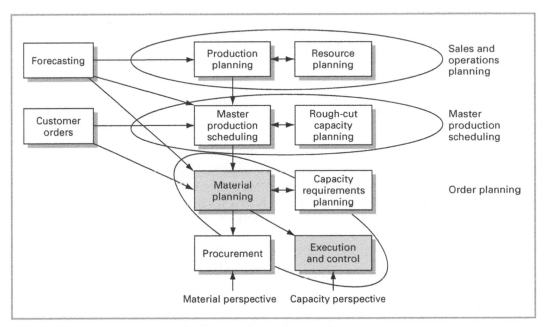

FIGURE 15.1 General relationship between execution and control in pull environments and other planning processes

15.1 What Is a Pull System?

The aim of a pull system is to create a smooth flow of materials through a production system that meets the demand, without requiring much inventory and work in process and without much information communication. The perfect pull system is thus simple at the same time as it results in low levels of tied-up capital and high customer response.

The pull production system was developed by Taiichi Ohno as part of the Toyota production system. The objective was to come up with a system that minimised the inventories and work in process and at the same time was responsive to the end demand. The idea to the final solution was adopted from the American supermarkets. Ohno's observation was that Americans made frequent visits to the supermarkets to buy the food they needed for just one or a couple of days, instead of doing weekly buys and storing the food at home. The responses by the supermarkets were to continuously order and replenish the shelves in small batches as they became empty. The customers consequently pulled the food stuff through the supermarkets. This way of pulling material downstream in a process is called a pull system.

In a pull system, materials movement and manufacturing only takes place on the initiative of and authorised by the consuming unit in the flow of materials. Production and procurement are initiated using some kind of physical card or signal, instead of issuing general production or purchasing orders. There is thus no need to process orders administratively in an ERP system. An example of a simple pull system is where the producer knows when to produce another product by looking at the downstream stock point. When the stock has been decreased to a minimum level, the producer is supposed to produce a batch and fill up the stock. In the optimum pull system the upstream producer produces a piece of the product as soon as the downstream unit has consumed a piece. There is never more than one piece of the product in the system but the setting-up, production, moving and waiting times are so short that no shortage occurs even though there are no safety mechanisms in the system.

A pull system in its simplest form consists of two operations with a stock point in between. The downstream operation withdraws items from the intermediate stock when it needs input material and the upstream operation produces and replenishes the stock in response to direct withdrawals from it or when the stock has decreased to a minimum level. If the operations are located geographically or time wise far away, it is necessary for two stock points between them, one outbound stock directly downstream from the first operation, holding the items that have been produced in the first operation, and one inbound stock point directly upstream from the second operation, holding the input items for the second operation. The second operation withdraws (pulls) items from its inbound stock. When the inbound stock needs to be replenished it withdraws items for the outbound stock of the first operation. When the outbound stock needs to be replenished the first operation produces items to deposit in the outbound stock. A process controlled by a pull system can consist of two operations with stock points in between or it can be a flow of successive pairs of operations and stocks.

A pull system works without traditional manufacturing and purchasing orders. Instead, production and materials movement is initiated by some kind of pull signal, coming from the directly downstream operation. The signal authorises what and when to produce and/or move. The signal normally takes the form of a card with information or a marked up empty pallet, box or container. To create as stable production environments as possible for the upstream operations, the aim is to keep the daily number of products to produce constant. If the final operation in a process is controlled by a production schedule and the upstream operations are controlled with a pull system, then the sequence and batch sizes of different products in the final operation's production schedule should also be designed in order to create as even demands and flows through the upstream operations as possible. Figure 15.2 illustrates how a pull signals control of a process of two operations with stocks in between.

A pull system is not stockless, because intermediate stocks are always necessary between each pair of operations. There are two reasons for having the stocks. The first is that the process throughput time would be very long if no stocks existed. Consider a pull system containing three successive operations. When a production order is issued at the last operation, there would be no input material available in the inbound stock. Therefore it is not possible to directly start producing according to the order. Instead, a signal to start producing must be sent to the upstream operation but, because this operation's inbound stock also is empty, another production signal to the first operation is needed. When the first operation has received input material it can then

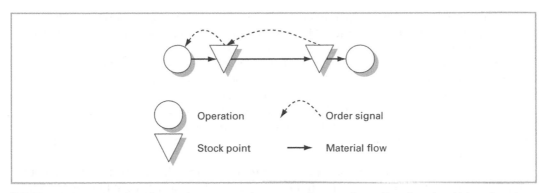

FIGURE 15.2 Illustration of a pull system in a process of two successive operations with outbound and inbound stocks in between

start to produce the item and deliver to the second operation. The second operation can then produce and deliver to the third operation which finally can produce and deliver according to the manufacturing order. The second reason for stocks is to even out small imbalances in production and demand rates between the successive pairs of operations.

Even though several stocks exist in a pull system, the aim is to level the production, minimise the set-up times, batch sizes and safety mechanisms and thereby keep each stock at a minimum level. When using a pull system in a stable environment and in an appropriate way the total amount of tied-up capital in the system could be very little, normally much lower than if not using a pull system. Execution and control in pull environments as described in this chapter, are never implemented as the only execution and control method in a company or production process. It order to make the pull system work it is necessary to first develop the planning environment, making sure the specific conditions outlined in the next section of the book exist. It will never be possible to develop such conditions for all production processes and material flows. More traditional planning and control methods also need to be used in parallel with a pull system, for example, to conduct capacity planning, generate delivery schedules, levelled master production schedules and detailed production schedules which may drive the last operation in the process.

15.2 Pull System Preconditions

A pull system is not appropriate in all situations but specific conditions are required in order to make the system work successfully. The main conditions are short set-up times and batch sizes, flow-oriented layout, **levelled production**, stable processes and responsible and involved personnel.

Short set-up times and batch sizes

Every operation in a pull system must be able to respond quickly to orders received from the downstream operations. To be able to do this, it is necessary to have short set-up times and batch sizes. A direct effect of this is short lead times.

Small batch sizes are not possible without short set-up times. Small batch sizes result in frequent changeovers and, if the set-up times are long, several machine hours will be lost in set-up and there may not be enough hours left for production. The costs for set-up will also be high. **Single minute exchange of die (SMED)** is a set-up time reduction approach developed by Toyota. The basic idea of SMED is to separate internal set-up (i.e. set-up work performed when a machine is stopped) from external set-up (i.e. work that can be done while the machine is still running). Focus is on reducing the internal set-up time to a minimum and converting internal set-ups to external set-ups. In this way set-up work is not reducing the available machine hours and is not increasing the total **cycle time**. In order to further decrease the set-up time and costs, all aspects of the set-up should be improved and, if possible, set-up should be completely abolished. There are several different techniques for set-up time reduction related to the SMED approach.

The package (pallet, box or container) size can also restrict the batch size. It may not be sufficient to store and move items in half-full packages. If the packages used to carry items in the factory or from suppliers are not small enough, then batch sizes may have to be large even though set-up times are short. Small package sizes are therefore necessary in a pull system.

Flow-oriented layout

A flow-oriented layout, for example using a cellular layout, is another important pull system requirement. Such a layout links different operations in a process together to a synchronous system. If every operation has about the same capacity and can produce with a similar cycle time (i.e. the same production time per item) it is possible to pull the items through the process in the sequence and pace dictated by the final operation in the process. A flow layout also results in short lead times which are important for effectively conducting pull-based execution and control.

Levelled production

In a pull system, the preceding operations work on order signals received from the respective subsequent operation. The entire process is driven by a production schedule for the most down-stream operation in the process. In order to create smooth production and flow of materials through the entire process it is therefore necessary that the production schedule for the most downstream operation is levelled, i.e. the same batch size of an item is produced every time and the production of every item is scheduled in regular intervals. The smaller the batch sizes the more levelled is the schedule and the smoother the material flow and the more even is the capacity utilisation. To create levelled production schedules, the demand needs to be continuous and even. The higher the demand volume and stability, the easier it is to develop levelled production schedules. As a rule of thumb, a pull system can only manage demand variation of +/− 10 per cent. The frequency and size of changes in the production schedules should also be kept to a minimum.

The starting point for levelled production is a levelled master production schedule as described in Chapter 10. When converting the master production schedule into daily equivalents, normally monthly volumes are translated into a daily **production rate**. The daily production rate is the number of products to produce per day during the master production scheduling period. The production rate is consequently the inverse of the cycle time. For example, a production rate of 10 products per hour corresponds to a cycle time per product of 6 minutes. The production rate may be changed between planning periods, and then the respective work centre has to adjust its cycle times by, for example, changing the number of operators or working hours per day. The cycle time is sometimes called **takt time**, because the same cycle time, or takt, is to be used consistently during a longer period. Levelled production based on daily production

Week		1	2	3	4
Outbound delivery requirement		200	200	200	200
Number of working days		5	3	4	5
Production rate per day		40	45	45	40
Stock on hand	90	90	25	5	5

FIGURE 15.3 The relationship between outbound delivery requirements, production times and stock in rate-based scheduling

rates, or a fixed production takt, is therefore also called **rate-based scheduling** or takt time planning.

The quantities that must be produced per period are planned on the basis of outbound delivery requirements, expressed in the master production schedule, and planned production time, meaning the quantity that is planned to be produced per day or per hour. The outbound delivery requirements per period are reconciled with assets in the form of quantities which can be produced with the current production times and closing stocks from the previous period. The relationship between outbound delivery requirements, production times and stock is illustrated in Figure 15.3.

In the example shown in Figure 15.3, the weeks have different numbers of working days. The loss of production caused by this is compensated by a higher production rate during weeks 2 and 3 so that the opening stock at the beginning of week 1 will be sufficient to be able to deliver in accordance with the requirements for all four weeks. If stocks are insufficient, either it must be accepted that deliveries will not correspond to demand or the production rate must be increased. Thus, rate-based scheduling is not only based on material requirements in the form of existing planned outbound deliveries. Consideration is also taken to manufacturing capacity in the form of the highest production rate that can be achieved.

Rate-based scheduling of this type is not only applied to standard products that are made to stock on a specific assembly line. Several similar standard products may be scheduled on the same assembly line in a corresponding fashion but with different production rates per product. In the same way, different order-specific variants of product models can be rate-based scheduled. Which variants of which products must be assembled and in what sequence will be determined by current stock levels and the backlog of orders. Producing multiple products on a repetitive basis, in a single machine centre, is called **mixed model production**, or **heijunka**, which is its Japanese term. Here, production of different products is evenly distributed over the day, week or month. In the smoothest schedule, each product is produced in single pieces or in several small batches every day. If for example products A, B and C are produced in the same machine centre and the demand for A is twice as high as the demand for B and C, then the optimum quantity and sequence of production would be to produce single pieces of the respective product in the following sequence: ABACABACABACABAC . . . etc. However, to produce in batch sizes of one requires very short set-up times and packages, and therefore smaller batches may be scheduled, instead of single pieces. Example 15.1 shows how a mixed model production schedule is developed for three products made in the same work centre.

Example 15.1

Products A, B and C are assembled in the same work cell. The monthly demand for product A is 1000 pieces, for product B it is 1500 pieces and for product C it is 2000 pieces. The actual month contains 20 working days. The mixed model production schedule is determined in three steps:

1 *Daily production requirement.* The daily production rate for each product would be:

Product A	1000/20 = 50 pieces
Product B	1500/20 = 75 pieces
Product C	2000/20 = 100 pieces

2 *Production sequence.* Every day a sequence of a number of products A, B and C is repeatedly produced. The optimum number of products in one sequence is determined by identifying the largest integer that divides the daily requirements into even figures. Here, the largest integer is 25. Dividing the daily requirements by 25 results in an optimum production sequence of 2 pieces of product A, 3 pieces of product B and 4 pieces of product C. This sequence (AABBBCCCC) should thus be repeated 25 times per day.

3 *Manufacturing order.* The final step is to determine the manufacturing order of the 9 products (2 product A, 3 product B and 4 product C) in the production sequence. The order which results in fewest numbers of set-ups is AABBBCCCC but this order is also the least smooth because it works with batches of 2, 3 and 4 products. An alternative, more smooth, order would be CCABBCCAB. However, it requires six set-ups instead of three.

Stable processes

The main idea with a pull system is to carry a minimum of inventory and create a smooth flow of material through the entire process and continuously decrease work in process and inventory levels. Stable processes without disruptions and production of high quality products are therefore important for maintaining such smooth flows. Maximising machine uptime, minimising breakdowns and improving process capabilities should therefore be secured through preventive maintenance and **total quality management (TQM)** strategies.

Responsible and involved personnel

In a pull system the responsibility of carrying out the execution and control is decentralised to the factory floor. The central planning organisation is not responsible for the daily planning-related activities. Instead, the foremen and direct production personnel have the responsibility. This is also the way for the continuous process improvement work necessary for creating short set-up times and stable processes. Therefore, there is a need for well-educated, trained and involved personnel. The management also has to give the shop floor personnel the responsibility and authority to conduct this work.

15.3 Authorisation Methods

Authorisation of materials movement in a pull system is conducted by a type of material planning method that is characterised by material requirements that arise in a consuming unit and more or less directly initiates manufacture and/or delivery from a material supply unit. The degree of direct initiation depends on the size of the inventory buffers which must be used for various reasons. These reasons include lead times being far too long to be able to order with immediate delivery, or that set-up times and ordering costs are far too high to justify manufacture and/or transportation of a direct demand. Restrictions of this type force companies to a certain extent to manufacture and/or deliver quantities that cover several direct needs.

Methods belonging to this category could be named direct call-off methods since in their very character they involve orders that are unlike those in other types of material planning methods, which are planned and registered in an administrative ERP system. Ordering takes place directly from the consuming to the supplying unit. The planning principle is the same as for the re-order point system and the methods may accordingly be considered as variants of the re-order point system. The most well known and widely used variant of such direct call-off methods is the one developed by Toyota in Japan, known as kanban. A physical form of the re-order point system, called the **two-bin system**, is also a common direct call-off method.

Two-bin system

Using physical re-order points is a special variant of the re-order point system described in Chapter 11. It is also called a bin system, or a two-bin system because normally two bins are used. In this variant of the re-order point system, the stock position or stock shelf of every item is divided into two physical parts: a larger bin from which withdrawals first take place, and a second smaller bin which contains a quantity corresponding to the expected consumption during the replenishment lead time plus a safety stock, i.e. the re-order point quantity. When stock in the first part has been consumed, a new order is made and consumption continues from the second part of the stock until stock in the first part has been replenished through a new delivery.

Using a two-bin system is an example of a pull system where a downstream part automatically initiates replenishment from an upstream part when there is a material need. A replenishment order is placed whenever stock falls to a critical level, in this case when the first bin is empty. The remaining stock equals the expected demand during the lead time and is calculated using the following formula that was introduced in Chapter 11:

$$ROP = SS + D \cdot L$$

where ROP = re-order point
        SS = safety stocks
        D = demand per time unit
        L = lead time

The lead time (L) is the total lead time to replenish the stock. It can be split into two lead time parts; a production lead time (P) and a moving lead time (M). The production lead time consists of the set-up time, manufacturing run time and waiting time. The moving lead time is the time for communicating the order to the upstream operation that will fill the order, and the time to move the materials to the downstream stock that initiated the replenishment.

Example 15.2

A two-bin system is used to replenish a stocked item from an upstream production operation. The lead time for set-up, running and waiting after production in the upstream operation is 2 days and the moving lead time is close to 0, because the stock level is inspected continuously and the stock and production operation are located physically close. The daily consumption is 100 pcs of the item. The safety stock equals one day's consumption. The re-order point bin should thus contain $(1 \cdot 100 + 2 \cdot 100) = 300$ pieces.

Kanban

When there is a need for replenishment, some form of communication has to be conducted between the downstream stock and the upstream operation responsible for replenishing the stock. The Japanese word "kanban" means card or sign and is the name of a method for authorising material movement or production based on a visual signal.

The kanban method is based on the use of kanban cards, **production kanban** and/or **move kanban**. The production kanban card is a signal to the manufacturing unit to start producing a batch of items and put them in stock. Production kanban circulates within a closed flow in the manufacturing unit. Move kanban is used to order movement or transport. It indicates that material can be withdrawn from stocks and transported to a predetermined destination, i.e. the consuming unit. Move kanban circulates within a closed flow between the supplying unit and the consuming unit. When the supplying unit is an external supplier, the card is usually called a supplier kanban instead of a move kanban.

Move kanban system

A move kanban, also called conveyance, withdrawal or transport kanban, is an authorisation to move a pallet, box or container from an upstream to a downstream stock. The kanban principle in stock transfer from an upstream to a local stock using move kanban cards is illustrated in three steps in Figure 15.4.

The basic principle of a kanban system is that when a pallet becomes empty or when removing the first item from a pallet at the consuming unit, the attached kanban card is released (step 1). This card is then sent to the supplying unit (step 2) which, via the card, is authorized to transport a full standard pallet to the consuming unit (step 3). The card received at the upstream unit is attached to the pallet before it is sent to the downstream consuming unit.

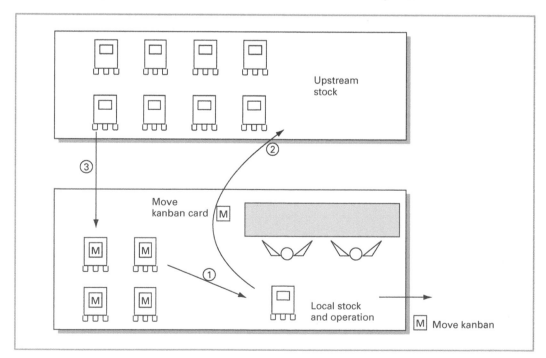

FIGURE 15.4 Illustration of the kanban principle in stock transfer

The example when using move kanbans to replenish a downstream stock from an upstream stock is called a one-card kanban system, because only one type of card is used. The upstream stock may be an outbound buffer of a production operation and the downstream local stock an inbound buffer of a local production operation.

The inventory in the local stock cannot exceed the number of full pallets that corresponds to the number of kanban cards in the system. The following formula is used to calculate the number of kanban cards, i.e.:

$$n = \frac{D \cdot L \cdot (1 + \alpha)}{a}$$

where n = number of kanban cards, standard pallets, boxes or containers (rounded up to integer figure)

D = demand per time unit

L = lead time

α = safety factor

a = capacity, number of pcs of the item, on one standard pallet, box or container

$D \cdot L$ is the demand during the lead time and $D \cdot L \cdot \alpha$ is the safety stock. The value of α is determined by the uncertainty in demand during the lead time. Figure 15.5 gives an example of software support for kanban card calculation. The goal at Toyota, where the kanban system was developed, was to keep the α value below 0.1. Another goal was to have pallets that held less than 10 per cent of one day's consumption. The number of kanban cards determines the

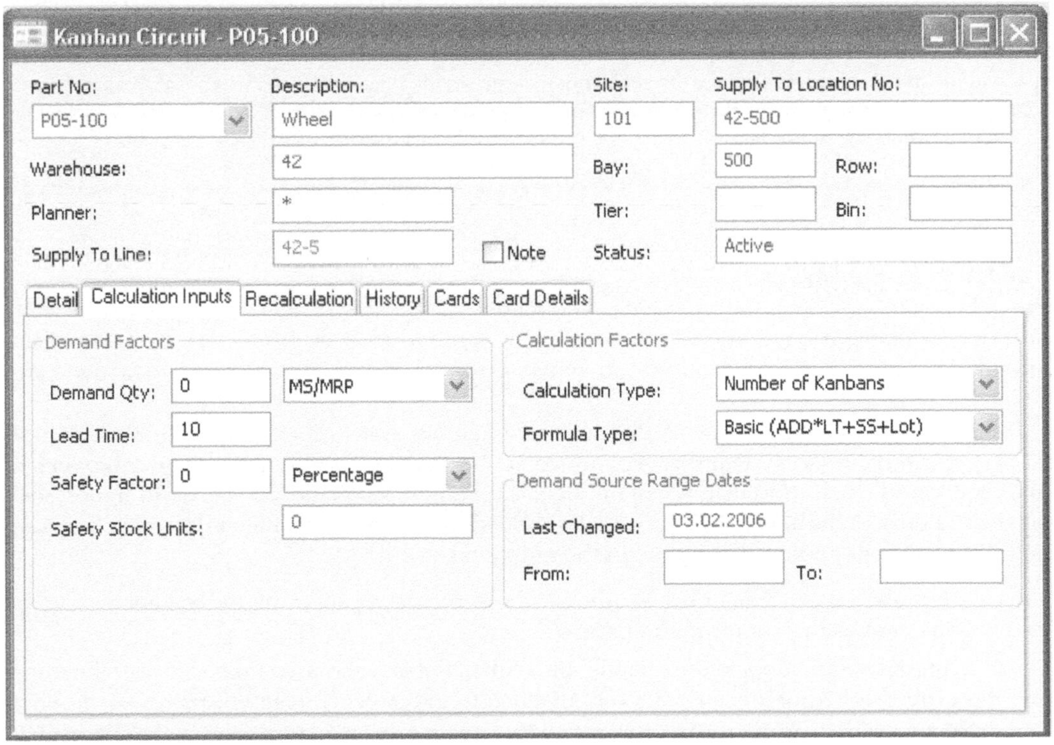

FIGURE 15.5 Example on screen of kanban card calculation

The picture shows one function of the ERP system IFS Application maximum stock in the system. The maximum is therefore $n \cdot a$ pcs tied up in the system. It can be seen from the above formula that the quantity $n \cdot a$ is equivalent to the demand during the lead time $(D \cdot L)$ plus a safety stock $(D \cdot L \cdot \alpha)$, which corresponds to the re-order point in the re-order point system. Determining the number of kanban cards in a kanban system is therefore the same as calculating the re-order point in a re-order point system. The kanban card triggers production or replenishment in the same way as in a re-order point system, i.e. depending on which reorder point and order quantity has been determined for the item. In a theoretically optimal kanban system, the pallet is the equivalent of one unit in terms of quantity, the lead time is zero and there is no demand uncertainty during the lead time (i.e. $\alpha = 0$).

The average lead time in a one-card kanban system using move kanbans is the sum of (1) the time from releasing a kanban to receiving the kanban at the upstream stock point, (2) the time from receiving the kanban at the upstream stock point to attaching the kanban on a full pallet and sending it back to the downstream point, and (3) the time from sending a full pallet from the upstream stock point to receiving the pallet at the downstream point.

Example 15.3

A one-card kanban system is used to replenish a stocked item from an upstream stocking point. The time from releasing a kanban to receiving the kanban at the upstream stock point is 15 minutes. The time from receiving the kanban at the upstream stock point to attaching the kanban on a full pallet and sending it back to the downstream point is 45 minutes. The time from sending a full pallet from the upstream stock point to receiving the pallet at the downstream point is 45 minutes. The daily consumption is 100 pcs of the item. One day contains 480 working minutes. The safety factor is 0.2 or one-fifth of a day's consumption. One pallet holds 10 pcs of the item. The number of pallets and kanban cards should thus be $100 \cdot ((15 + 45 + 45) / 480) \cdot 1.2 / 10 = 2.62$ pallets, rounded up to 3 pallets and kanbans. The maximum size of the local stock is thus 3 full pallets of items, i.e. 30 pieces of the item.

Production kanban system

Production kanbans are used to initiate production or assembly of items. In the same way as for move kanbans, no action is allowed without a kanban card. A production kanban system normally includes both production and move kanbans and is therefore called a two-card kanban system.

The following six steps illustrate the two-card kanban system in Figure 15.6. Steps 1 to 3 define a closed kanban system in which move kanban initiates withdrawals from a producing unit's outbound stocks and delivery of full pallets to the consuming unit. Steps 4 to 6 represent the second closed kanban system, in which production kanban cards initiate manufacturing in the producing unit and replenishment in the outbound stock.

Step 1: when the consuming unit has used all the items on a pallet, the move kanban card is released and sent to the supplying unit, the stores.

Step 2: a quantity equivalent to the quantity on a full pallet is withdrawn from stock. At the same time the production kanban card which is attached to the pallet is released. For every production kanban card that is removed from the pallet, a move kanban card is attached to the pallet.

Step 3: the full pallet with the move kanban card is then delivered to the consuming unit. When the items delivered have been used at the consuming unit, step 1 is repeated.

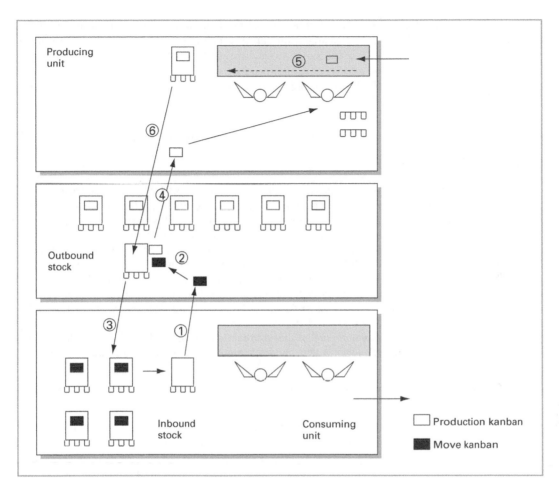

FIGURE 15.6 Illustration of the two-card kanban system

Step 4: the production kanban card that was released from the stores in step 2 is then sent to the unit that produces the items.

Step 5: the production of the items takes place according to the specification on the incoming production kanban card. More than one kanban is required to start production if the pallet size is smaller than the batch size. When a stated number of production kanban cards has been received, production is allowed to start.

Step 6: when the items have been produced they are placed on an empty pallet and the associated production kanban card is fixed on the pallet. The pallet, the physical items and the production kanban card are put in the stores for subsequent delivery to the consuming unit as described in step 2.

Compared to the one-card kanban system, the two-card kanban system controls two instead of one stock point. The number of move kanbans decides the maximum number of full pallets in the consuming unit's inbound stock and the number of production kanbans decides the maximum number of full pallets in the producing unit's outbound stock. The number of production kanbans and pallets in the closed production loop is calculated with the same formula as for move kanbans and pallets. The only difference is that the lead time is the production lead time, i.e. the sum of (1) the time from releasing a kanban to receiving the kanban at the first station in the work cell, (2) the time from receiving the kanban at the first station to finished making (i.e. time for setting up, running and waiting after production) filling a full pallet with pieces and (3) time from sending a full pallet from the producing unit to receiving the pallet at the outbound stock point.

Example 15.4

The one-card kanban system in Example 15.3 is changed to a two-card kanban system. The move kanban system of Example 15.3 is still remaining in order to replenish the downstream stock from the upstream stock. A production kanban system is added to initiate production and replenish the upstream stock (outbound production stock). Three move kanban cards and pallets are still appropriate for the move kanban system. The time from releasing a kanban to receiving the kanban at the first station in the work cell is 15 minutes. The set-up time is 20 minutes, the production run time 3 minutes per piece and the average waiting time after production and moving time to the outbound stock is 15 minutes. The daily consumption is still 100 pcs of the item and one day is 480 minutes. The safety factor for the production kanban system is 0.2 or one-fifth of a day's consumption. One pallet holds 10 pcs of the item. The number of pallets and kanban cards in the production kanban system should thus be $100 \cdot ((15 + 20 + (10 \cdot 3) + 15) / 480) \cdot 1.2 / 10 = 2$ pallets, rounded up to 5 pallets and kanbans. The maximum size of the upstream outbound stock is 5 full pallets of items, i.e. 50 pieces of the item. The system should thus be able to work with 3 move kanban cards, 2 production kanban cards and 5 pallets.

When there is only one stock point between the upstream producing unit and the downstream consuming unit, it is possible to use production kanbans in a single-kanban system, i.e. without combination with move kanbans in a two-kanban system. This is, for example, the case when the two operations are located physically close together. It could also be the case when the units are not adjacent but the move time is short.

Kanban lot sizing
When calculating the required number of kanban cards, consideration is given to current lead times for replenishment and to desired safety stocks to safeguard against variations in demand and other uncertainties. In the ideal case, the manufacturing order quantity and the transportation quantity are equal to the quantity of a standard package and what is stated on the kanban card. If this is not possible due to high ordering costs or large package sizes, multiples of these standard quantities are either produced or transported. The smallest number of kanban cards for starting production or transportation is stated on the kanban card. A move card contains information about the item, number of cards in the loop and the number of the specific card, consuming unit, supplying unit, unit load capacity, etc. A production kanban card normally also contains information about input material, tools, least and/or largest number of cards to start manufacturing, etc. Figure 15.7 shows an example of a kanban card.

Sometimes the changeover time between production of two items may be so long that the total available time is not enough for several changeovers. It is then necessary to order production batches that are larger than one pallet, box or container. The maximum time used for changeover is the total production time per day minus the total run time to make one day of consumption. The maximum number of changeovers per day is then calculated as the maximum

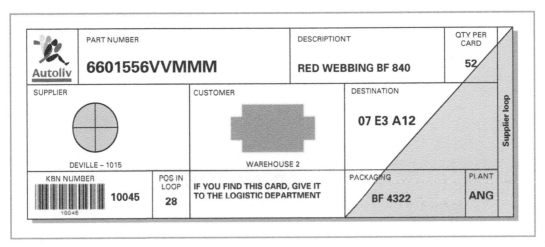

FIGURE 15.7 Example of kanban card used at Autoliv Inc.

time used for changeover divided by the time per changeover. The minimum batch size for an item is determined by dividing the daily production volume by the maximum number of changeovers per day; see Example 15.5.

Example 15.5

The minimum production batch size for the production in Example 15.4 is calculated as follows: There are 480 available working minutes per day. 300 (100 · 3 = 300) minutes are required to make the 100 pieces and 180 (480 − 300) minutes are available for changeover per day. The set-up time is 20 minutes. The maximum number of changeovers per day are thus 9 (180 / 20 = 9) and the minimum number of pieces per batch is 12 (100 / 9 = 11.1). The batch size should be expressed in full pallets and kanban cards. One pallet and kanban card equals 10 pieces, and then the production batch must be 20 pieces, i.e. two pallets and kanban cards.

There are some different ways of controlling for kanban batch sizes in practice. A common way is to combine a kanban run line with a blocker card on a kanban board, as illustrated in Figure 15.8. If the production batch size equals two kanbans, then there is one empty spot for a kanban between the blocker card and the run line. If the batch size equals three kanbans, then there are two empty spots for kanbans between the blocker card and the run line, etc. When receiving an empty production kanban it is attached to the board. The first card is put directly above the blocker card. The next card is attached over the first kanban, etc. Attaching a kanban above the run line means that enough kanbans for starting production are received. All kanban cards are then removed from the board and production is started. The blocker card should be easy to remove in order to make it possible to change production batch size, for example as a result of changed demand or set-up times.

Another way of dealing with production batch sizes in a kanban system is to use so-called **signal kanbans**. A signal kanban is a special card tagged onto the edge of the pallet or container positioned as the re-order point in the store room. If there are, for example, six pallets in the

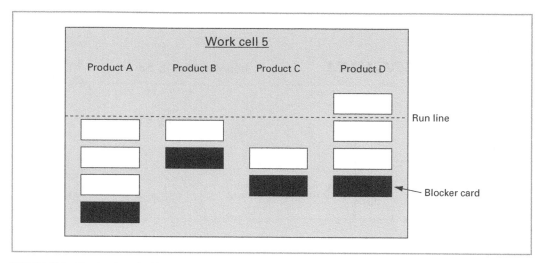

FIGURE 15.8 A kanban board with blocker card and run line. Product D is due for production

kanban system and the re-order point equals two full pallets, then the signal kanban is tagged on the second pallet in the store room. The signal kanban is moved from the pallet when only two full pallets remain in the store. When the upstream operation receives the signal kanban it is informed that the downstream stock has reached the re-order point and it is thus authorised to produce a batch size equal to the number of empty production kanbans it has received, i.e. four full pallets.

Types of kanban signals

There are two main types of kanban methods: those based on some form of physical and visual initiation of re-ordering, and those based on some form of administrative initiation. The traditional kanban system using printed kanban cards is an example of the first type. In the second type, initiation takes place through an administrative control system. Terms such as electronic kanban (see Figure 15.9) and faxban are used for this type. Transfer of re-ordering to the supply unit may take place physically when a kanban card is printed out from the system, or it may take place through a "kanban message" being sent by fax, email or over the Internet to the supplying unit.

A large number of variants of visual kanbans are in use. Generally it is the physical medium that is the carrier of re-ordering which distinguishes these variants. These could be the pallets, boxes or containers used to carry the items. Returning an empty pallet, box or container to an upstream operation then means that a batch has been consumed and should be replenished. If using empty pallets, boxes or containers as signals, they need to be marked in order to specify what particular item should be put on it and where to send it. The number of packages to produce and store can be restricted by marking up squares on the floor or designated space in a shelf rack. An empty space authorises production or replenishment of the required quantities that will fill the empty places. In highly repetitive environments, such signals as lights or sounds can be used to inform the feeding operation when it is time to produce or replenish a new batch. Small balls of various colours and markings that represent different items are sometimes also used. When a downstream unit has consumed a piece it rolls a ball along a gravity chute upstream to the feeding operation. A received ball is an order signal and authorises production or stock

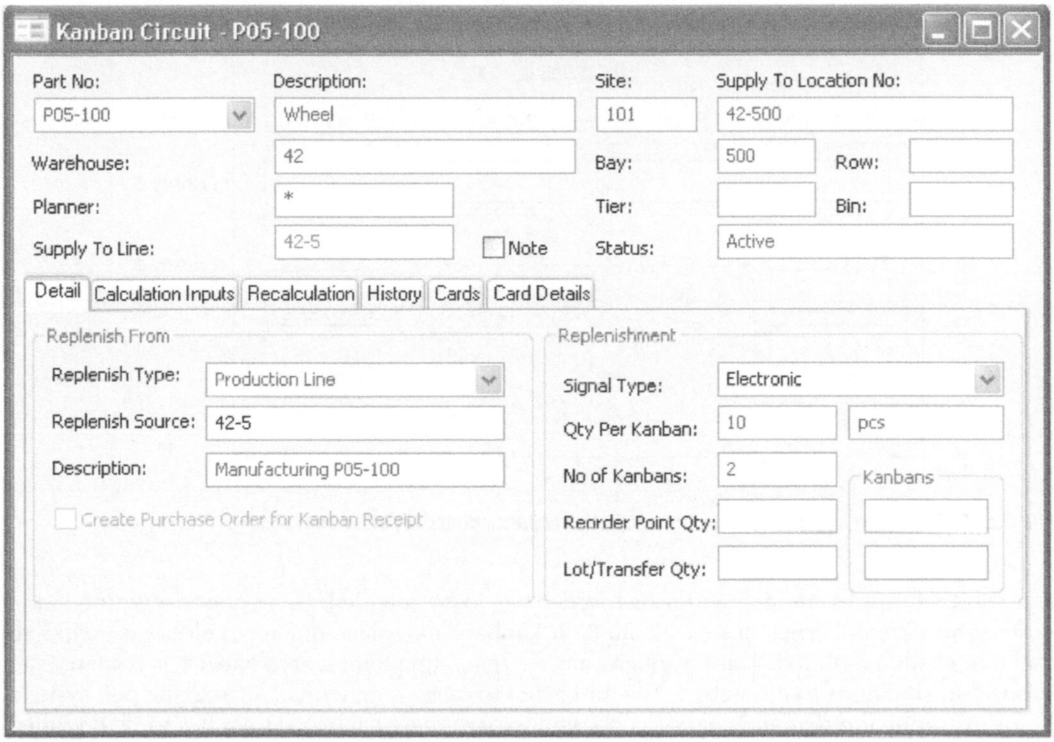

FIGURE 15.9 Example on screen of electronic kanban details
The illustration shows one function of the ERP system IFS Application

replenishment. The sequence of balls of various colours and marks gives the sequence of production or replenishment. Pull signals can also be generated at the consuming unit and announced on electronic monitors at the upstream feeding unit.

Kanbans are processed in the sequence they arrive to the upstream operation. However, from time to time a backlog of different kanbans can be built up. To prevent shortages of certain items and to achieve a balanced workload, production of the different types of items then has to be prioritized and sequenced. This can be done, for example, using a kanban sequence board (see Figure 15.10). When kanbans are received they are put on the board, starting from the top. The board is partitioned into different segments indicating various priorities. Cards with priority 1 have priority over those with priority 2, etc. The kanban boards of Figures 15.8 and 15.10 could also be combined into one common board. Then the priority lines should be drawn from the bottom and up, because the empty cards are attached from bottom-up when using blocker cards.

Special situation kanban

A pull system requires a stable planning environment, in terms of lead times, demand, quality levels, etc. But short-term situations may occur that require deviations from an originally designed kanban system. Special situation kanbans can then be released in order to deal with the changed situation. Temporary kanbans can be issued when a sudden demand increase occurs or when the defect rate has been higher than normal and there is a temporary need for stock build up. Another situation when special situation kanbans may be released is when there

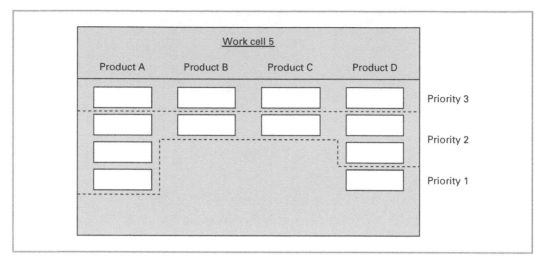

FIGURE 15.10 Kanban sequence boards. Product D has highest priority and should be produced first

is a shortage risk of an item and a rush order has to be released, i.e. a prioritised production order. The different types of special situation kanbans may have different colours or marks so they are easily identified. These kanbans aim at generating temporary changes as responses to short-term situations to the system. The aim is not to make long-term change to the pull system. If, for example, the demand change is not temporary a new kanban system has to be designed and a new set of kanbans need to be printed and issued. In order not to affect the originally designed kanban system, a special situation kanban has to be taken out from the system directly after production and stock replenishment.

Generic kanban

Pull production is characterised by high repetitiveness, and the more frequently a specific item is produced and moved the better the system normally works. High repetition does not necessarily occur only because there is high-volume production. There may be a large number of products and variants produced, and some variants may be required quite infrequently. In such make-to-order or assemble-to-order situations, a traditional kanban system will not be very successful. Instead, a kanban variant called **generic kanbans** can be used.

A generic kanban system is designed and works in the same way as the traditional kanban system, but with one exception. The generic kanban is an authorisation to start to produce or move an item but it does not specify exactly what item to produce or move. It often specifies a product family but not the exact model, variant or option. A dispatch list (see Chapter 16) specifying the exact item configurations and sequence of items to produce and move during a day therefore has to be generated and displayed at each work centre. Using a generic kanban system is thus a mixed model pull technique that combines the pull principles of the kanban system with the push principles of centralised dispatch lists.

Kanban rules

A number of rules are necessary to follow in order to make a kanban system work. The following five kanban rules were put forward by Toyota:[1]

---

1   Monden, Y. (1993) *Toyota production system*, Institute of Industrial Engineers, Norcross (GA).

Rule 1: the first rule says that "*The subsequent process should withdraw the necessary products from the preceding process in the necessary quantities at the necessary point in time*". This rule means that nothing but what is stated on a kanban card can be withdrawn and moved downstream in the process. It also means that a kanban card or equivalent information must be attached to the pallet, box or container moving in a closed kanban loop. A necessary requirement for rule 1 to work is to plan a periodic, for example monthly, production rate in advance, and communicate this to all processes and suppliers in the system. This will give the process proactive information about the capacity and material needed during the coming period. It is also necessary with a levelled daily production plan at item level, with small lot production and material withdrawal, in the subsequent production. Without a more or less even demand, a kanban system will result in high inventory levels and require flexible capacity in order to respond to the fluctuating demand.

Rule 2: the second rule says that "*The preceding process should produce its products in the quantities withdrawn by the subsequent process*". The subsequent operation will require items in small lot sizes in order to attain smoothed production. Therefore, the preceding operations need to make frequent and quick set-ups when changing between producing different items. Larger quantities than stated on the received kanban cards cannot be produced and the production sequence must follow the sequence in which the cards were received.

Rule 3: the third rule is that "*Defective products should never be conveyed to the subsequent process*". When discovering a defective item, it should be returned upstream where the causes of the defects are analysed. A defective part or operation should be corrected at once.

Rule 4: the fourth rule says that "*The number of kanbans should be minimised*". The number of kanbans in the system decides the work in process and inventory levels. It should thus be kept as small as possible. The long-term goal is to improve processes and, for example, decrease lot sizes and shorten lead times. Such improvements result in a smaller number of kanbans and thus also lower inventory levels. A common improvement strategy is to continuously reduce the number of kanbans in order to expose areas which are problematic and need to improve. In the short run, the total number of kanbans is however kept constant. When daily demand is changing, the production capacity should change correspondingly in order to increase or decrease the cycle time and thereby manage deliveries to the subsequent process in time with the receipt of empty kanbans. Special situation kanbans can, however, be used for temporary demand peaks and extra safety stock can be necessary if demand variations occur frequently.

Rule 5: The fifth kanban rule says that "*Kanban should be used to adapt to small fluctuations in demand*". This means that no detailed production schedule needs to be developed in advance for the upstream operations and no replanning is necessary when the daily demand is changing. When the production is levelled and there exists some production flexibility it is possible to react to market changes by producing a few more or a few less than originally scheduled. According to Toyota, demand variations around 10 per cent are possible to handle by changing only the frequency of kanban transfer. But if the demand changes more, the production processes must be re-arranged in order to adjust the cycle times or the number of kanbans must be increased or decreased.

CONWIP

CONWIP stands for constant work in process. It is similar to the generic kanban method in the sense that it generates and sends a signal upstream when it is time to produce items but it does not specify what to produce. It is thus an alternative method to apply in a mixed model production environment. In a CONWIP system, the last unit in a process is the only unit sending a

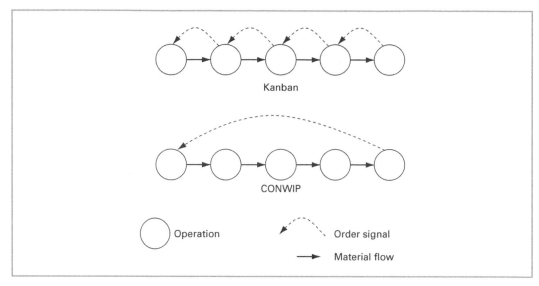

FIGURE 15.11 Kanban and CONWIP comparison

signal and the first unit in the process is the only unit receiving a signal, no matter how many individual operations are included in the process. This procedure secures a constant level of work in process in the system at the same time as it allows for mix variations over time. Sending signals directly from the last to the first operation in a process that may contain several individual operations is the major difference between the kanban and CONWIP systems. If using a kanban system, cards would instead have been sent upstream between each successive pair of operations. If the process consists of only two operations, the CONWIP and generic kanban systems would thus be the same. Cards similar to kanban cards could be used as order signals also in a CONWIP system. The number of cards in the system is calculated using the same formula as for kanban calculations and using the entire process cycle lead time. Figure 15.11 illustrates the principle difference between the CONWIP and kanban systems.

What to make when receiving an empty CONWIP card at the first operation is determined by a centrally generated dispatch list (i.e. an order list specifying what item to produce when receiving the first card, second card, third card, etc.). This means that an item is only produced if there is an expected future demand for it and not just because there has been a historical consumption of it, as is the case when using a kanban system. A first-come-first-served (FCFS) principle is used to push the items through the operations in the process. When an item is produced in the first operation it is pushed to the second operation. The second operation produces items in the sequence they are received at the inbound stock. When finishing producing an item in the second operation it is pushed to the third operation, etc. When finishing producing an item in the last operation of the process, the CONWIP card or signal is transferred to the first operation. The first operation is then authorised to start producing the next item on the dispatch list. Because the items are pushed through the process and sequenced on a first-come-first-served basis in the individual operations it is required that all items in the process follow the same sequence and routings of operations. The CONWIP method is not applicable if that is not the case.

15.4 Considering Bottlenecks

In the most optimal pull-based system, a levelled master production schedule sets the pace of the system and final assembly drives the entire system by sending pull signals upstream. A prerequisite for achieving smooth and uniform material flow in such a pull system is that all upstream operations have enough capacity to meet the demand-driven pull requirement from final assembly. This is, however, often not the case. Instead, physical constraints, in terms of manufacturing resources which have less capacity than what is needed to fulfil the demand from a downstream resource, may occur in upstream resources. Such a constraint is called a bottleneck. An approach to consider capacity limitations in bottlenecks when executing and controlling material flows is using the **drum–buffer–rope** (DBR) principle, which is a part of the theory of constraint (TOC), outlined in Chapter 3.

The basic idea of TOC is that the bottleneck constrains the entire system and the only way to increase capacity and throughput in the system is to increase the bottleneck throughput. This is achieved by maximising the capacity utilisation at the bottleneck, by minimising set-up times and running large batches. To reduce the risks of long lead times due to the large batch sizes, the manufacturing orders should be split and transfer batches from the bottleneck to the downstream operations should be as small as possible. The DBR system contains three principles (the drum, the buffer, the rope) which determines how materials are pulled through a bottleneck constrained production system.

The drum

The production rate, or throughput pace, of the process is determined by the capacity of the bottleneck. The production rate should thus be determined by the bottleneck capacity. The bottleneck can be described as a drum, where its drumbeat (production rate) sets the production rate and throughput pace of the entire process.

The buffer

An hour lost in a bottleneck is an hour lost in the entire system and it can never be regained. A bottleneck should consequently never be idle and therefore an inbound buffer should be established at the bottleneck resource. The objective of the buffer is to decouple the bottleneck from upstream operations, and minimise the risk that disruptions in an upstream operation affect the bottleneck.

The rope

When applying the DBR system, the bottleneck sets the pace of the system and it drives the material through the system, by pulling materials from upstream operations and into the entire process, using a so-called "rope". The rope can be a kanban or CONWIP card but is normally a schedule determining the time for orders to be released in order to arrive to the bottleneck resources when needed (plus a safety time, equivalent to the inbound buffer).

15.5 Order-less Production and Reporting

When batch sizes are reduced in the pull system, the number of transactions is increased. However, manufacturing orders are eliminated in the system and thereby several transactions and related documentation are reduced. A pull system also reduces the need for feedback reporting when releasing and conducting production. Detailed routing information, order due dates, operation start dates and finish dates for an order are for example not necessary. The routing steps are easily communicated and obvious in a cell or flow layout. Various manufacturing documents specifying what to do, therefore, do not have to be carried with the item. The kanban is the authorisation to produce in accordance with the information on the kanban. The necessary count points along the production process are few and normally it is enough to register when a product is finished. Much of the status and feedback reporting is carried out by the kanbans themselves.

When using pull systems that are based on some form of physical initiation of production or procurement such as kanban cards, there is no need to process orders administratively in an ERP system. The physical signal given by the kanban card is a type of order, and the card is sufficient to achieve a flow of materials. From the viewpoint of material planning, no stock accounting is required for visual pull systems to work. This is necessary, however, for administrative systems. Here, stock transactions trigger new kanban cards via the administrative pull system. Companies that use pull-based execution and control systems sometimes use the kanban card or other physical ordering mechanisms in parallel to create orders in the administrative system. Likewise, there may be orders in the administrative system when different types of electronic kanbans are used. However, there are also companies that do not use orders administratively in ERP systems in their material planning. The function of the manufacturing order as an object for reporting and following up can therefore not be used. For example, reporting work cannot be carried out to order and costing in the traditional sense cannot be carried out. Neither can withdrawal reporting to order be performed. In this type of planning environment, material withdrawals for stock accounting purposes are usually reported in the form of backflushing with the help of bills of material, as is described in Chapter 18. Deductions of stock on hand take place at the same rate as products are reported completed at the end of the production process. For example, if a fridge is reported as fully manufactured, two hinges are deducted from stock on hand. The number of count points for reporting and feedback is thereby minimised.

15.6 Characteristics and Planning Environments

Pull-based execution and control can be a very effective system if used it in an appropriate way and in an appropriate planning environment. But using a pull system in an environment without fairly stable and continuous demand, smooth material flows and the other preconditions outlined in Section 15.2 could be very defective. Characteristics and primary planning environments of a pull-based execution and control system are summarised below.

Characteristics of pull-based execution and control

A pull-based execution and control system is based on the same principles as the re-order point system and has the following characteristic properties:

- The methods are easy to understand and transparent for personnel involved, especially the methods based on visual direct call-off.
- The pull authorisation methods do not require computer-supported systems. However, they are usually supplemented by some form of material requirements planning system as support for overall and long-term material planning.

- Data for the calculation of the number of kanban cards required can also be obtained with the aid of gross requirements calculations from the master production schedules for raw materials and purchased components in manufacturing companies.

- In the ideal levelled production environment that kanban and corresponding methods are really intended for, there is no need for priorities. Manufacturing or transportation is normally intended to be carried out in the order in which kanban cards are received.

Primary application environments

For pull-based execution and control to be used in an efficient manner a number of environment requirements must be fulfilled:

- *Stable demand and levelled production.* In order to create smooth material flows without high stock levels and low service levels it is necessary to create a levelled production schedule, with consistent daily production rate and a repetitive production sequence of various products. Such levelled and repetitive production is not possible without a somewhat stable and continuous demand. It is not necessary with highly frequent production of all products but it is necessary with a levelled daily production rate and a repetitive production pattern of the different products. The more standardised an item is the more even is normally its demand. Therefore, standard items such as raw material and modules may be appropriate items to control with pull systems.

- *Few engineering changes.* Somewhat similar manufacturing processes and stable products with few engineering changes and a limited number of product variants are other important preconditions for creating a smooth flow.

- *Small batch sizes and short lead times.* Small batch sizes are necessary for creating short and stable throughput times.

- *Stable processes and continuous flow.* Despite a stable demand and levelled production, the flow will not be smooth unless all processes are stable and cycle times and quality are consistent. The production process therefore has to be characterised by few equipment disturbances and high quality output. In order to keep constant cycle times and flow in situations of small demand variations, the processes need to have some degree of flexibility. Flexibility with respect to machine capacity and working hours, for example, should therefore be built into the system.

15.7 Summary

This chapter described how a pull-based execution and control system is designed and can be used. In a pull system, materials movement and manufacturing only take place on the initiative of and authorised by the consuming unit in the flow of materials. Five types of preconditions for pull-based execution and control were presented. These were short set-up times and batch sizes, flow-oriented layout, levelled production, stable processes and responsible and involved personnel. Without first making sure these conditions exist, it will not be possible to effectively apply a pull system. Three methods for initiating and authorising production and materials movement were described. All methods are variants of the re-order point method described in Chapter 11. The two-bin system is a physical re-order point system using two bins.

The most common pull systems are the one-card kanban and two-card kanban systems. It was described how the kanban systems are based on move and production kanbans, how to lot size and use alternative kanban signals and kanban cards within a kanban system. Generic kanbans and CONWIP were described as alternative methods to use in situations characterized of several item or product variants with infrequent requirements. The chapter ended with a discussion about the drum–buffer–rope principle for bottleneck considerations, feedback reporting and characteristics and planning environments for pull systems.

 Key concepts

CONWIP 335	Production kanban 326
Cycle time 322	Production rate 322
Drum–buffer–rope 337	Pull system 318
Generic kanban 334	Rate-based scheduling 323
Heijunka 323	Signal kanban 331
Kanban 318	Single minute exchange of die (SMED) 322
Levelled production 321	Takt time 322
Mixed model production 323	Total quality management (TQM) 324
Move kanban 326	Two-bin system 325

 Discussion tasks

1 Describe under what conditions and in what planning environments a pull-based execution and control system is not appropriate.

2 Discuss how MRP can support pull-based execution and control.

3 Compare the one-card, two-card and generic kanban systems and explain in what situations the respective system is most appropriate.

Problems[2]

Problem 15.1
Calculate the necessary number of kanbans for production of a product with a demand of 1000 pieces per week, 40 working hours per week, cycle time for a kanban of 5 hours, a package size equal to 10 and a safety factor of 20 per cent.

Problem 15.2
A one card kanban system is used for controlling the material flow of item XC78 from an upstream machining work centre to an assembly station. The average daily demand for item XC78 in the assembly station is 240 pieces and one day contains 8 working hours. The time

---

2 Solutions to problems are in Appendix A.

from releasing a kanban to receiving the kanban at the upstream stock point is on average 20 minutes. The time from receiving the kanban at the upstream stock point to attaching the kanban on a full pallet and sending it back to the downstream point is on average 20 minutes. The time from sending a full pallet from the upstream stock point to receiving the pallet at the downstream point is on average 20 minutes and the average waiting time in the downstream stock point until the pallet is accessed and the kanban is released is on average 4 hours. A pallet holds 30 pieces of XC78 and there are currently 8 kanbans and pallets in the kanban loop of item XC78 between the machining and assembly cells.

a) How large is the safety stock in the system, measured in pieces and as a percentage safety factor?

b) How much does the cycle time need to be decreased in order to keep the same percentage safety factor if removing one kanban and pallet?

c) What would the safety stock in pieces and percentage safety factor be if the demand increases to 300 pieces per day and 7 kanbans and pallets are used?

Problem 15.3
The waiting time for a specific item in front of a machining work cell is 6 hours; it takes 10 minutes to set up the machine and 6 minutes to make one piece in the machine. A pallet holds 10 pieces of the item. What daily demand does this correspond to if there are 10 kanbans in the kanban system and the safety factor is 10 per cent? The number of working hours per day is 8.

Further reading

Chausse, S., Landry, S., Pasin, F. and Fortier, S. (2000) "Anatomy of a kanban: a case study", *Production and Inventory Management Journal*, Vol. 41, No. 4, pp. 11–16.

Gupta, S., Al-Turki, Y. and Perry, R. (1999) "Flexible kanban systems", *International Journal of Operations and Production Management*, Vol. 19, No. 10, pp. 1065–1093.

Haslett, T. and Osborne, C. (2000) "Local rules: their application in a kanban system", *International Journal of Operations and Production Management*, Vol. 20, No. 9, pp. 1078–1092.

Hopp, W. and Spearman, M. (1996) *Factory physics*. Irwin, Chicago (IL).

Hyer, N. and Wemmerlov, U. (2002) *Reorganizing the factory: competing through cellular manufacturing*. Productivity Press, Portland (OR).

Likert, J. (2004) *The Toyota way*. McGraw-Hill, New York.

Monden, Y. (1993) *Toyota production system*. Institute of Industrial Engineers, Norcross (GA).

Nicholas, J. (1998) *Competitive manufacturing management*. McGraw-Hill, New York (NY).

Ohno, T. (1988) *Toyota production system: beyond management of large-scale production*. Productivity Press, Cambridge (MA).

Sanders, W. (1989) *Just-in-time: making it happen: unleashing the power of continuous improvement*. John Wiley & Sons, New York.

Autoliv Inc. develops and makes airbags, seatbelts, safety electronics, steering wheels, anti-whiplash systems, seat components, child seats and other safety related systems for the automotive industry. It supplies nearly all automotive original equipment manufacturers (OEMs) in the world with these products, from its about 80 manufacturing plants over the world. This case study deals with one of Autoliv's major manufacturing plants.

Demand, production and supply
The short-term demand from the OEMs is quite stable. New products are developed for new car models. The average product life cycle varies between three and eight years, i.e. equal to that of a car model. Most products are standard components in the final vehicle but there also exist options, with more uncertain demand.

The manufacturing is organised according to cellular manufacturing principles. Different products are made in different production cells. There may be about 50 production cells in a typical Autoliv plant. Cells with similar products and production processes are grouped into a smaller number of production groups. The cells are designed for specific products and are therefore rebuilt when old products are phased out and new are phased in. Autoliv bases its production on monthly levelled production plans, and uses inventory buffers to even out short-term demand variations. Still, extra capacity may be required, for example due to higher than forecasted customer demand or disruptions. The number of produced items per time unit in a cell could be varied by increasing the number of operators in a cell, buying capacity from another cell in the same production group or manufacturing during unplanned hours, for example, weekends.

Autoliv sends weekly updated delivery schedules for the expected future of component demand to its suppliers, generated by MRP calculation. A typical delivery schedule contains 20 weeks of planned demand. As a result of the levelled production, the weekly Autoliv demand on its suppliers is quite stable, with exceptions for some non-standard items. The daily demand, however, can vary considerably. The supplier lead times normally vary between 0.5 and 5 days for most suppliers. The average suppliers make daily deliveries but some deliver twice a day and some once a week.

When a production cell has consumed components and needs more material from the component store, a kanban system is used to authorise material movement. In the same way, supplier kanbans are used to trigger replenishment from suppliers. There are consequently two sequential kanban loops. Each loop has a certain amount of kanban cards. The figure illustrates the material flows and kanban systems at Autoliv.

The move kanban system
The move kanban system controls the material flow between the component store and the respective production cell. The figure illustrates how a loop of kanban cards pulls material from the store to the cell. The loop contains the following steps M1 to M5:

---

3   Based on Veeremäe, D. and Schneider, M. (2007) "Aligning material planning methods with planning environment – focus on supplier kanban at Autoliv Vårgårda", master thesis E 2007:117, Chalmers University of Technology, Gothenburg, Sweden.

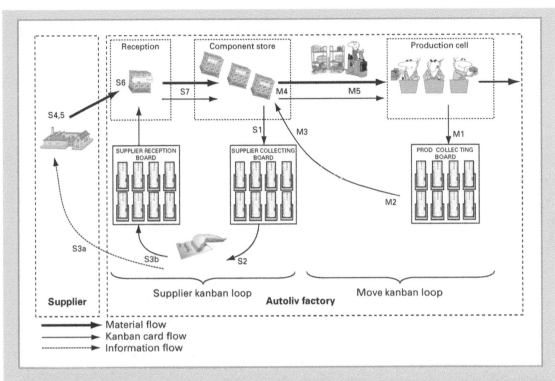

- When the last piece of a component is taken from a box at the production cell, the attached move kanban card is removed from the box and put on the production collection aboard next to the cell (M1).

- A so-called pull flow vehicle is continuously circulating in the factory. When it passes by a production collection board it takes all empty move kanban cards from the board (M2). Each picked-up move kanban card is a pull signal from the production cell to the component store to replenish another full box of components.

- When the pull flow vehicle has visited all production collection boards in the factory it returns to the component store, where it withdraws boxes of components that correspond to the collected move kanban cards (M3).

- The move kanban cards brought from the production cells are attached to the boxes with-drawn from the store (M4). At the same time the supplier kanban cards which were attached to the stored boxes are taken from the boxes and put in the supplier collecting board (S1). Thereafter, the pull flow vehicle drives to the production cells and delivers the ordered boxes, and collects move kanban cards from the production collection boards, i.e. the production kanban process starts again (M5).

The supplier kanban system

Traditionally, the supply of material from suppliers has been controlled by MRP, but Autoliv has started to also pull some material from suppliers using a supplier kanban system. The figure illustrates how a loop of kanban cards, pulls material from the supplier to the component store. The loop contains the following steps S1 to S7:

- Step M4 in the move kanban loop initiates removal of the supplier kanban card from the stored boxes to the supplier collecting board (S1). The collection board is structured according to the existing kanban suppliers.

- Every day at 11:00, the goods reception staff take the supplier kanban cards from the collection board and scan them (S2). This information enters a software system called M.A.S.K (Move Autoliv Supplier Kanban), which was developed as an external kanban module to the ERP system. For suppliers with two deliveries per day, this scanning procedure is also done twice a day, with the second scanning at 19:30. Thereafter, M.A.S.K creates a kanban generated purchase order in the ERP system.

- The purchase order is converted to a delivery just in time EDI message and is sent to suppliers (S3a). Since some suppliers are not able to receive the EDI message, there are different ways of sending it. For some suppliers the EDI message goes directly from Autoliv's ERP system to the ERP system of the supplier. Another possibility is to use a web-based system, from where Autoliv's EDI message is converted to an application that is readable with a normal web browser. The supplier can print the message and manually register the data into its own ERP system. Another solution for some suppliers is fax, with all the information of an EDI order.

- After scanning the cards and printing the kanban generated purchase order, both cards and printed papers are placed on a reception board (S3b). That board is divided vertically into weekdays and horizontally into time laps of two hours (8:00–10:00, 10:00–12:00, etc.). The cards are placed to an intersection of day and time the shipment is expected to arrive in order to have a visual control over the reception of the goods.

- The order to delivery activities at the suppliers' site depend on the manufacturing strategy and are not described here (S4).

- When the goods are shipped from the suppliers, the suppliers also send a dispatch note (S5). It is either a dispatch advice EDI message, which is sent via EDI or web and registered in the ERP system, or fax is used for those suppliers that do not have EDI.

- After the goods have arrived at the reception area at Autoliv, the reception staff take the previously printed kanban generated purchase order, which is still with the supplier kanban cards on the reception board, and scan the order sheet. When there is a digital dispatch advice message, the order pops up on the screen in M.A.S.K and the worker compares if the shipment is the same and accepts the delivery by pushing a button. When there is no digital dispatch advice message, the reception worker has to enter the quantities of sent items, batch numbers and dispatch number in order to accept the delivery. Information about received goods enters the ERP system and the inventory balance is updated automatically (S6).

- Thereafter, supplier kanban cards are put on the dedicated boxes (S7). The larger the shipment, the more time it takes for the operators to match supplier kanban cards with the right box or pallet. Sometimes some parts from the delivery are sent to the Quality department in order be inspected. The rest of the delivery is stored. Now the next supplier kanban loop can start again.

Temporary kanban cards
In case of a higher demand than forecasted from the customer, material planners have to confirm the increase with the customer. Thereafter, the planner evaluates if there is a need to

release a temporary card, called one-shot card. The one-shot card is a kanban card that can be scanned only once as an additional card without increasing the previously calculated number of cards in the loop. If the increase in demand is spontaneous, the one-shot card is an option. However, if the demand seems to stay on a higher level, then the number of cards in the supplier kanban loop has to be increased. A new set of kanban cards is then printed and used.

When the supplier plants are not producing due to, for example, national holidays, the supplier cannot receive any kanban order, and cannot prepare and ship the goods to Autoliv with the usual lead time. Hence, there is a risk that Autoliv runs out of inbound stock. In those cases, the company either releases one-shot cards some days ahead according to the regular lead time for increasing the stock before the supplier's holiday, or uses the safety stock and replenishes it after the weekend. Normally, Autoliv has a safety stock of two days which is included in the parameters to calculate the number of kanban cards.

Discussion questions

1 The product life cycles of most products made by Autoliv are several years. But this does not mean that no change is made to a product during this time. Engineering changes of components included in a product are continuously taking place. Some components may never be changed but others may be changed several times a year. How well do the kanban systems manage engineering change situations? What negative effects could you identify? How would you suggest Autoliv reduces the negative effects of the kanban systems during engineering change?

2 Some of the present kanban suppliers have informed Autoliv that the delivery schedule accuracy is not high enough, and that high accuracy is a necessity for efficient material control when using supplier kanbans. How would you suggest Autoliv reacts to that comment?

Execution and Control in a Traditional Planning Environment

Manufacturing in many companies is characterised by set-up times requiring order quantities bigger than the immediate needs, lead times being several weeks long, discrete and non-repetitive material flows, and varying capacity requirements. In such planning environments execution and control of manufacturing orders has to be carried out in a much more extensive way than in cases where manufacturing is rate based and the material flow more or less is pulled through the company by kanban cards or similar means. The manufacturing orders may be a direct result of customer orders at the master production scheduling level, such as in planning environments with order-based products and customer-oriented production. A manufacturing order may well be a direct result of material planning; for example, based on a forecast or calculated material requirements or to replenish stocks. Execution and control of manufacturing orders has traditionally been called **production activity control** or shop floor control.

Irrespective of the planning level at which the manufacturing order was created, it should specify the earliest start time, the latest due time and the quantity. Without these details it will not be possible to ensure its links to higher planning levels. Giving consideration to these links also means that planning margins when executing manufacturing orders will lie within the frame of the start and completion times of the order as well as its quantities. In other words, the **operations** that a manufacturing order consists of will be carried out between these two points in time. To try to start before the planned start time for the order may not be possible since the availability of materials at lower structural levels has been planned for this time and not earlier. Starting earlier than necessary will also involve unnecessary tied-up capital in work in process. If the last operation is not completed before the planned due time for the order, there may be a delay in delivery to the customer or shortages of material for manufacturing orders requiring the corresponding item.

To achieve efficient production and utilisation of capacity as well as obtaining short through-put times, a balance is necessary between available manufacturing capacity and capacity requirements for current orders and operations. To release more orders than **available capacity** in the workshop can cover will only create long queues and tied-up capital in work in process. Start-up materials must also be available for orders to be carried out cost-effectively and without disruptions. Releasing manufacturing orders without all materials being available will mean unnecessary administrative work and loss of efficiency in production. Different sequences may be selected for released orders. These sequences may be more or less efficient with regard to throughput times and delivery precision. Suitable procedures, sets of rules applied and methods for establishing sequences will all be required to achieve efficient execution and control. Consequently, execution and control may be said to have three main purposes:

1 To release orders at the rate that capacity conditions will allow them to be executed with reasonable throughput times

2 To ensure that start-up materials are available when each order is planned to start

3 To ensure that orders released for manufacturing in the workshop are completed in a suitable sequence with respect to delivery precision and throughput times

The first two main purposes concern **order release control** and material availability checks. Together with the printout of the **shop packet**, they precede the release of the manufacturing order to the workshop. The third main purpose concerns **priority control**, involving detailed control of released orders to the relevant work centres. In addition to dealing with these three basic issues, execution and control also includes **labour reporting** from the manufacturing process and final reporting of completed manufacturing orders. The relationship between execution and control and other planning processes is illustrated in Figure 16.1.

This chapter describes the execution and control process and the fundamental principles behind different methods of controlling order release, material availability checks and priority control.

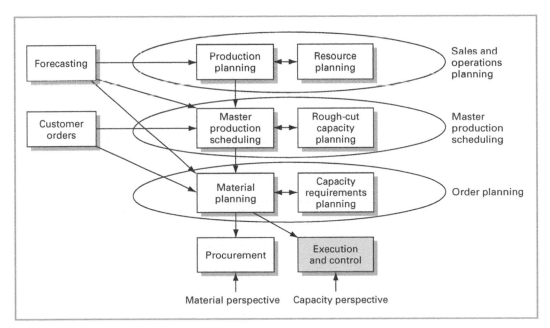

FIGURE 16.1 General relationships between execution and control and other planning processes

16.1 The Execution and Control Process

Activities in the execution and control process

The execution and control process consists of a number of activities as described below and illustrated in Figure 16.2. The exact structure of the different activities differs from company to company, depending on the planning environment in question, but in general the following apply:

FIGURE 16.2 The main components of the execution and control process

1 *Order initiation.* A manufacturing order is initiated at upper planning levels, most often by material planning. The order normally contains information on start time, due time and quantities.

2 *Order release control.* For an order release to the workshop to be meaningful, there must be available capacity to carry it out. The number of orders released must be controlled to avoid overload situations.

3 *Material availability check.* Before an order can be started, the materials necessary for its manufacture must be available in stock. The process of ensuring that this is the case is called material availability checking.

4 *Generating the shop packet.* The information required for executing a manufacturing order must be communicated to the workshop. This often takes place in the form of a shop packet, which contains information on the order number, sequence of operations, standard times, material involved, tool requirements and so on. The shop packet is either a paper-based document that accompanies the order, or information communicated via a terminal in the workshop.

5 *Releasing manufacturing orders.* When a manufacturing order is scheduled, the material availability check carried out and the shop packet printed, the order is released to the workshop.

6 *Priority control.* When the order is released to the workshop its operations must be executed in an appropriate sequence. In job workshops, this step means assigning priority to orders that are issued to the same work centre during the same time period.

7 *Labour reporting.* To be able to monitor how the execution of manufacturing orders is progressing and if necessary take corrective measures, labour reporting from the workshop is necessary. It is also required for follow-up of real operation times, checking of workloads and follow-up of capacity utilisation in work centres.

8 *Final reporting.* When the last operation of the order has been carried out, the finished products are delivered to stock and reported.

Execution and control in various types of manufacturing

The basic activities involved in the execution and control process are generally carried out in one form or another, although the scope of each activity may differ between companies and different planning environments. The primary differences are between manufacturing cells and line processes compared to job shop processes.

The manufacturing cell or production line is characterised by high volumes of few or similar products and short **set-up times** and lead times. In such environments, control of order release and priority control take place almost simultaneously. Priority control is considerably easier in comparison with manufacturing in a job shop, since the ordering point is only the cell/line as a whole, and not individual machines or work stations that are parts of the work centre. The **routings** for all manufacturing orders are normally identical and are governed to a large extent by the physical manufacturing layout. The sequence of execution of issued orders is not decided on the shop floor, but by the order in which manufacturing orders are released, i.e. according to the **priority rule** first in, first out. Queues within the cell/line are normally non-existent or very short.

In contrast to manufacturing in cells or lines, where the focus is on the overall manufacturing process rate, every single operation in a manufacturing order in a job shop is given special treatment. All activities involved in execution and control in a job shop are roughly equal in importance and scope. The difference between various activities is also more clear-cut than in manufacturing cells or line processes. After order release, material availability check and printing the shop packet, the manufacturing order is released to the first operation. Since queues normally arise at work centres, detailed priority control will be necessary for sequencing released orders. To be able to apply advanced priority rules in this process, some form of system support will be necessary and labour reporting of the order status must be carried out continuously and in more detail.

16.2 Order Release Control

The control of order release involves determining which manufacturing orders will be released to the workshop to be executed. An important starting point is the start time and due time established at the higher planning level, but current workload must also be taken into consideration.

There are two general aspects on releasing or not releasing an order for manufacturing to the workshop. The first is related to available capacity: an order is released as and when there is free capacity, even if there are no material requirement reasons for doing this immediately. This can contribute to better utilisation of capacity, but at the cost of more tied-up capital. The second concerns the material requirements that lie behind the order being initiated. The due time for the order then has higher priority than available capacity in controlling order release, and the order is back scheduled from the due time without consideration to capacity.

In many planning environments the control of order release is a relatively complex issue. To be able to take into consideration capacity and material requirements and at the same time fulfil the efficiency goals for execution and control, it is sometimes necessary in the short term to utilise the margin of action that exists.

When planning with consideration to available capacity there is a certain margin for short-term increases in capacity, as discussed in Chapter 14 on capacity planning. In many cases the use of overtime work can rapidly increase available capacity, such as temporarily planning work on a Saturday. Shift work and hiring more labour are other capacity variables of a more long-term character which provide margins of action at upper planning levels. Subcontracting may also be an option, but may involve some uncertainty with respect to delivery precision. If work is handed out to new subcontractors it may be necessary to plan extra slack time after subcontracted operations.

Another margin of action is to influence the manufacturing order itself instead of increasing capacity within existing resources. If it is a case of make to stock with large order quantities in relation to short-term demand, changing the order quantity could be an alternative. A decrease in the order quantity may, for example, be sufficient to cover the short-term material requirements at the same time as it decreases the demand on manufacturing capacity. There is no such margin of action in make-to-order environments since every deviation in delivered quantities means there will be a shortage situation if the quantity made is too small, and a surplus for which there is no consumption if the quantity made is too large. Other courses of action which are sometimes available include processing an order in an alternative work centre or temporarily changing operation sequences, i.e. the sequence of operations the item must go through. In many cases there are two or more identical or similar machines in a work centre. A division of the order quantity can then be made and manufacturing can take place in parallel using several machines – a measure known as **order splitting**. This will bring about a considerable decrease in production time but will increase the total set-up time. Consequently, this option is only suitable in environments with relatively short set-up times.

One further order-related option to decrease throughput times in overload situations to enable deliveries on time is **overlapping** operations. This means that a completed part of the order quantity is moved to the next operation before the entire order quantity is complete. The total throughput time for the order quantity will then be considerably less than the sum of throughput time in the operations involved in the undivided order quantity. However, transportation costs will be greater and rational flow layouts will be required together with proximity between stages in refinement if overlapping is to be suitable.

There are different methods for supporting the control of order release. They have different characteristics and vary in usefulness in different planning environments. The following three general methods of controlling order release are described: (1) order release from planned start times, (2) **regulated order release** and (3) **input/output control**.

Order release from planned start times

Releasing orders from planned start times means that manufacturing orders are released to the workshop more or less continuously at the rate at which start times are assigned to orders by material planning. There is no real short-term modification of release for cases of capacity shortage or surplus in the workshop. The material requirement situation governs the process, and start times established by material planning are not adapted to available capacity. In the case of more permanent overload or surplus capacity the flow of new orders may be modified. This takes place at the master scheduling level through changes to master production schedules.

Variants of the method

Order release to the workshop may take place in two different ways: per order or per operation. Release per order involves the release of all operations belonging to a manufacturing order together with the order. A planned start time for the order then governs all releases of operations. Release per operation means that the first operation is released together with the order and other operations are released with the desired time margins based on the start times planned for each operation relative to the equivalent start time for the order.

Start times and due times for orders are obtained directly from material planning. The corresponding times for operations involved are obtained through scheduling on the basis of the start time and due time for each order in generally the same way as for capacity requirements planning, as described in Chapter 14. There is one difference, however, relating to where queuing times should be placed between operations.

To be able to calculate capacity requirements as accurately as possible standard queue times should be located in the work centre in question before every operation, as in the upper part of Figure 16.3. Scheduling then reflects the most probable outcome for when the operation will be carried out in reality, and thus the most probable time for the capacity requirements it brings about.

When scheduling operations for release per operation, the situation is somewhat different. Scheduling should then be based on when an operation can be planned to start earliest, and the normal queue time is thus placed after the operation as in the lower part of Figure 16.3.

Characteristic properties of the method

The method of controlling release of manufacturing orders with the aid of scheduled start times by material planning is characterised by the following:

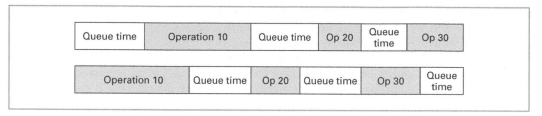

FIGURE 16.3 Alternative scheduling principles for operations

- The method is simple and requires very small planning resources.

- Since the start times and due times are not modified as a result of available capacity, they always represent real demand points from the material perspective. They can thus be used as a basis for prioritising in the workshop.

- Since capacity is not considered, the method gives rise to uneven workloads, especially when there are large order quantities and long operation times.

- Since overloading may arise, the method may give rise to queuing problems that are difficult to control, and in consequence uncertain and varying throughput times.

- The method means that execution and control is completely subordinate to material planning.

Primary environment requirements for the method

The method of controlling order release through scheduled start times is useful in the following environments:

- *Smoothed work load.* It is primarily of use together with material planning methods that have workload smoothing characteristics.

- *Stable demand and high volume flexibility.* Other important environment requirements are stable demand and relatively uniform manufacturing processes if the method is to function well. The method may also work reasonably well in planning environments with high volume flexibility, since available capacity can be varied to a larger extent on the basis of demand in such an environment.

- *Short throughput times and small order quantities.* The method has its relative strengths in planning environments with short throughput times and small ordering costs. This environmental requirement is related to the fact that large order quantities tend to result in uneven workloads.

Regulated order release

When orders are released through scheduled start times, they are released entirely on the basis of timing by material planning in which only material requirements are considered, and not available capacity. Regulated order release, on the other hand, means that scheduling of orders is adapted to capacity supply through checks made on short-term workload. In the case of overload or surplus capacity, rescheduling is carried out to achieve a more even workload. The principal for rescheduling through bringing forward and/or postponing workload that exceeds the available capacity is illustrated in Figure 16.4. When rescheduling, checks are made that it is possible to carry out measures from the perspective of material supply.

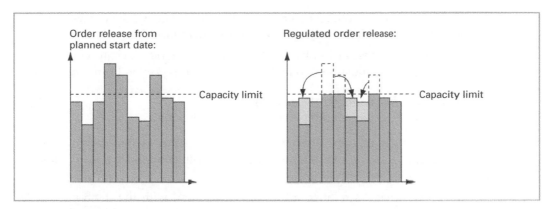

FIGURE 16.4 Rescheduling with regulated order release

Order release to the workshop takes place on the basis of start times of rescheduled orders, in the same way as order release from scheduled start times. The difference between the two methods is thus completely a question of capacity adaptation of start times for orders before release.

The work involved in executing rescheduling may be rather complex. This generally does not apply to the rescheduling itself from the capacity perspective, but the analyses and rectification of material planning consequences that must be carried out. For example, **forward scheduling** a manufacturing order may force another manufacturing order to be brought forward in order to ensure the supply of start-up materials if quantities in stock are insufficient. **Backward scheduling** of an order may also have repercussions through manufacturing orders for items at a higher structural level being postponed as a result. As a consequence there may be reasons to postpone orders for items at the same structural level since the start-up materials related to these will not be used until the postponed manufacturing order at the same structural level is delivered. In other words, rescheduling single orders can lead to chain reactions and new requirements for rescheduling.

Variants of the method

One type of method variant that exists relates to order release per order or per operation in a corresponding fashion to the previous method. The other method variants are related to different ways of rescheduling. There are three options. The first is when only start times are changed, the second when due times are changed and the third when both start and due times are changed. Only changing the start times or due times assumes that rescheduling is minor and that lead time can be compressed or expanded as much as the rescheduling corresponds to. If not, the timing of order will not have any significance as an information carrier to the workshop about when an order is expected to be started or completed.

Characteristic properties of the method

The method of regulating release of manufacturing orders with the aid of capacity-adapted order start times can be characterised as follows:

- The method is considerably more time-consuming and difficult to master than the previously described methods. Advanced system support that allows material supply analyses and shortage controls during rescheduling is a prerequisite.

- Since it results in some smoothing of workload, the method gives rise to fewer queuing problems and fewer problems with uncertain throughput times.

- Since rescheduling of orders takes place with respect to available capacity, current capacity requirements levels must be updated through labour reporting from production.

Primary application environments for the method
Regulated order release is primarily used in the following environments:

- *Uneven capacity requirements.* Since the method takes capacity into consideration, it is mainly suitable for use in planning environments with short-term, uneven capacity requirements and limited options for varying capacity.

- *Long throughput times.* Its relative advantages compared with the method of order release from scheduled start times are greatest in environments with long throughput times, since the other methods mostly control through start times for the first operation of the order. It may therefore be described as especially suited to job shops.

- *Products with simple bills of material and few operations.* Using the method is difficult in environments containing products with complex bills of material and where the number of operations is large. This is related to the fact that difficulties in planning increase at a more than linear rate with increasing complexity of structure and operation sequences. Problems involved in analysing the consequences of executed scheduling also increase. This applies to the analysis of material supply consequences for other orders and items as a result of forward and backward scheduling. It also relates to the extent to which remedies for capacity supply problems in one work centre give rise to capacity supply problems in another work centre.

Input/output control

Input/output controlled order release means regulating order release on the basis of available capacity in the workshop, as in the previous method. However, regulation does not take place with the aid of rescheduled orders. The number of orders released is regulated instead by how many can be dealt with at a certain point in time. The basic principle is to control the release of orders and through this the number of orders and queue sizes in the workshop, and the amount of work in process. The start and due times calculated by material planning are in principle not rescheduled, but remain unchanged and correspond entirely to current material requirements and material supplies.

The principle of input/output control is illustrated with the aid of an example in Figure 16.5. Planned output from the work centre in the example is 120 hours per week and corresponds to net capacity that can be planned. For various reasons real output deviates from planned output during the first four weeks. This may be due to lower productivity than normal in the work centre caused by machine breakdowns, losses due to stoppages and slowdowns, poor quality results, absence of personnel, etc. It may also depend on disruptions in deliveries from supplying work centres. Planned input is the number of orders planned to be released to the work centre. The number of orders released is determined by the size of the accumulated order queue. The basic principle is not to release new orders to a work centre that already has a long order queue, except in situations where a current order has a higher priority than those already released and in the order queue. Start times generated by material planning are normally used for setting priorities for the orders when selecting their release. In the example in Figure 16.5, in weeks 3

Week		1	2	3	4	5
INPUT						
Planned		120	120	110	110	110
Actual		110	130	110	115	
Cum. deviation		−10	0	0	5	
OUTPUT						
Planned		120	120	120	120	120
Actual		80	110	115	110	
Cum. deviation		−40	−50	−55	−65	
Queue						
Planned	200	200	200	190	180	170
Actual	200	230	250	245	250	

FIGURE 16.5 Example of calculations when using input/output control of order release

and 4 fewer orders were released than the corresponding planned available capacity in order to decrease the order queue. Despite this measure the order queue did not decrease, primarily as a result of lower output than planned, but also because the real input of the work centre during this time exceeded planned levels. This may have been caused by deviations between planned and real lead times for the current order. Changes to the real queue are the difference between actual input and output.

The approach used to apply the method means that for each planning period and work centre/department/foreman area, manufacturing orders are released corresponding to the capacity requirements which each work centre/department/foreman area has the capacity to manage during the same period of time. Start times generated by material planning are normally used to set priorities for orders when selecting their release.

Variants of the method

There are two main variants of input/output control. In the first variant, the number of released orders is regulated so that the amount of released capacity requirements per work centre and time period are at the most equivalent to planned capacity adjusted by the difference between planned queue size and real queue size.

In the second variant, the number of orders released is regulated so that the capacity requirement per work centre and time period is equivalent to capacity performed during the previous period. This variant may therefore be characterised as performance-based release.

It is also possible to distinguish two variants with respect to what is released to the workshop. In the first variant, complete orders are released with all their respective operations. In this case, the work centres for the first operations will regulate order release. In the second variant, release takes place per operation. In this case, planned operation start times are used instead of planned order start times as a basis for prioritising release of orders.

Characteristic properties of the method

Input/output control of manufacturing orders can be characterised as follows:

- The method is simpler to handle than regulated order release. It does not demand access to advanced system support in the same way to enable analysis of rescheduling effects.

- By controlling the number of manufacturing orders in the workshop, the method is a powerful aid for reducing problems with bottlenecks and uncertain throughput times.

■ Input/output control may prevent the release of orders that should have been released from the material planning perspective due to shortage of capacity supply. The consequences of this in terms of material planning are not included in the framework of the method.

■ Allowing the order start and delivery times established by material planning to remain fixed, especially the delivery times, means that they are assigned priorities in the workshop since they represent when the orders should be able to be delivered from the material planning perspective.

Primary application environments for the method

Input/output control is useful in the following environments:

■ *Manufacturing cells and line processes.* The method is primarily suitable for use in planning environments with manufacturing cells or line processes. It is also useful in job shops if throughput times are reasonably short and the number of operations is limited. If these environment requirements are not fulfilled it is difficult to plan how much capacity requirement should be released due to mutual influence between many work centres. This applies especially to the variant in which complete orders are released separately. It may then happen that a release motivated on the grounds of low workload in a work centre causes other operations in the order to make the queuing situation in another work centre even worse.

■ *Detailed production preparations.* The method requires detailed production preparations and a satisfactory quality level of operation times in order to function well.

Comparison of methods for controlling order release

The methods for controlling order release can be characterised according to whether they take capacity into consideration, whether release takes place periodically or sequentially, whether the methods are dependent on labour reporting and whether consideration is given to material requirements. A comparison of the methods described in the chapter with respect to these variables is summarised in Table 16.1.

Property variables	From planned start times	Regulated order release	Input/output control
Capacity considerations	Not adapted to available capacity	Adapted to available capacity	Adapted to available capacity
Type of release	Sequential release	Sequential release	Periodic release
Labour reporting	Release not dependent on labour reporting	Release dependent on labour reporting	Release dependent on labour reporting
Material requirement considerations	Yes, consideration given to material requirements situation	Yes, consideration given to material requirements situation	No consideration given to material requirements situation

TABLE 16.1 Comparison of methods and their order release characteristics

16.3 Material Availability Checking

An operation for a manufacturing order cannot start unless materials, tools, manufacturing instructions and so on are available. It is therefore necessary to schedule so that all resources required to start operations are available simultaneously at the start time. This chapter is limited to the process of checking the availability of necessary starting materials for operations. The process is called materials availability checking. Its purpose is to avoid orders with material shortages being released for manufacture in the workshop.

Shortages of materials may depend on disruptions in an earlier work centre, delays or faulty delivery from a supplier, or the real stock on hand differing from the accounted balance according to the ERP system. If the material availability check shows that materials are not available in stock and will not be available in time, corrective measures will be necessary such as forcing materials from suppliers or from internal manufacturing, postponing the due time of the order, compressing its lead time or reducing the order quantity.

In principle, there are three different methods of performing material availability checks. These methods are (1) through physical allocations, (2) accounted stock on hand and (3) available stock on hand. They are described briefly below.

Material availability checks in the form of physical allocations

Material availability checks in the form of physical allocations mean that the physical access to material is controlled and that material is pre-picked from stocks and stored at a special storage place or in the workshop until the order is released to the workshop. Alternatively, pre-picking is limited to the material normally directly picked for an order, while materials in shop floor stocks at workstations and assembly stations are not pre-picked. One reason for using the method may be that stock on hand information in the ERP system is uncertain and verification is required that material really is available. Since material is picked in advance, the method results in more tied-up capital. This also means increased material handling, since material must be moved twice. Furthermore, conditions are made worse for establishing new priorities in the case of material shortages, since it is difficult or even impossible in practice to re-pick material that has already been picked.

Material availability checking through accounted stock on hand

Material availability checking through stock on hand means that the quantity of a certain item required to start a manufacturing order is compared with the accounted stock on hand quantity. If stock on hand of the item in question exceeds the quantities required, the order is cleared for materials. Material availability checking through accounted stock on hand will require access to a computerised stock accounting system. If the items are missing, the ERP system can indicate when they will be available at the earliest. Decisions on start of manufacturing can be taken even if no items are available, provided that it is known that the item will be available when required. There is a certain amount of uncertainty associated with the method since a check of physically available quantities is not carried out. This is particularly relevant when material availability checks must be carried out in good time before picking and starting an order.

Material availability checking through accounted available stock on hand

The method of material availability checking through accounted available stock on hand means that the quantity of a certain item required to start a manufacturing order is compared with the quantity available in stock according to stock accounts including consideration to allocated quantities, i.e. the quantity in stock minus allocated quantities for other orders. If the available quantity in stock is greater than the quantity required, the order is cleared for materials with respect to this item. In the same way as for the previous method, this is an administrative material availability method that requires a computerised stock accounting system with allocations processing in order to work. Accurate information can normally be achieved if the stock accounting system has sufficient accuracy. This is also the case if material availability checks are carried out in good time before physical picking and order start.

16.4 Generating the Shop Packet

On release, the planned order is given the status of released to the workshop. In addition to carrying out detailed control of the release and ensuring that the required materials, tools and capacity are available, shop packets are printed before the order can be released to the workshop. The shop packet contains detailed information on each order, routings, tools required, required materials and so on. The purpose of this is to convey information on how manufacturing will be carried out and to provide information for labour reporting. In certain companies the shop packet is only on paper. In others most of the shop packet information is transferred to terminals close to the work centres. In some companies documentation has been eliminated since its purpose has disappeared to a large extent due to automation or the method of organising the manufacturing process. For example, necessary information can be identified with the aid of a bar-code reader when the order arrives from an earlier operation. Information may also be linked with an operator's employment number, and then all necessary information can be obtained from a workshop computer terminal through entering the employment number. It is also possible to completely automate reporting by equipping load carriers with identity badges which are automatically registered when they pass a reader.

Exactly which documents follow a manufacturing order will vary from company to company. Below are some examples of information that is either conveyed with accompanying documents or can be made available through computer terminals:

- *Shop traveller*. The purpose of the shop traveller, or operation card, is to identify unique orders. It follows the order from release to final reporting. The operation list on the card contains detailed information about what operational steps and work centres the order is to be processed in. Start and due times for every operation may also be stated, as well as information on what tools are to be used in the different operational steps.

- *Labour card*. For every operation there is one or more **labour cards** for reporting. These contain detailed information on the operation, including set-up and run times. It is used to authorise start of operation, but also for reporting actual labour time.

- *Tool card*. The tool card states what specific tools will be used in the different operations. This information is used to order necessary tools or other manufacturing aids.

- *Pick list/material requisition*. This list states what start-up materials are required for manufacturing and in what quantities. It is used to order picking of items from stock.

■ *Report card.* For every operation there is at least one **report card**. The aim of the report card is, among other things, to report that operations have been started and completed.

16.5 Priority Control

The aim of priority control is to ensure that those orders that have been released for manufacturing in the workshop are executed in an appropriate sequence with respect to delivery precision and throughput times. The time schedules established before order release may be too approximate, such as stated per day or per week, or manufacturing conditions behind the schedules may have changed. Several orders may then be in a queue ahead of a common resource. Priority control will determine in which sequence the queued orders are to be executed as efficiently as possible.

There are a number of methods that are available as support for performing priority control. They differ in a number of respects and are therefore more or less appropriate for use in different scheduling environments. However, in different ways they all tackle the basic problem of priority control, namely that orders released for manufacturing in the workshop are executed in the most appropriate sequence with respect to delivery precision and throughput times. The following three different priority control methods are described below: (1) supervisor-managed priority control, (2) priority control by priority rules and (3) priority control by **dispatch lists**.

Supervisor-managed priority control

Supervisor-managed priority control means that supervisors or operators in the workshop set priorities and choose in what sequence released orders and operations will be executed. Workshop personnel do not really use any formal rules for prioritising sequences.

The larger the number of orders released, the greater will be the influence by workshop personnel over the sequence in which manufacturing takes place. The remaining control over the successive execution of operations by planners is then closely connected to how far ahead manufacturing orders are released. The further in advance, the less control is exercised by the planning department. Potential control decreases also as throughput times become longer.

Variants of the method

Release may take place per order or per operation, and for this reason there are two different variants of supervisor-managed priority control. When the release takes place per order, all operations on the order are released at the start time of the order. In the case of release per operation, operations are released separately. Start times and finish times of operations involved are obtained by scheduling based on each order's scheduled start time and finish time. Since priorities in this case are completely based on decisions in the workshop, there are no differences in principle between the two variants with respect to priorities generated. The difference lies in the fact that releasing operation by operation will involve considerably fewer operations released at the same time, and consequently there will be fewer operations to choose between. The potential for the planning department to influence ordering sequences in the workshop will therefore be greater when release is based on operations, especially in situations with long throughput times and many operations.

Characteristic properties for the method

Supervisor-managed priority control can be characterised as follows:

- The method is very simple to use and has a positive effect on the motivation of workshop personnel.

- The method often results in factors related to set-up times, incentive payments and similar workshop factors being decisive for the sequence of manufacturing.

- Since priorities are strongly influenced by conditions in the workshop, manufacturing sequences have a tendency to become random from the material planning perspective, i.e. they have little connection with the timing of the real material requirements that exist.

- The extent to which consideration can be given to capacity is related to how well individual supervisors or order/operation distributors know the capacity situation in different parts of the workshop, either through workload reports or through contacts with other foremen/order distributors at joint planning meetings. There is nothing inbuilt in the approach that ensures that consideration is given to the current capacity situation.

Primary application environments for the method

The primary application environments for this method are as follows:

- *No production preparations.* Since the method has no real demands on production preparation, its relative advantages lie in environments without detailed production preparation and existence of routing data.

- *Small workshops.* Supervisor-managed priority control is inferior to other priority control methods in most planning environments. Its primary area of use is in small workshops where it is easier to gain an overview of the total material supply situation and thus take necessary consideration to material planning in the selection of priorities. The method may also work reasonably well in planning environments where workshop personnel and material planning personnel work closely together.

- *Short throughput times and few operations.* The relative weaknesses of the method are greatest in planning environments with long throughput times and a large number of operations per order. It is generally less satisfactory in job shop environments.

- *Sequence dependent set-up times.* The greater the importance of manufacturing sequence from a manufacturing efficiency point of view, the greater are the relative advantages of the method. Such importance of manufacturing sequence occurs, for example, when set-up times between consecutive manufacturing orders vary with the sequence in which they are carried out. When using supervisor-managed priority control, workshop personnel have greater opportunities to select suitable manufacturing sequences for released orders.

Priority control by priority rules

Priority control with the aid of priority rules means that supervisors and operators in the workshop themselves choose the sequence in which released orders and operations are executed. In contrast to the previous method, chosen sequences are governed by priority rules intended to be applied by the workshop personnel.

Types of priority rules

There are basically two types of priority rules: (1) general priority rules and (2) scheduling-based priority rules. General priority rules means rules that can be formulated and applied without using information related to the specific planning situation. This means that only order-based information is required in the workshop for the method to be applied. Scheduling-based priority rules, on the other hand, means rules that are based on information about the current scheduling situation for released manufacturing orders and their associated operations. This information refers to scheduled start and finish times for orders and operations. The priority information is conveyed through some form of manufacturing order document, for example an operation card. Information may also be displayed on terminals in the workshop.

Below are some general priority rules which can be used by workshop personnel in the application of the current priority control method:

- *Shortest operation time rule* means that orders with short operation times are given priority. The average throughput time and tied-up capital is thereby kept at a minimum. The risk involved in using this rule is that orders with short operation times are sometimes finished so rapidly that they may lie waiting to be delivered. Another disadvantage of this rule is that orders with long operation times may have to wait very long for manufacture, and thus risk being late. For this reason it is common to combine the rule with a maximum waiting time. Orders which have exceeded the maximum waiting time are automatically given high priority.

- *Highest number of remaining operations first* means that orders with many operations are given priority. The more operations an order has, the greater is the risk of it being delivered late. Delivery precision is given priority using this rule, although it might take place at the cost of throughput times and tied-up capital.

- *First in, first out* means that the order with the earliest arrival time is also processed first. In situations with few orders in the queue, for example in manufacturing cells and line processes, this is a very simple and useful rule.

- *Largest order value first* is a method that keeps tied-up capital at low levels. Some calculation work will be necessary and access to standard cost information is required.

Example 16.1

For manufacturing orders with arrival times, operation times and remaining number of operations as stated below have been assigned to a work centre. Arrival times are stated in running workdays.

Order	Arrival time	Operation time	Number of remaining operations
A	116	2	4
B	112	3	5
C	115	1	3
D	120	4	2

Priorities and average queue times per order with the application of the above described general priority control methods are stated below. It can be seen that the priority rule shortest operation time gives the shortest average queue time per order: $((0 + 1 + 3 + 6) / 4 = 2.5)$.

Priority rule	Sequence	Average queue time
Shortest operation time	C-A-B-D	2.5
Highest number remaining operations	B-A-C-D	3.5
First in, first out	B-C-A-D	3.25

Several scheduling based priority rules are available of which the following two are commonly used.

■ *Earliest finished time first* prioritises orders with early finished times. It is a priority rule that focuses on keeping delivery times. The rule can be based on either the due time for the order or due time for the individual operation. It is also called EDD (earliest due date).

■ *Earliest start time first* means that orders scheduled to start first are indeed started first. As above, the rule may be based on the order start time or the start time for each operation.

Start and finish times for orders are obtained directly from material planning. The corresponding start and finish times for operations involved are obtained through scheduling based on the start time and finish time of each order, in the same way as in capacity requirements planning.

Example 16.2

The four manufacturing orders below have arrived at a work centre. Start and finish times, expressed as running working days for each order and operations, are as in the table.

Order	Start time order	Start time operation	Finish time order	Finish time operation
A	110	119	122	121
B	115	115	127	118
C	111	112	120	113
D	120	120	132	124

When using the priority rule earliest finish time for orders, priority between orders will be C-A-B-D, i.e. those with highest order priority have shortest remaining time to delivery. If the earliest start time for order is used, the sequence will be A-C-B-D instead.

Variants of the method

In the same way as in supervisor-managed priority control, we can distinguish between two different variants with respect to release per order or release per operation. In the case of release per order, scheduling-based priority information for all operations is based on the scheduling situation prevailing at the time of order release. For release per operation, there are two subvariants. In one, there is no updating of priority control information after release of the order and its first operation. This variant will have the same priority character as the variant release per order. The second sub-variant uses successive rescheduling, and scheduling data for remaining operations not released is updated as rescheduling takes place. The priority-based information for each operation reflects the situation prevailing when the operation was released, not the situation prevailing when the order was released. Using this variant, decisions on priorities can be made to a larger extent on the basis of the current scheduling situation.

Within the framework of this method, you can also distinguish between cases where start and finish times are only scheduled at the order level and consequently do not give consideration to operations involved, and cases in which start and finish times are also scheduled for each operation. The latter variant is a prerequisite for using release per operation in combination with successively rescheduled operations.

Characteristic properties of the method

Priority control with the aid of priority rules can be characterised in the following way:

- The method is simple to use and puts, in the case of general priority rules, no demands on transfer of scheduling information to the workshop.

- Since priorities in the case of scheduling-based priority rules are based on information on the current scheduling situation, this method compared with the previous method can obtain closer correspondence between what is carried out in the workshop and what is expected to be achieved through material planning.

- Since the selection of priorities is based on manufacturing order documents and the scheduling situation prevailing on release, the method is rather conservative from a rescheduling viewpoint. The further in advance release takes place, the more noticeable this disadvantage becomes relative to methods based on priority control by dispatch lists.

- Consideration of the requirements of material planning is greater when using the variant with release per operation.

Primary application environments for the method

Priority control with the aid of priority rules is primarily useful in the following environments:

- *Manufacturing cells and line processes*. The method based on general priority rules is most useful in planning environments with manufacturing cells or line processes. The need for detailed control of manufacturing sequences in these environments is less critical since to a large extent it is determined by the physical layout of the work centres.

- *Job shops*. The method based on scheduling-based priority rules is useful in most manufacturing plants, including job shops.

- *Short throughput times and few operations*. The method has its greatest limitations in environments with long throughput times and many operations in combination with high change rates and large requirements for rescheduling, especially the variant in which release per order is used.

- *Detailed production preparations*. The demand for detailed production preparations depends on which type of priority rule is used. When using general priority rules the method only puts small demands on production preparations, while more detailed production preparations are required if scheduling-based rules are used. This is because start times and finish times have to be derived from routings and operation times.

Order number	Operation number	Operation time	Start time	Finish time
13687	10	3	081220	090120
25467	20	6	090103	090130
38794	10	1	090109	090124
46751	30	5	090108	090122

FIGURE 16.6 Example of a dispatch list

Priority control by dispatch lists

A dispatch list is a list of operations for each work centre in the sequence in which the operations are expected to be carried out. One example of a dispatch list is shown in Figure 16.6. Otherwise expressed, a dispatch list may be said to be a scheduled manufacturing sequence for each work centre. If a dispatch list is generated with the aid of priority rules, then we call it a priority-based dispatch list. The priority rules may be of the same type as the scheduling-based priority rules in the previous section, but there are also possibilities to use considerably more advanced rules with closer links to material planning.

The following three dynamic priority rules constitute examples of more advanced priority rules that can be used when generating dispatch lists:

■ *Least difference between remaining time to due time and accumulated remaining operation time (shortest slack time)* is a priority rule that in principle has the same aim as earliest finish time, i.e. it tries to minimise the risk of late deliveries. The difference between remaining time to due time and accumulated remaining set-up time and run time is calculated. Controlling priorities with respect to the least difference between remaining time to due time and accumulated remaining operation time means that the order with the greatest risk of late delivery, i.e. the smallest time margin to due time, is carried out first. The rule is primarily intended for make-to-order environments. An order with 30 hours to scheduled due time and an accumulated remaining operation time of 22 hours would be given priority over an order with 26 hours to scheduled due time and an accumulated remaining operation time of 15 hours. The time margin for the first order is only 8 hours, as compared with 11 hours for the second order. One variant of the method is to divide the total slack time by the number of remaining operations.

■ *Critical ratio* is a commonly used variant of shortest slack time. The ratio between remaining time to due time and accumulated remaining set-up and run times are calculated. The order with the lowest ratio is given the highest priority.

■ *Least difference between stock run-out time and remaining throughput time* is a rule that focuses on the risk of shortages in stock. An order with a stock level having a short run-out time, i.e. how long the stock for the corresponding item is expected to last, and long remaining throughput time will be given high priority since the risk of shortage is high in this situation. The rule is primarily intended for make-to-stock environments. Using this rule, an order for an item with a stock level equivalent to 100 hours' demand and a remaining throughput time of 35 hours would be given priority over an order for an item with a stock level equivalent to 90 hours' demand and a remaining throughput time of 5 hours. The time margin for the first order is 65 hours, compared with 85 hours for the second order. In the same way as for critical ratio, there is a variant of the rule that expresses the relationship as a ratio instead of as a difference.

Example 16.3

Four manufacturing orders have arrived at a work centre. The data in the table applies to each order.

Order	Remaining operation time, total	Remaining time to due time	Remaining Lead time, total	Run-out time in stock
A	12	25	20	0
B	12	30	23	70
C	15	26	25	75
D	22	30	31	100

The sequence of order completion will be different depending on which priority rule is used. If the rule "least difference between remaining time to due time and accumulated remaining operation time" (shortest slack time) is used, the sequence for the orders will be D-C-A-B, since the slack times are 8, 11, 13 and 18 hours respectively. The greatest risk of delay will be for order D, with 8 hours' difference between the due time and remaining operation time. If the rule "least difference between stock run-out time and remaining throughput time" is used, the sequence will be A-B-C-D instead, since the least difference between run-out time and throughput time for orders is –20, 47, 50 and 69 hours respectively. The greatest risk of shortage is for the item to be manufactured in order A, and that order is thus assigned highest priority.

Dispatch lists may also be generated by scheduling released operations to the finite capacity of each work centre. Advanced operations research methods and algorithms are used to accomplish this. These dispatch lists, which to a greater or lesser degree are optimal, are called dispatch lists from **finite capacity scheduling**, and have in principle the same appearance and function as dispatch plans generated on the basis of priorities. Printouts or terminals connected to an ERP system may be used to communicate current dispatch lists to work centres and departments/foremen areas in the workshop. The information flow in the use of dispatch lists based on finite capacity scheduling has the same starting point as priority-based dispatch lists, but includes also information on the capacity of work centres.

Variants of the method

Variants of the method of priority control using priority-based dispatch lists occur in the sense that different priority rules are used when generating the dispatch lists. It is also possible to distinguish between two different categories of priority rules in this context. The first category is based on start times and finish times for orders and operations as scheduled by material planning, which are thus indirectly an expression of underlying supply and demand of materials.

This category of rules was called scheduling-based priority rules earlier in this chapter. The second category is directly based on the current situation with respect to supply of materials in relation to demand for materials, and may therefore be characterised as dynamic priority rules. Which priority rule to use will depend on the current environment and the properties of each rule. It is also possible to use combinations of rules, i.e. to have one main rule and other rules for distinguishing between orders that are not differentiated by the main rule.

There are also method variants with respect to whether the dispatch list includes all released operations or only those ready for start. In principle, an operation is ready for start if the required start-up materials are available and if the previous operation is completed. Conditions are more complex in the case of overlapping operations.

Characteristic properties of the method

Priority control by dispatch lists may be characterised in the following way:

- More system support is required for generating dispatch lists than the other methods described earlier in the chapter. Apart from this, the method is simple to use and places relatively small demands on transfer of scheduling information from the scheduling unit to the workshop.

- Since the selection of a sequence is based on dispatch lists that can be generated and distributed daily, and in theory for every transaction if presented on a terminal, this method enables manufacturing to be run in a sequence that almost always reflects the current material planning situation.

- Since the timing of operations is based on how corresponding orders are scheduled, and the fact that this scheduling can be updated almost in real time, the method provides an effective link between the requirements of material planning and activities carried out in the workshop.

- Reporting of completed operations is required for the method to work. To fully utilise its properties, successive work carried out on operations must also be reported. This is particularly the case if operation times are long, such as more than half a day.

Primary application environments for the method

Priority control based on dispatch lists is primarily useful in the following environments:

- *All types of manufacturing layouts.* The method is useful in all types of manufacturing layouts. In most scheduling environments it is superior to the previous alternatives, even though its relative advantages are less in manufacturing cells and line processes.

- *High integration of material flow and manufacturing activities desired.* Since priority control using dispatch list has stronger connections to material planning compared to other priority control methods, it is superior in job shop layouts. It contributes to the integration of the flow of materials from master production scheduling to material planning and all the way down to manufacturing activities in the workshop.

- *Detailed production preparations.* Planning environments with detailed production preparation are required for the method to be used, i.e. operation sequences and operation times must already have been prepared. Requirements for high-quality operation data are also greater than previous methods if its relative advantages are to be exploited from the planning viewpoint.

Property variables	Supervisor-managed priority control	Priority control by priority rules	Priority control by dispatch lists
Capacity considerations	Capacity considerations in own work centre	No capacity considerations	Capacity considerations if based on finite capacity scheduling
Priority control of operations	No influence by scheduling unit	Indirect influence by scheduling unit	Direct influence by scheduling unit
Timing of sequences	Priorities not defined by start/finish times	Depending on type of priority rule used	Priorities defined by start/finish times
Links with material planning	No consideration taken of materials situation	Depending on type of priority rule used	Consideration taken of materials situation
Dependency on job reporting	Job reporting not required	Job reporting not required	Times and final reporting required

TABLE 16.2 Comparison of methods for priority control

■ *Sequence independent set-up times.* Since it is difficult with this method to take into consideration sequence dependent set-up times, environments with such conditions are not suitable. Necessary consideration to sequence dependent set-up times in the workshop may force the selection and use of manufacturing sequences other than those on the dispatch list. The greater the deviations made in relation to the dispatch list drawn up, the more it loses value as an instrument for synchronising activities in the workshop in terms of material planning.

Comparison of priority control methods

The priority control methods of supervisor-managed priority control, priority control with general priority rules and scheduling-based priority rules are all decentralised methods, meaning that priority control is carried out to a larger or smaller degree by supervisors in the workshop or by operators themselves. Indirect influence by a central scheduling unit may be achieved through the use of priority rules. Using the simplest method, supervisor-managed priority control, some consideration to capacity may be taken directly by workshop personnel when sequencing decisions are made. Consideration to capacity is otherwise only given in the use of dispatch lists based on finite capacity scheduling. Through the use of priority-based or finite capacity scheduling dispatch lists, priority control is carried out centrally and communicated through printouts or terminals connected to the ERP system. Other aspects which differ between the methods are how priority rules are formulated and used. Table 16.2 shows a summary and comparison of the three priority control methods and their characteristic properties.

16.6 Job Reporting

The fact that manufacturing orders and operations have start and finish times does not automatically mean that they will be started or completed at the times stated. Consideration given at the planning stage to supply of materials and capacity can never be completely accurate. Changes to scheduled orders are therefore inevitable. Neither is it possible to completely avoid disruptions of various types, such as stoppages in production or delays in promised deliveries of start-up materials. Thus, there are good reasons to monitor and follow up the completion of released orders in the workshop. For this to be possible it is necessary to receive job reports describing how the planned orders have been executed in practice. There are three purposes of job reporting: (1) following up released orders, (2) updating resource availability and (3) supplying information on real resource consumption.

One of the purposes of job reporting is to transfer information for monitoring and follow-up of released manufacturing orders. In the case of deviations it will then be possible at an early stage to introduce rectifying measures that will minimise any negative consequences.

To be able to schedule a manufacturing order it is necessary at all times to have information on what resources are available, primarily in the form of start-up materials and manufacturing capacity. Since these resources are consumed when executing manufacturing orders it is desirable that information on their consumption is reported so that current resources available can be updated. Examples are reporting of material withdrawals from stock and work executed so that workload can be checked off in the relevant work centres. This is the second purpose of job reporting.

The consumption of resources that takes place when a manufacturing order is executed is seldom exactly as large as when it was pre-calculated and planned, and for which there is data in the bills of material and routing files. More or less material may have been consumed than calculated, and likewise for man-hours. To be able to carry out post-calculations for the manufacturing cost of a product or item it is necessary to have information on real consumption. It is equally necessary to have information on real resource consumption to be able to update the pre-calculated information in basic data files. If a company applies some form of performance-based salary system, information on real resource consumption will also be required. This may be the actual time taken for work executed as compared with the planned time consumption, which will be the basis of bonuses paid on top of salaries. Supplying information on real resource consumption is the third purpose of job reporting.

There are three different levels involved in job reporting: reporting of the entire order, reporting of operations and reporting materials withdrawn for the order or delivered when the order has been completed. The materials aspect is treated in Chapter 18.

Order reporting

The different reporting items that may be relevant for a manufacturing order are illustrated in Figure 16.7. Of these, only orders delivered are completely necessary. This is also the most common scope of reporting. Reporting takes place in conjunction with inbound deliveries to stocks or on direct delivery to another manufacturing order, and includes not only the fact that the order has been completed but also the total manufactured quantity. Reporting may also be supplemented by quantities not quality approved and order closure. Closing an order may also

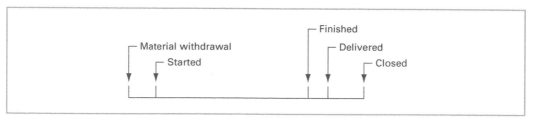

FIGURE 16.7 Reporting items for a manufacturing order

be carried out as a separate reporting occasion, and means that no further reports such as man-hours spent on an operation or withdrawals of materials are expected. When an order has been closed it is ready for post-calculation.

In planning environments in which material withdrawals are reported, there is generally a simultaneous and automatic updating of start of order. If there is no withdrawal reporting there may be reasons for arranging separate start reports to obtain information on the status of the order for monitoring purposes. Separate start reporting is also desirable in planning environments with long throughput.

Order reporting may take place with the aid of different types of cards in the shop packet or via workshop terminals or special order reporting systems connected to the ERP system. The reporting cards used may have barcodes or similar machine readable codes to rationalise reporting work and minimise errors.

Operation reporting

For different reasons there may also be motives for **operations reporting**, i.e. reporting operations involved in orders. The most basic of these reports is that the operation has been completed. This may be supplemented by reports of manufactured quantities and scrapped quantities, and deducting the operation time from the workload in the work centre that carried out the operation.

If post-calculations are required, or data on which to update planned operation times in routing files in the ERP system, actual operation times must also be reported. This may take place in two main ways: either actual, measured operation times are reported directly to the ERP system, or start reports and finish reports on operations are entered in a job reporting system. In the latter case, actual operation times are automatically calculated as the difference between start times and finish times. For operations that extend over stoppages in work, such as from one workday to another or through breaks of different types, reports of interruption and resumption of work will be required.

In planning environments with long operation times and priority control through finite capacity scheduling, there may be reasons for continuous reports of actual operation times. These reported times are deducted from workloads so that scheduling of operations to finite capacities can better take into consideration real capacity requirements. For other priority control methods, the same problems do not arise since they are only based on priorities between released operations.

Reporting of operations may take place with the aid of a work card from the shop packet, or a special report card, through workshop terminals and special job reporting systems connected to the ERP system. In this case, too, the different types of card used for reporting may have bar codes or similar machine readable codes.

16.7 Summary

This chapter has discussed the three main purposes of execution and control: to release orders at the same rate as capacity allows them to be executed with reasonable throughput times, to ensure that start-up materials are available when each order is scheduled to start and to execute orders released for manufacturing in the workshop in an appropriate sequence with respect to delivery precision and throughput times. The first two main purposes relate to the control of release and material availability checks. Together with the printing of the shop packet, they precede the release of the manufacturing order to the workshop. The third main purpose relates to priority control and provides detailed control of released orders in the current work centres.

In addition to solving these three basic problems, execution and control also includes job reporting from the manufacturing process and final reporting of processed manufacturing orders.

This chapter has accounted for the purposes and the design of execution and control processes for different types of company and manufacturing situations. Furthermore, detailed descriptions and comparisons were made of three methods for the control of order release, three methods for material availability checks, three methods for priority control and the principles behind shop packets and job reporting. Variants, characteristic properties and primary application environments for the different methods were also examined.

🔓 Key concepts

Available capacity 346	Order release control 347
Backward scheduling 353	Order splitting 351
Dispatch list 359	Overlapping 351
Finite capacity scheduling 365	Priority control 347
Forward scheduling 353	Priority rule 349
Input/output control 351	Production activity control 346
Job reporting 348	Regulated order release 351
Labour card 358	Report card 359
Labour reporting 347	Routing 349
Material availability checking 348	Set-up time 349
Operation 346	Shop packet 347
Operation reporting 369	

Discussion tasks

1 The release of manufacturing orders is not just an issue of available capacity. It also has material planning implications. Which are the major material planning consequences in cases when the manufacturing order is a result of a customer order compared to the consequences when it is a result of a stock replenishment order?

2 Different alternatives are available to control that released manufacturing orders are executed in an appropriate sequence. Why is supervisor-managed priority control more favourable than the other alternatives in cases with sequence dependent set-up times and what are the consequences for the lead time if sequence dependent set-up times are considered when establishing priorities?

3 Operation reporting can refer to just reporting start times and/or finish times for operations but it can also include reporting of actual spent operation time when an operation is in process. Under what circumstances is this important and for which of the various priority control alternatives is it most important?

Problems[1]

Problem 16.1

A company called Critical Ltd has one critical resource, the work centre WC 600. On one occasion, workday 100, there were six manufacturing orders handed to the work centre. The information on these manufacturing orders is shown in the table. All times are expressed in days. Start and finish times are expressed in running working days. Make a dispatch list for WC 600 based on priority rules (a) shortest operation time, (b) earliest start time for operation and (c) critical ratio.

Order	Operation	Operation start time	Operation finish time	Operation time	Remaining operation time	Order finish time
101	20	103	104	1	9	112
103	10	102	105	3	24	125
105	30	108	111	2.5	8	118
107	20	100	100	0.5	10	110
109	40	95	97	2	8	105
110	20	101	103	1.5	9.5	110

Problem 16.2

A planner at the company Mechanics Ltd is ready to release a manufacturing order TO4321 for item A11, as in the table. The planned delivery time is Monday in week 6.

Operation	Work centre	Operation time (hours)
10	WC 100	9
20	WC 200	5
30	WC 300	14
40	WC 400	5

The company applies regulated order release, i.e. orders are scheduled according to available capacity. The company counts on 2 days (16 hours) of slack time (transportation time and queue time) between operations and 1 day (8 hours) of administration time before delivery. Working hours are daytime, 40 hours a week, and no overtime is planned. The second table shows capacity requirements for released orders and available capacity in the coming weeks for four of the company's work centres before the release of TO4321. Is it possible to deliver TO4321 on time without delaying already released orders?

Work centre	Available capacity (hours/week)	Capacity requirements (hours) Week						
		1	2	3	4	5	6	7
WC 100	70	35	26	18	9	4	0	0
WC 200	35	35	35	26	35	13	9	5
WC 300	70	70	70	70	70	61	52	70
WC 400	35	35	35	35	35	31	30	26

1 Solutions to problems are in Appendix A.

Problem 16.3

The company Mechanics Ltd has problems with item B12. Due to a misunderstanding, an order for 40 B12s was not entered into the ERP system and the customer is demanding delivery within three weeks. It is now Monday morning in week 1 and a manufacturing order for 40 B12s has been created and given highest priority. The planner, Jonas Valfisk, has already ordered a new set-up of WC 1, the first work centre which the item is processed in. Operation data for item B12 and data for the three work centres are shown in the tables.

Operation	Work centre	Set-up time (hours)	Unit run time (hours)
10	WC 1	1.0	1.0
20	WC 2	2.0	1.5
30	WC 3	2.0	1.0

Work centre	No. of machines	Available capacity (hours/week)
WC 1	1	40
WC 2	2 equivalent	80
WC 3	2 equivalent	80

Transportation time between work centres is one hour. One working day is 8 hours.

a) Will it be possible to deliver the complete order within three weeks, i.e. before Monday in week 4, without ordering overtime and without splitting the order?

b) Can the delivery be managed through order splitting?

c) Is there any way of making the delivery without utilising both machines in WC 2 and WC 3 in parallel for operations 20 and 30?

Further reading

Bauer, A., Bowden, R., Browne, J., Duggan, J. and Lyons, G. (1991) *Shop floor control systems*. Chapman & Hall, London.

Fogarty, D., Blackstone, J. and Hoffmann, T. (1991) *Production and inventory management*. South-Western Publishing, Cincinnati (OH).

Hyer, N. and Wemmerlöv, U. (2002) *Reorganizing the factory*. Productivity Press, Cambridge (MA).

Melnyk, S. and Carter, P. (1987) *Production activity control*. The Oliver Wight Companies, Essex Junction (VT).

Nicholas, J. (1998) *Competitive manufacturing management*. Irwin/McGraw-Hill, New York (NY).

Oden, H., Langenwalter, G. and Lucier, R. (1993) *Handbook of material & capacity requirements planning*. McGraw-Hill, New York (NY).

Philipoom, P., Markland, R. and Fry, T. (1989) "Sequencing rules, progress milestones and product structure in a multistage work shop", *Journal of Operations Management*, Vol. 8, No. 3, pp. 209–229.

Plossl, G. (1985) *Production and inventory control – principles and techniques.* Prentice-Hall, Englewood Cliffs (NJ).

Taylor, S. (2001) "Finite capacity scheduling alternatives", *Production and Inventory Management Journal*, Vol. 42. No. 3/4, pp. 70–74.

Vollmann, T., Berry, W., Whybark, C. and Jacobs, R. (2005) *Manufacturing planning and control for supply chain management.* Irwin/McGraw-Hill, New York (NY).

CASE STUDY 16.1: PRIORITY CONTROL AT MARINE ENGINE LTD

Marine Engine Ltd develops and manufactures marine engine and gas turbine components. It also carries out some engine maintenance. Most of Marine Engine's products have very deep and wide bills of material and are made in very low volumes. A high volume product at Marine Engine is made in about 100 pieces per year and the average product is made in 20–30 pieces per year. There exist several products made only in one or two pieces per year and most maintenance work results in unique one piece production.

The production process and layout
The production is capital intensive; both raw material and machine-hours are quite expensive. The product variety is high and customer order sizes small. In order to manage production under these circumstances, the factory has a functional production layout in accordance with job shop processes. Machines and other work centres are grouped and located in the factory according to their production functions. The materials to be used in the manufacturing are moved from work centre to work centre, and in principle there are one or a few manufacturing operations conducted per work centre. Marine Engine's main plant contains four factories (Factory A, B, C and D). Factory A contains eight larger production areas, within which there are several work centres and machines. A typical product is moved between 10 to 15 work centres before it is finished. A couple of times during its way from raw material to finished product, work is normally carried out in an external production area, by work centres used also by Factories B, C and D.

Thermal spraying – a common resource
One of the external production areas is thermal spraying. It is located in the same plant as Factories A–D, but in an own building. Thermal spraying normally contains the following three operations: manual masking, blasting and plasma spraying. Some components are prepared with masking tape when arriving at thermal spraying but most are not. If already masked, the first operation is blasting, otherwise the first operation is manual masking in WC 9109. The blasting work centre (WC 9100) consists of six blasting machines. Most components can be worked in any of the blasting machines but some are only approved to be worked in specific machines. The operation after blasting is plasma spraying. To achieve high spraying quality, there cannot be longer than a three-hour time span between blasting and spraying. Therefore, the same operator carries out both these operations for a specific component. There are five plasma spraying machines in the work centre (WC 9093). Most components are approved for spraying in at least two different machines. After spraying, the component is put in the outbound store and a test piece on which the same operations as on the actual component are carried out, is sent to the internal laboratory for analysis. After receiving positive laboratory analysis results, the component is available for moving to its downstream operation.

▶

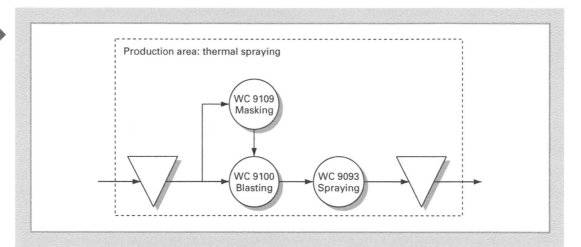

The masking, blasting and spraying operation times for a component vary between some minutes to several hours, and the average total lead time varies from about a day to a couple of days (see table with data for six components). The operation lead times are quite stable but can vary, for example, because of not properly washed components, unavailability of correct masking, blasting or spraying tools, machine disruptions, inexperienced personnel, etc. Even though the operation times are stable, the total lead time variation is high. When analysing the total lead times of the last 200 components in thermal spraying, the average was 1.5 days but it varied between 1 and 12 days.

Component	Planned process time (hours)			Average lead time (days)
	Masking	**Blasting**	**Spraying**	
Shaft front	0.7	0.7	2.0	2.5
Spoolar	No masking	1.0	2.0	0.8
Diffuser case	No masking	0.8	2.5	0.7
Sumpar	0.2	0.3	0.7	1.4
Transply	No masking	0.4	0.7	2.0
AC-liner	0.5	1.3	5.7	1.3

Thermal spraying has on average overcapacity, both in machine- and man-hours. During the last year, the available net capacity was 21 400 machine-hours and the machine-hours used were 14 400 hours. Only 3 per cent of the nominal capacity was lost due to disruptions, maintenance and quality defects. The average available man-hours were 5.2 per day and the capacity used was 3.6 hours per day. The work teams are highly educated and most operators are authorised to work on all machines in the three work centres.

The work centre carrying out the preceding operation is responsible for the transportation of components to thermal spraying. There are two transport alternatives to thermal spraying. The first is an internal truck system which is scheduled on daily work hours according to a timetable. It picks up goods at about every second hour at the respective work centre. Direct transportation could be ordered if transport is required during, for example, shift hours or other pick-up times than in the timetable. Sometimes there is a time lag between finishing an operation and ordering transportation, which may delay the transport.

The queuing time in the inbound store of thermal spraying is sometimes zero and the capacity is idle and sometimes the queue is several days. One reason for the varying queuing time is the uneven flow of components arriving to the production area. The table shows the number of components arriving at thermal spraying during a two-week period. The distribution of arrival times during the day is also quite random.

Arrival date	Weekday	Number of components
1015	Tuesday	7
1016	Wednesday	4
1017	Thursday	8
1018	Friday	4
1019	Saturday	0
1020	Sunday	0
1021	Monday	6
1022	Tuesday	10
1023	Wednesday	12
1024	Thursday	6
1025	Friday	3
1026	Saturday	0
1027	Sunday	0
1028	Monday	12

The time for process planning which precedes the actual work depends on the number of available authorised process planning personnel and the quality of the documents in the shop package. For prioritised orders, process planning is normally carried out before the actual component arrives at thermal spraying. For normal orders, process planning is carried out when the component and shop package are available at the production area. The time for this varies from a half to several hours.

Quality defects may be identified before or during the work in thermal spraying. If the defect is severe it is analysed together with personnel from the responsible cost centre. It is then either corrected at thermal spraying or sent to another work centre for rework and correction. Defects can consequently delay a component for several days.

The priority control process

Weekly production and capacity planning are conducted using MRP and CRP calculations. Based on the weekly plans, orders are released in other work centres in the four factories and when finishing the preceding operations the orders and semi-finished components arrive at thermal spraying. The planning conducted in thermal spraying is a type of supervisor-managed priority control. The goal is to determine an appropriate execution sequence of the queued orders, in order to optimise the delivery precision and minimise the average lead time. No software support is used for the priority control but it is a result of the experience and work of the local planner. The masking, blasting and spraying operations are planned as a common operation, i.e. the planning point is when starting masking (or blasting if the component is already masked). The planner announces the priority orders by writing the order sequence on a whiteboard. Normally, the planner has no information about forthcoming orders, unless the production manager or a responsible cost centre has told her that a critical order is about to arrive.

The following are examples of considerations taken by the planner:

- Orders with short masking time are prioritised in order to decrease the average queuing time.
- Orders using the same powder in the plasma spraying machines are batched together in order to decrease set-up times.
- Orders with available operators authorised to run machines and machines in which the actual components are approved to be worked are prioritised.
- Orders flagged as rush orders by the production manager are prioritised.
- Orders with long accumulated queue times can be prioritised.
- Orders which are behind schedule can be prioritised. The planner, however, is seldom aware of order due dates, unless the responsible cost centre or production manager has informed about it.

Discussion questions

1 Evaluate and characterise the execution and control environment of thermal spraying. What makes the priority control difficult?

2 Evaluate the current priority planning process. Would you suggest any changes?

CHAPTER 17

Procurement of Materials

Just as manufacturing orders generated at the **material planning** level are executed in the workshop, the purpose of materials **procurement** is to execute the procurement of **purchase items** required as identified at the level of material planning. Such requirements may stem from the need to replenish stocks, or a direct requirement for start-up materials from a manufacturing order, or a requirement resulting from an incoming customer order. Irrespective of the origins of the requirement, it must be filled in some way through a **purchase order** and a delivery from a supplier. The purpose of the materials procurement process is to ensure that the requirements of purchase items are satisfied in an efficient manner. The relationship between materials procurement and other planning processes is described in general terms in Figure 17.1.

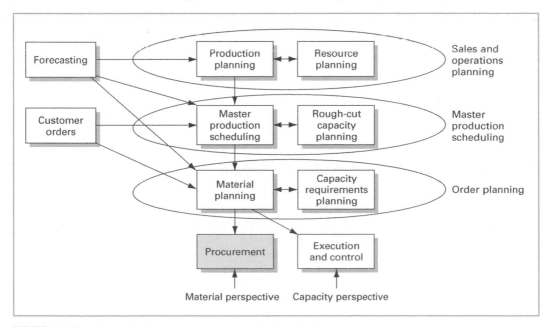

FIGURE 17.1 The relationships between procurement and other planning processes

The procurement process varies in different situations but contains always a number of more or less common activities, which are treated in this chapter. Some principles and alternative paths of action for increasing the efficiency of the materials procurement process through simplification, automation and reconfiguration of the activities in the process are also described in the chapter.

Procurement is only discussed here from operative viewpoints. The various aspects and issues related to strategic purchasing lie outside the framework of this book. The chapter is also limited to the procurement of direct materials, i.e. the procurement of items used as start-up materials for manufacturing of products.

LEARNING OBJECTIVES

After studying this chapter, you should be able to:

- Outline a general procurement process from a logistics perspective
- Outline and explain strategies for simplification, automation and reconfiguration of the materials procurement process
- Explain the principles of **customer-managed ordering** and **vendor-managed inventory (VMI)**

17.1 The General Procurement Process from a Logistics Perspective

The structure of the materials procurement process varies depending on the type of item involved, how supplier relations are structured and what IT support is available. It is also influenced by whether the supplier manufactures the item in question to stock or schedule, or if it is assembled, manufactured or designed to order. Irrespective of what unique conditions exist and how the procurement function is organised, the materials procurement process always involves certain general activities. Figure 17.2 illustrates a general model of the material procurement process. Activities related to invoicing, invoice control and payment have not been included.

Material requirements and purchase requisitions

Anticipated material requirements for the future always form the basis of materials procurement. These requirements are identified at the material planning level through the generation of planned orders or order proposals. This may take place from an MRP system, a re-order point system or through the release of a kanban card. Order proposals in the form of **purchase requisitions** may also be created manually. This is the case for one-off purchases of an order-based product.

An order proposal or an order requisition is a request to the purchasing unit to place a purchase order to a supplier. A purchase requisition is thus the document or signal for procurement that provides the purchase unit with the task of creating the purchase order and executing the purchase.

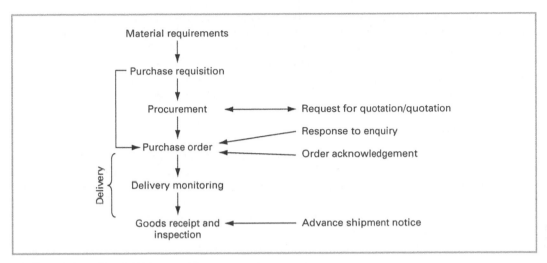

FIGURE 17.2 General model of the materials procurement process

Procurement

If the purchase requisition applies to an item for which there is no established relation to a particular supplier, enquiries must be made to potential suppliers and formal **quotations** will possibly be requested. The aim of a **request for quotation** is to make a comparison between potential suppliers in terms of the factors affecting total costs, quality and delivery service for the materials procured. However, many purchases are relatively simple and do not require an entire quotation process, but can be quickly decided at the time of purchase. This is generally the case when purchasing standard items in small quantities once only. For the procurement of more complex items, a clear specification of requirements will be needed and a more exhaustive quotation procedure will be carried out. This is also the case when procurements are made for large values or recurrent deliveries of large volumes over long time periods.

Procurement activities may also include drawing up agreements with selected suppliers. These agreements may relate to product performance, quality levels, quantities, prices, volume discounts and delivery conditions. These conditions may also include times for the delivery of goods, delivery dates, the location to which the goods are to be delivered, stipulations for who pays freight costs, who pays costs for loading at the place of delivery, for how long the vendor is responsible for the goods, when the purchaser takes over responsibility for the goods, who is responsible for customs clearance and the payment of any customs duties required, etc. If the parties have not made an agreement on delivery conditions, the normal basic rule applied is "Ex-works or warehouse of vendor". This means that the vendor, at the agreed point in time, makes the goods available to the purchaser at the vendor's factory or warehouse. The purchaser is thus responsible for transportation costs and transportation risks.

In the case of recurrent purchases from the same supplier, a long-term **delivery agreement** will normally be signed with the purpose of securing future deliveries, but also to negotiate a better price and delivery conditions. A delivery agreement may be drawn up in many different ways, but in general it is an agreement to purchase a certain minimum quantity of a specific item or group of items at a certain price during a certain time period, normally six months or one year. Minimum and maximum quantities ordered during a sub-period such as one month may also be specified in the agreement. What volumes will be called off and when they are called off will be decided later and successively as needs arise at the customer company.

If a supplier relationship is already established for an item that needs to be procured, it is not often necessary to go through a quotation procedure or other type of procurement process. A purchase requisition can be directly changed into a purchase order.

Purchase orders

A purchase order is a document which authorises the supplier to deliver. It defines the item in question, what quantities are to be delivered and when, and the price of the item. Current delivery conditions are also specified in the purchase order. Figure 17.3 shows an example of purchase order information.

If there is a delivery agreement, the parties have already negotiated and agreed on delivery lead times, quantities, quality levels, prices, etc. A simplified purchase order procedure in the form of **call-off** from the current agreement can then be applied. In such cases the supplier normally has access to a **delivery schedule** as a basis for short-term deliveries as well as a forecast for more long-term material planning. The required quantities of the item are stated on the delivery schedule for future time periods. The delivery schedule is generally divided into three parts. The first part consists of call-offs for delivery and may be considered as purchase orders to the supplier. This part normally covers a time period of between a number of days and a couple of weeks in the future. The subsequent part consists of quantities that are expected to be called off but which do not constitute a full and fixed commitment to call off. The third part is a forecast of probable requirements for future deliveries. The use of delivery schedules is described in more detail in Chapter 8.

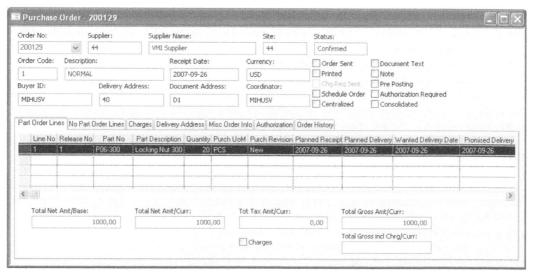

FIGURE 17.3 Example of purchase order list and details

The illustration shows purchase order number 200129 to supplier number 44. The order contains one order line of 20 pieces of part number P06-300. The picture represents a function from the ERP system IFS Application

Order acknowledgement

In order to ensure that the conditions for a purchase order have been approved by a supplier, the order is often acknowledged. An **order acknowledgement** from a supplier means that the order has been received and that quantities and delivery times have been accepted. However, an order acknowledgement may also contain changes relative to the customer's order, such as a new delivery time.

As a rule there are no order acknowledgements made when calling off from a delivery schedule. In other contexts, too, companies may sometimes not demand or send order acknowledgements.

Delivery monitoring

Delivery monitoring is aimed at ensuring that deliveries take place at the agreed point in time. Information is sent to the supplier about when delivery is expected to take place in accordance with the agreement made at the time of ordering. Delayed deliveries may result in disruptions to production, dissatisfied customers and loss of sales. Deliveries made too early cause unnecessary tied-up capital, disrupt the normal flow of materials and require unnecessary storage space, among other things.

Monitoring may take place before or after the agreed delivery time. Pre-monitoring means that a company wishes to make the supplier aware that delivery is expected soon in accordance with the current purchase order. The aim is to improve the probability that the delivery will take place as agreed. Delivery monitoring before the promised delivery date is generally only carried out selectively. Monitoring may be limited to include items for which it is especially important to avoid shortage situations, or it may be limited to suppliers that have shown poor **delivery precision** in the past.

Post-monitoring means that delivery monitoring takes place after the agreed point in time of delivery as a measure to accelerate an already delayed delivery. In this case, delivery monitoring has the character of a delivery reminder. The aim is to minimise the effects of an already delayed delivery. This process may also be carried out selectively. Selective monitoring in this context means that reminders are only sent if the delay has already led to a shortage or is estimated to lead to a shortage within the near future. A company may choose not to send a reminder if there are sufficient quantities in stock to cover requirements at present and in the near future.

Advance shipment notice

In some contexts a supplier will inform customers that their delivery is on the way. This is done by sending out an **advance shipment notice**. The purpose of the information is to alert the customer in advance that delivery is imminent so that goods reception and quality control can be prepared. The aim is also to transfer information about loading and packing so that identification, goods receipt control and possibly onward transport can be facilitated. Advance shipment notice is above all given by suppliers to customers with repetitive manufacturing, such as the automobile industry and the white goods industry.

Goods receipt and inspection

Goods receipt and inspection means that goods are received and checked to ensure that the delivery is correct and in accordance with the corresponding order. The inspection may be visual to check for any exterior damage, quantity inspection and/or quality inspection. Any deviations are noted and will form the basis of possible complaints.

How extensive receiving inspection is and how it is carried out varies a great deal from one situation to another. In some cases quality inspection may take place during the manufacturing process at the supplier, meaning that this part of receiving inspection may be eliminated. If quality inspection is carried out at goods receipt, it may either be 100 per cent inspection, i.e. all pieces of a delivered item are checked, or it may be through acceptance sampling, meaning that a small quantity of items from a large batch are taken, and conclusions are drawn for the entire batch on the basis of the results.

When a delivery has been approved for quantity and quality it is put into stock or moved directly to the place of usage in the workshop.

17.2 The Materials Procurement Process from a Supply Chain Perspective

From a supply chain perspective, the materials procurement process interacts with the supplier's **order-to-delivery process**. Some common ways of simplifying, automating and reconfiguring the materials procurement process from this starting point are described below.

Simplification of the materials procurement process

When structuring and rationalising processes, it is necessary to take into consideration the specific conditions that prevail. An important starting point in rationalisation work is to eliminate as far as possible any activities that are unnecessary and do not add any value, and to create and simplify procedures for value-adding activities.

The materials procurement process can be simplified by separating the commercial part of the process from the part related to material flows. The commercial part, negotiating and entering into purchasing agreements and delivery agreements, can then be handled as an activity performed less frequently, such as once a year. The agreements established make it possible to simplify the ordering process for repetitive, recurrent orders and deliveries.

Three other options for simplifying the common materials procurement and order-to-delivery process are shown in Figure 17.4. Initially, information on requirements is conveyed via material planning to the purchasing unit, which then contacts the sales or order entry units at a supplying company, possibly after evaluation and selection of supplier. When an agreement on delivery has been made, the purchase order is sent from the customer company and becomes a customer order at the supplier. If it is for a stocked item, the order is processed by a unit for delivery planning, which plans and prepares outbound delivery. The order is then processed by a dispatch unit that executes picking, packing and shipment. If it is an order that is made to order, it will be processed by the material and production planning unit that plans the corresponding manufacturing order, and will subsequently be processed by the manufacturing and shipment units.

One alternative for simplifying the process described above is where the purchase order is placed directly with the supplier by the materials planner within the framework of agreements already made. This alternative means that the involvement of the purchasing unit in the parts of the materials procurement process related to operative material flows is eliminated or radically reduced, but that there is no major impact on the order-to-delivery process at the supplying company.

One further step in simplifying processes is to integrate the material planning activities in the two companies and arrange for letting the suppliers' and customers' material planning processes

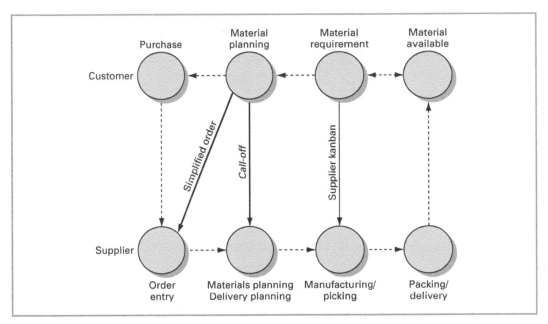

FIGURE 17.4 Different measures for simplifying the materials procurement process

respectively also execute ordering and order entry. This means that order processing in the traditional sense by the supplier is not necessary. The order represents then a call-off from a current agreement or delivery schedule. If there are any delivery schedules these schedules will be the basis for call-offs and inputs to production planning at the supplier. Even if it is possible to initiate large parts of purchasing without demanding direct authorisation from the purchasing department, in many situations it is desirable that the order is approved before ordering takes place. One possibility to differentiate the work involved in creating or authorising purchase orders is to allow a purchase requisition for low value items to automatically generate purchase orders, whereas purchase requisitions for high value items will require authorisation by the purchasing department before a purchase takes place.

The third course of action, illustrated in Figure 17.4, involves the department in the customer company which has a requirement for materials directly calling off from the supplier company's production department or stores. This type of procedure can for example be arranged by using a **supplier kanban** system. Such a system is described in Chapter 15.

The procurement process can also be streamlined by not demanding an order acknowledgement when a purchase order has been placed. Tacit acceptance is used instead. This means that the supplier accepts all conditions and preferences in the purchase order if no contact is made within an agreed time, such as a couple of days. In principle, then, action is only taken in the case of deviations and order acknowledgements are eliminated when the two parties are in agreement on conditions.

Automation of the materials procurement process

Automation means that activities previously carried out by personnel are transferred to an automatic process such as an ERP system, and using methods of communication such as EDI, XML and the **Internet**. Automation of the materials procurement process may be applied to

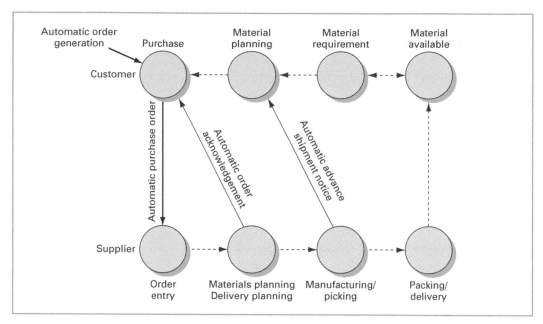

FIGURE 17.5 Different measures for automating the materials procurement process

order generation, transfer of purchase orders, order acknowledgements and advance shipment notices, as in Figure 17.5.

The application of **electronic data interchange (EDI)** is a common approach to automating the transfer of recurrent information between customers and suppliers. For the system that creates and sends information to be able to communicate with the receiving system and interpret and process the message, both systems must use the same message protocol. An EDI message may relate to purchase orders, delivery notes, invoices, call-offs or delivery schedules, and may include the receiving company's order files being automatically updated when the message is received. One disadvantage of EDI is that it is a relatively complex method of communication that requires large investments and has high running costs. As a result, it is mainly large and medium-sized companies that use this method.

To be able to rationalise communication with smaller companies, system solutions based on a combination of EDI and Internet have been developed, called web-EDI systems. Solutions of this type make it possible for larger companies to achieve full systems integration even though some of the suppliers do not have a complete EDI system.

The transfer of purchase orders from customer companies to supplier companies can be automated by communicating system to system, i.e. sending the order directly from the customer's system to the supplier's system by, for example, EDI transactions. The degree of automation depends to a large extent on the capability of the companies' systems to create, send, convert, receive and process orders digitally. A partially automatic solution is for the order to be automatically sent to the supplier, who prints the order and then enters it into his own customer order system. In a fully automatic system, the purchase order is transferred directly to the customer order system without any manual work involved.

Automatic order generation will involve further steps of automation. If generated stock replenishment orders fulfil certain set criteria with regard to upper limits of item prices and order value, for example, purchase orders are automatically created for predetermined suppliers. These automatically generated purchase orders are stored in the purchase order file and are automatically transferred to the respective suppliers using EDI or other information transfer technology.

The transfer and reception of order acknowledgements can be automated in a similar fashion as for purchase orders. Further automation is possible by order acknowledgements being automatically created in the supplier's system and transferred to the customer when the order has been fully entered or changed. It is also possible that order acknowledgements received are automatically compared with the corresponding purchase orders in the customer's ERP system and that any deviations within accepted tolerance limits are automatically updated in the system.

Advance shipment notices may be carried in the same way as order acknowledgements. The notices are automatically created in the supplier's ERP system when the packing and dispatch is reported for the customer order. The transfer takes place with the aid of EDI messages and is automatically processed in the customer company's system.

Reconfiguration of the materials procurement process

According to conventional wisdom, the procurement process is a customer company process and the order-to-delivery process a supplier company process in a dyad. Looking at these two processes more carefully, it becomes however apparent that from a supply chain perspective they are rather two sub-processes that together represent one common cross-company business process. This common process starts for example when there is a need to re-supply stock in the customer company, continues with fulfilment activities in the supplier company and ends when the needed goods are received from the supplier and available to put away in stock.

Reconfiguration of a business process refers to improving performance through reallocation and consolidation of activities along the process. Reconfiguration may take place through reallocation of activities between departments within a company, such as between material planning and purchasing in a procurement process. Looking at the procurement process and the order-to-delivery process as sub-processes in a common process, however, also allows reconfiguration to be carried out between departments across company boundaries. Two basic strategies are then available when trying to improve performance by reconfiguration. One option is to let the customer be responsible for as much of the common process as suitable. This is called customer-managed ordering and means that parts of the order-to-delivery sub-process in the supplier company are added to the customer's procurement sub-process. A second option is to let the supplier to be responsible for as much of the common process as suitable. This option is called vendor-managed inventory and means that parts of the procurement sub-process in the customer company are added to the order-to-delivery sub-process in the supplier company.

Customer-managed ordering

Customer-managed ordering means basically that the common cross-company process is cut short as illustrated in Figure 17.6 and that individuals in the customer company carry out order entry activities that used to be carried out by individuals in the supplier company. The circles in the figure refer to activities in the process. Short-cutting the process like this can be accomplished by connecting workstations at the customer company to the supplier's ERP system via fixed or dialled-up telecommunication lines. It can also be accomplished by using the Internet and web-shop type of system integrated with the supplier's ERP system.

Two options of customer-managed ordering are typically used. One option means that the customer is allowed to enter orders into the supplier's ERP system. The entered customer orders are considered preliminary and must be checked and possibly modified by personnel at the supplier company before being approved and confirmed. By letting the order entry activities be

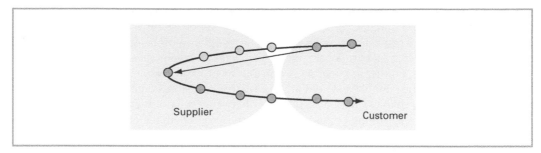

FIGURE 17.6 Customer-managed ordering as an integrated cross-company business process

carried out by the personnel at the customer company, the total order entry work in the common process is reduced and the data quality increased. The customer can also to a larger extent carry out his or her procurement work on his or her own since he or she can get direct access to various types of information such as price, available in stock, current lead times, etc. and not have to ask the supplier's personnel over the phone or by mail.

A second and a more far-reaching option is to allow the personnel in the customer company to not only enter orders but also execute and finalise them in the supplier's ERP system, including allocating material and establishing delivery dates on his or her own. As a consequence of this option, there is no need for any activities to be carried out by the customer company personnel to finalise the ordering part of the process. Since the customer company personnel finalise the orders themselves, there is no need for the supplier to confirm the order. Accordingly the whole order confirmation process can be eliminated which reduces the administrative work involved even further.

Vendor-managed inventory

As stated above, customer-managed ordering means allocating some of the supplier's work and responsibilities for the common cross-company process to the customer in order to reduce the total administrative work included and to reduce the process lead time. Opposite to this, vendor-managed inventory means allocating activities in the common process from the customer to the supplier to carry out and the supplier assumes responsibility for managing the customer's inventory. Accordingly, vendor-managed inventory means basically that the supplier executes the customer's procurement and material planning activities and enters replenishment orders into the customer's ERP system. The principle is illustrated in Figure 17.7.

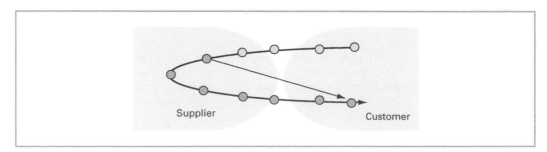

FIGURE 17.7 Vendor-managed inventory as a reconfiguration of the common cross-company process

To apply vendor-managed inventory, individuals at the supplier company must have access to information such as stock on hand and estimated future demand from the customer company, i.e. in principle the same type of information as the personnel at the customer company need when they manage their inventory themselves. Within the concept of vendor-managed inventory a number of dimensions can be identified, each with different options. One dimension is who owns the inventory in stock at the customer's facility, with the options being customer owned or supplier owned. If the customer owns the inventory, the supplier invoices the customer when he or she ships the replenishment orders, while if the supplier owns the inventory, the customer is invoiced when he or she withdraws quantities from stock. In the case of supplier-owned inventory, the inventory can be considered as a consignment type of inventory.

Another dimension concerns the existence of certain limits for the supplier to replenish the customer's inventory. In the case of customer-owned inventory, agreements between the customer and supplier about limits of this kind are of special importance. Typically such agreements state a minimum and a maximum level for allowed stock on hand. Within these limits the supplier is allowed to deliver as much or as little as he wants and when he wants. They represent, in other words, a window for the delivery flexibility allowed by the customer. The bigger the difference between these limits, the bigger the freedom for the supplier to utilise his resources and to allocate quantities to customers who most urgently need replenishments.

A third important dimension within the concept of vendor-managed inventory concerns access to information and in whose ERP system the management of the customer's inventory is supposed to take place. Access to information refers to in what way the supplier gets access to information necessary to manage the customer's inventory. One option is to let the supplier have online access to the customer's ERP system and, as a consequence, also carry out his or her inventory management responsibility in the customer's ERP system or in a web-based system connected to the customer's ERP system. The supplier then also enters new replenishment orders directly into the customer's ERP system.

Information needed to carry out the inventory management responsibilities can also be communicated by batch transactions on a daily basis from the customer's to the supplier's ERP system, e.g. by EDI. In this case the supplier can manage the customer's inventory in his or her own ERP system. When having planned new replenishments orders, information about these orders is communicated by batch transactions back to the customer's ERP system. An advantage with this alternative is that the supplier does not have to learn how to use various customers' ERP systems as he or she can manage the inventories for all of them in his or her own system. A major advantage is also that by managing all customers' inventories in his or her own system, the opportunities to consider the total demand and integral inventory for all customers when planning new replenishments are much bigger than in the case when the supplier carries out his or her inventory management responsibilities in each of the customers' own ERP system and with one at a time.

17.3 Summary

The purpose of materials procurement is to execute the procurement of purchased items required as identified at the level of material planning. Such requirements may stem from the need to replenish stocks, or a direct requirement for start-up materials from a manufacturing order, or from a requirement resulting from an incoming customer order. Irrespective of the origins of the requirement, it must be filled in some way through a purchase order process and a delivery from a supplier. The purpose of the materials procurement process is to ensure that the requirements of purchase items are satisfied in an efficient manner.

This chapter has described the content of the procurement process from the perspective of manufacturing planning and control. The procurement process varies in different situations but always contains a number of more or less common activities, which have been treated in this chapter. Some principles and alternative paths of action for increasing the efficiency of the materials procurement process through simplification, automation and reconfiguration of the activities in the process were also described. Procurement was only discussed from an operative perspective. The various aspects and issues related to strategic purchasing lie outside the framework of this book. The chapter is also limited to the procurement of direct materials, i.e. the procurement of items used as start-up materials for manufacturing of products.

🔐 Key concepts

Advance shipment notice 381	Order-to-delivery process 382
Call-off 380	Procurement 377
Customer-managed ordering 378	Purchase item 377
Delivery agreement 379	Purchase order 377
Delivery precision 381	Purchase requisition 378
Delivery schedule 380	Quotation 379
Electronic data interchange (EDI) 384	Request for quotation 379
Internet 383	Supplier kanban 383
Material planning 377	Vendor-managed inventory (VMI) 378
Order acknowledgement 381	

Discussion tasks

1 Delivery monitoring may take place before or after the agreed delivery time. Discuss possible policies that can be applied in the two cases to avoid spending unnecessary administrative time while still securing reasonable delivery precision from suppliers.

2 To improve the efficiency of the materials procurement process it is possible to generate and send replenishment orders for standard items automatically. Why is it reasonable to primarily use this approach for low value items? What other prerequisites should be considered to use this approach successfully?

3 When applying vendor-managed inventory, customers and suppliers often agree upon certain maximum and minimum levels within which stock on hand is allowed. Why is the size of the maximum level of importance for the customer and the supplier respectively? Is there any difference depending on whether the customer or supplier owns the inventory?

Further reading

De Toni, A. and Zamolo, E. (2005) "From a traditional replenishment system to vendor managed inventory", *International Journal of Production Economics*, Vol. 96, No. 1, pp. 63–79.

Dobler, D., Burt, D. and Lee, L. (1990) *Purchasing and materials management – text and cases*. McGraw-Hill, New York (NY).

Elvander, M., Sarpola, S. and Mattsson, S.-A. (2007) "Framework for characterizing the design of VMI systems", *International Journal of Physical Distribution & Logistics Management*, Vol. 37, No. 10, pp. 782–798.

Gerald, A., Giunipero, L. and Sawchuck, C. (2002) *ePurchasingPlus*. JGC Enterprises, Goshen (NY).

Riggs, D. and Robbins, S. (1998) *Supply management strategies*. AMACOM, New York (NY).

Schorr, J. (1998) *Purchasing in the 21st century*. John Wiley & Sons, New York (NY).

Van Weele, A. (2005) *Purchasing and supply chain management*. Thomson Learning, London.

CASE STUDY 17.1: VENDOR-MANAGED INVENTORY (VMI) AT HEAT EXCHANGE PRODUCTS LTD

Heat Exchange Products is a globally operating OEM manufacturer of heat exchange products with a turnover of about €1.1 million. For quite a few purchased items the procurement process has been replaced by letting the corresponding suppliers be responsible for managing the inventory at the company, i.e. the company is using the concept vendor-managed inventory for these items. One of the suppliers with which Heat Exchange Products has a VMI relationship is a rather small manufacturer of turned metal products. Sales to Heat Exchange Products represents some 20 per cent of its total turnover. The initiative to introduce a VMI relationship was the customer's, i.e. Heat Exchange Products'.

Production and material flow

Heat Exchange Products is an assemble-to-order type of company. As soon as a customer order arrives, a corresponding manufacturing order is released and purchased components allocated. All these purchased components are stocked items. The release of manufacturing orders is on average made a couple of weeks before the order is planned to start but in some cases much less in advance.

The supplier applies a combination of a make-to-stock and make-to-order type of manufacturing. All products are customer specific. The reason for still making some of the products to stock is that the requested lead times from customers are shorter than the company's own accumulated manufacturing lead time. Fifteen different products are supplied to Heat Exchange Products within the frame of the VMI relationship. All these products are only stored at the customer.

System used and information exchanged

The system used for making it possible for the supplier to manage Heat Exchange Products' component inventory was designed solely by the customer and is owned, operated and maintained by the customer. It is a web-based solution where the information needed is continuously retrieved from the customer's ERP system and presented for the supplier in real time through a web-interface. See the figure.

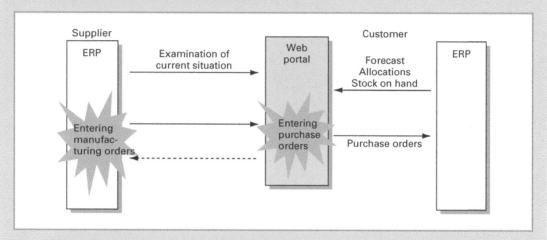

Through the web, the supplier gets information regarding stock on hand, allocations to manufacturing orders made by the customer company and forecasted future usage. It is also possible for the supplier to get access to the drawings regarding for products.

Working procedures at the supplying company

Based on the information received and rules concerning maximum and minimum allowed available inventory in stock that has been agreed between the companies, a production planner at the supplying company determines how much to manufacture and when to deliver. In this application available inventory means stock on hand less the sum of allocations within the manufacturing lead time for the supplier. Based on these decisions, the planner enters purchase orders into the web system. These purchase orders are then automatically downloaded to the customer's ERP system. After having entered a purchase order, he enters a corresponding manufacturing order in his own ERP system.

The planner is allowed to later change manufacturing order quantities and delivery dates if needed due to capacity reasons provided that the available inventory remains within the maximum and minimum limits. He can, for example, increase the order quantity if he has free capacity available and by such type of action smooth the load in his manufacturing resources. The production planner handles one customer at a time and the manufacturing capacity is then allocated on a first come, first served basis. Changing and prioritising among manufacturing orders are accordingly not possible until all VMI customers have been considered.

When a manufacturing order is ready to ship to the customer, the planner makes a last update of the purchase order and enters the final delivery date and order quantity in the web system. The status of the purchase order in the customer's ERP system is updated so that the material planners at Heat Exchange Products know that a shipment is on the way.

Ownership and payments

In VMI business relations the customer or the supplier owns the inventory. If the customer owns the inventory, the supplier invoices him when he delivers, while if the supplier owns the inventory, the supplier invoices when the customer withdraw quantities from stock. Heat Exchange Products has chosen a third solution. They own the inventory of delivered components and the supplier invoices when he delivers. However, Heat Exchange Products does not pay the supplier until the time according to agreed terms of payment has passed after quantities have been withdrawn from stock.

Discussion questions

1 How do chosen maximum and minimum limits for allowed availability in stock affect the supplier's possibilities to plan their capacity utilisation and abilities to prioritise between different customers?

2 Which are the consequences of the principles of ownership and payments applied by Heat Exchange Products?

3 What are the benefits and drawbacks respectively for the supplier in this type of VMI relationship?

CHAPTER

18

Inventory Accounting

In this book **material planning** is described as the planning function that ensures efficient flows of material from a supplier to a company, through production and out from the company to customers. This is achieved by identifying imbalances between requirements and supply of materials, and if imbalances occur, by initiating new **purchase orders** or **manufacturing orders**. Supply of materials is related to quantities in stock at the company, among other things. To take these stocks into consideration their quantities must be known, and for this reason there must be some form of **inventory accounting** at the company, that is, a system that keeps track of **stock on hand**.

There are two different methods of recognising current stock on hand in a company: transactional inventory accounting and periodic inventory accounting. In transactional inventory accounting, also called perpetual accounting, stock on hand is updated as each transaction affecting stock takes place. Inbound deliveries are added and **withdrawals** are subtracted. Periodic inventory accounting means that the balance is only updated when there is a need or a plan to physically check how large remaining stocks are, and possibly to enter this quantity in the inventory accounting system. It may be considered as a form of **physical inventory counting**.

Periodic inventory accounting is generally used for stock valuation and as a part of the financial accounting in a company. There is a formal auditing requirement that stock must be physically counted at least once a year. For the purposes of material planning, transactional inventory accounting is by far the most common approach in industrial companies. Periodic inventory accounting is used when it is less expensive to check stock as required or planned than reporting every stock withdrawal and delivery. It is also used when it is difficult to correctly count or measure withdrawals, such as when tapping off liquid from a tank in the chemicals/technical industry. Only transactional inventory accounting is treated in this chapter.

As described in Chapters 11 and 15, for some material planning methods it is not necessary to know current stock on hand. This is the case, for example, in the variant of the re-order point system known as the two-bin system, and when using **kanban** cards. In these cases no inventory accounting is required for material planning purposes.

In this chapter different aspects of inventory accounting for inbound delivery transactions and withdrawal transactions will be described. Physical inventory counting is also presented as a method of correcting inaccuracies in stock on hand that have occurred. Inventory accounting will only be discussed with respect to quantities in stock. Different methods of pricing items in stock and valuation of inventories are not included. For studies of these issues refer to textbooks on product calculation and financial accounting.

LEARNING OBJECTIVES

After studying this chapter, you should be able to:

- Describe different considerations and strategies of inventory accounting for inbound deliveries
- Describe different considerations and strategies of inventory accounting for stock withdrawals
- Explain the process and strategies of physical inventory counting

---

18.1 Inbound Deliveries

From the viewpoint of inventory accounting, inbound deliveries mean that the quantity of stock increases. These deliveries may be of different types. There may be a delivery from an external supplier that stems from a purchase order, or a delivery from an internal workshop that stems from a manufacturing order, depending on whether the item is purchased or internally manufactured. It may also consist of a transfer from another warehouse, or a transfer from one storage point to another in the same warehouse. Finally, it may be a return delivery from an external customer that has made a complaint or from an internal workshop as a result of excessive quantities withdrawn.

Irrespective of the type of delivery, it is essential that the quantity delivered goes through quality inspection before it is placed in stock. Quality inspection is carried out in conjunction with goods reception, or in direct connection with production in an internal workshop or a supplier's workshop. The purpose of this inspection is to ensure as far as possible that the quantities in stock are available for production or delivery to customers. In other cases, the available-to-promise analyses as described in Chapter 8 will give incorrect information on the capability to deliver. Other problems will also arise as a consequence of shortage situations.

When deliveries take place from purchase orders or manufacturing orders, the quantity delivered must be matched with the corresponding order. In the case of purchase orders, the quantity delivered may be larger, equal to or smaller than the quantity in the order. If the quantity delivered is larger than the quantity ordered and the surplus delivery can be accepted, stock will be increased by the delivered quantity in the same way as if the delivered quantity was the same as the order quantity. In other cases, the surplus quantity will be returned and stock on hand will be updated by the order quantity. If, on the other hand, the quantity delivered is less than the order quantity and short delivery is accepted, stock on hand will be increased by the delivered quantity and the remaining quantity to be delivered is set at 0. If short delivery is not accepted, the delivered quantity on the purchase order is changed and stock on hand is updated by the delivered quantity. The remaining quantity on the purchase order is expected to be delivered on a later occasion and the purchase order is retained until this takes place. It is vital

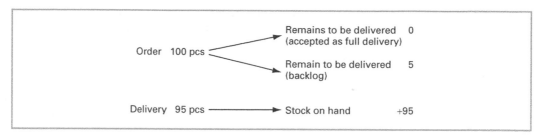

FIGURE 18.1 Example of updating stock on hand and purchase order for short delivery.

that updating of stock on hand and remaining quantities to be delivered from the purchase order take place at the same time, otherwise availability calculations will be incorrect. Figure 18.1 illustrates an example of updating stock on hand and purchase orders in the case of a short delivery.

For manufacturing orders, there are corresponding conditions. In general, however, surplus or short deliveries are not always accepted. If split deliveries take place, stock on hand and manufacturing orders are updated in the same way as described above for purchase orders. Final updating of the quantity of the manufacturing order takes place when the order is closed. For the same reasons as for purchase orders, it is essential that updating of stock on hand and manufacturing orders takes place simultaneously. As will be made clear in Section 18.2, reporting of manufacturing orders often includes updating stock on hand for items incorporated in the delivered product as well as the removal of corresponding reservations.

Other information that is required in addition to quantities when reporting deliveries will depend on the stock location system used. There are two main systems: **fixed position stock location system** and **random stock location system**. The fixed position system means that every item number has its own place reserved in stock. Quantities delivered are always placed in this location and there is no need to enter stock location addresses. The stock location in question is fixed information in the inventory files in the ERP system.

The random stock location system means that quantities delivered are placed in the first available free location in stock. The address of the stock location where the delivery has been put away must then be entered in the inventory files. In general the delivered quantity is not put in the location where remaining quantities of the same item are located. Using this system, the same item may be located in many different places in stock. The quantity delivered may also need to be divided between several stock locations for reasons of available space.

In addition to these specific stock location systems, there are different forms of mixed systems. In one such commonly occurring mixed system, picking stock, or the stock from which withdrawals are normally made, has fixed stock locations while larger quantities are stored in a random location system, where inbound deliveries are also placed. When the picking stock is exhausted, it is replenished from the random location stock. Irrespective of which mixed system is used, the stock address must be stated in reports. Random location systems and mixed systems require that the inventory accounting system is able to keep track of stock on hand at several different stock addresses.

The requirement to report on delivery also depends on whether there is a requirement for batch reporting, meaning that every batch delivered is kept separate from other delivered batches. This requirement is relevant when a company must be able to trace from which batch and supplier certain materials originate, such as when products delivered must be recalled due to material or manufacturing defects. **Batch accounting** is also required if batches delivered have different quality levels, and may be classified as first or second quality goods. If batch accounting is required, delivery reports must be supplemented with order numbers, batch numbers or other identities which can distinguish between batches. The inventory accounting system must be able to keep track of current stock on hand per batch, as in the case of a random location system.

When transfers are made between warehouses and stock locations in the same warehouse, stock on hand is increased at the receiving stock location at the same time as stock on hand is decreased at the stock location it is delivered from. The total stock on hand is not changed.

In the case of returns, stock on hand is increased by the quantity returned.

18.2 Withdrawals

Withdrawals from stock are the opposite of inbound deliveries, and stock on hand decreases after withdrawals. Normal withdrawals fall into four main categories. A withdrawal may take place as a result of a customer order or a manufacturing order. Withdrawals can also take place in conjunction with transfers of quantities from one warehouse to another, and from one stock location to another in the same warehouse. Finally, withdrawals may be of an indirect type, as in withdrawals for unspecified usage.

Withdrawals for customer orders are reported either by order line or for the entire order at once. Updating of stock on hand must take place at the same time as the removal of corresponding allocations for the order so that available-to-promise calculations will be correct. If the quantities picked are greater than those stated on the order, the entire quantity allocated is removed. If the quantities picked are smaller, the entire quantity allocated is removed if each order line, despite this, is considered as fully delivered. Otherwise the allocated quantity is decreased by the quantity picked and delivered.

There are three main options for reporting withdrawals for manufacturing orders. One option is that every withdrawal of material made for a manufacturing order is reported and stock on hand is updated as for withdrawals for customer orders. Withdrawals are often made using a pick list printed out from the ERP system. This pick list contains information on which items are to be picked for the manufacturing order, including information on the quantities of each item to be picked. When the quantities picked are reported in the system, stock on hand is decreased for each item and the corresponding allocations are removed. Any differences between the originally allocated quantity and the quantity picked are eliminated, either when picking is reported or when the manufacturing order is closed after reporting the quantity manufactured as finished. In terms of accounting, this option means that the inventory account for the items picked will be decreased by the value of the quantities withdrawn and that a work-in-process account for the manufacturing order will be increased by the same amount.

Another option for reporting withdrawals updates stock automatically when a manufacturing order is reported as finished. This means that stock on hand is decreased by allocated quantities in proportion to the quantities reported as finished. If a manufacturing order is for 100 four-legged chairs, then 400 legs have been allocated for the order. When 25 chairs are reported as finished, the stock on hand of legs will be decreased by 100. The approach is illustrated in Figure 18.2. At the same time as the stock account is updated, corresponding allocations are removed. When the order is reported as finished, the remaining stock on hand updating and elimination of allocations will be carried out. Any remaining allocations due to the number of finished products being fewer than the quantity stated on the manufacturing order will also be removed. **Backflushing** is a commonly used term for this kind of automatic stock on hand updating.

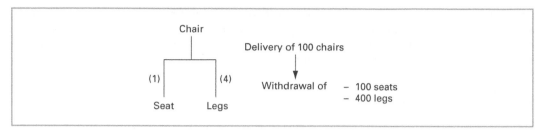

FIGURE 18.2 Backflushing by means of allocations.

Backflushing in conjunction with reporting of manufacturing orders means that stock on hand of incorporated items is not updated when real withdrawals are made by picking. Consequently, the physical stock on hand will be lower than the administrative stock on hand, as in the inventory accounting system. The available stock on hand will remain correct, however, since allocations will remain until stock on hand is updated. This approach also means that the work-in-process account will not contain tied-up capital from material, which is the equivalent of the value of items incorporated, but will only consist of direct salaries and manufacturing overheads for labour time reported. If the quantities actually withdrawn deviate from what has been reserved, such as in the case of scrapping, a manual withdrawal report and updating of stock on hand must be carried out in addition to backflushing.

When kanban methods for material planning are used, there are no manufacturing orders. Backflushing cannot be used in the same way as in the case above. The information in the bill-of-material file is used instead to automatically calculate stock withdrawals and to update stock on hand. When the quantities of the item on the kanban card are reported, the quantities of material used are calculated in the ERP system automatically with the aid of explosion quantities in the bill-of-materials file. Stock on hand is updated using these quantities. The difference compared with backflushing based on allocations for manufacturing orders is that there are no allocations. What is available in stock according to the system may therefore be more than what is the case in reality.

Backflushing as an option for reporting withdrawals is useful above all in cases where lead times in production are relatively short so that the delay between real withdrawals from stock and stock updating is not too long. This option also assumes high bills-of-material quality, otherwise automatic deduction from stock will be incorrect since it is based on explosion quantities in these bills of material, either directly or indirectly through allocations being automatically created from the bills-of-material file on order release. In order for this to work satisfactorily the amount of scrapping must be small, otherwise there will be a need for extensive supplementary manual reporting of withdrawals that must take place to cover extra requirements as a result of scrapping.

One variant of the above procedure is to report the start of a manufacturing order and to carry out backflushing on the basis of this report. This approach means that the work in process account will also contain tied-up capital from withdrawals of material and will therefore more closely represent the real value of ongoing manufacturing orders. In addition, when used with the kanban method it will mean that information on incorporated materials available in stock will be more correct than if reporting is only carried out on inbound delivery of finished products. It will, though, require one extra reporting occasion. One further variant exists for cases in which manufacturing orders and allocations are used. This lies somewhere between start and final

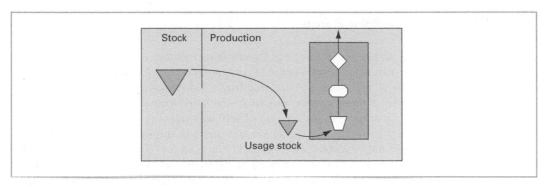

FIGURE 18.3 Withdrawals to usage stocks in production.

reporting, and means that quantity reporting takes place at each operation, or at selected operations during the process of manufacture. This form of automatic updating is called **synchroflushing**.

With the exception of kanban, the options above will mean that withdrawals are reported for manufacturing orders. Thus, the value of materials withdrawn will be transferred from an inventory account to a work-in-process account, either at the start of a manufacturing order or on completion of a manufacturing order. The value will subsequently be transferred to an inventory account for the manufactured item on reporting of inbound delivery. Provided that scrapping or other forms of usage beyond that envisaged do not occur, no materials costs will arise during the manufacturing process from withdrawal to inbound delivery. Costs for usage of materials do not arise until the product has been sold and delivered to a customer. The third main option with regard to reporting withdrawals for manufacturing orders is when withdrawals are made for usage in general, without this being reported to or charged to any particular manufacturing order. Withdrawals from stock are reported and accounted for in the overhead account when the stock on hand is updated. Items handled in this way are not generally allocated.

When this third option is utilised, quantities are withdrawn from stock to cover current needs during a certain period, such as one week or one month, irrespective of which manufacturing order is in question. The quantities are "stocked" in the factory close to production as a general usage stock, often called floor stock, weekly usage stock or similar. The principle is illustrated in Figure 18.3. Withdrawals from usage stocks are not reported, the stock is not included in inventory accounting, and it does not represent any book value. This option is primarily used for low value items which would have to be picked and reported frequently if they were handled in accordance with the two previous options.

The above three options can also be used in combination with each other. For example, in the case of high value items every withdrawal for manufacturing orders may be picked and reported, whereas low value items may be removed from usage stock without being associated with any specific manufacturing order. There are occasions when usage stocks are accounted for and withdrawals from real stock to usage stock constitute a stock transfer. Stock balance in usage stock is updated using backflushing of manufacturing orders or kanban cards. In this way it is easier to maintain high accuracy of stock on hand since all real withdrawals are reported. At the same time, reporting costs for many small withdrawals for each individual manufacturing order are kept low through backflushing in the usage stock.

18.3 Physical Inventory Count

One condition for the efficient functioning of the planning methods presented in Chapters 11 and 15 is that stock on hand as shown in the ERP system is correct, i.e. that reported stock on hand is equal to the physical, real stock on hand. Kanban and the two-bin system are exceptions in this respect since neither of these systems requires access to information on current stock on hand to operate. Other planning methods are completely dependent on information on current stock on hand to function correctly. Generally we can say that the more advanced planning methods are, the greater is the importance of the quality of stock on hand figures. For example, material requirements planning places greater demands on high stock on hand quality than does the re-order point system if its relative advantages are to be exploited. A correct stock on hand is also required to achieve an accurate inventory evaluation. Inventory value is part of a company's balance sheet and any inaccuracies regarding inventory value will influence the company's financial position and its profit or loss.

For various reasons there is always a risk that accounted stock on hand is incorrect. These reasons may include incorrect quantities being reported on inbound delivery or withdrawal of material, wastage, or personnel forgetting to report stock movements of various types. Inaccuracies may also arise due to incorrect bills of material if backflushing is used for reporting withdrawals. Physical inventory counting is an administrative procedure aimed at correcting any inaccuracies that may have arisen and ensuring that the stock on hand in the ERP system files is equivalent to the real, physical stock on hand.

The process of physical inventory counting

Physical inventory counting is carried out by counting the quantity of an item in stock and correcting the accounted stock on hand so that it corresponds to this quantity. A typical physical inventory counting process may include the following steps:

1 *Printing of request for physical inventory counting.* A request for physical inventory counting is a printed list from the ERP system. The list includes all the items to be counted at the time. A general example of a request for physical inventory counting is shown in Figure 18.4.

2 *Counting real quantities in stock.* The quantities counted are noted on the request for physical inventory counting.

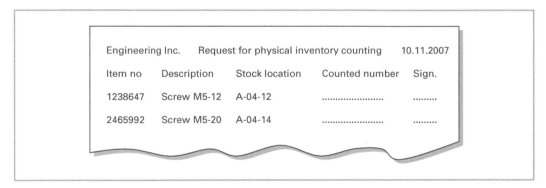

FIGURE 18.4 Illustration of a request for physical inventory counting

3 *Reporting quantities counted.* The results of the physical inventory counting are reported to the ERP system. This can take place either by reporting the counted quantities, or by reporting the difference between the counted quantity and the quantity according to the inventory accounting system. The latter is called differential physical inventory counting.

4 *Calculation of physical inventory counting difference.* The physical inventory counting difference, or the difference between the quantity of stock counted and the quantity as stated in the ERP system, is calculated in the ERP system.

5 *Printing a list of physical inventory counting differences.* A list is printed out that includes items with physical inventory counting differences and is used as a basis for approval.

6 *Approval of physical inventory counting differences.* All physical inventory counting differences that have arisen must be approved by authorised personnel since they influence a company's profit or loss. Different authorisation levels are often applied for different values of physical inventory counting differences, so that differences involving small values may be approved by the person carrying out the physical inventory counting, whereas larger values of counting differences must be approved by the financial manager or equivalent. When required and when the differences are large, it may also be appropriate to carry out a new physical inventory counting to check the result of the previous one.

7 *Updating stock on hand.* When differences in physical inventory counting have been approved, stock on hand is updated in the ERP system so that it corresponds to the quantity counted. The date of the physical inventory counting is stored in the inventory files. At the same time accounting transactions are created for financial accounting to update the inventory account and cost account.

Three different types of physical inventory counting are usually encountered in companies: periodic physical inventory counting, **cycle counting** and **direct physical inventory counting**. Each of these types is described in overall terms below.

Periodic physical inventory counting

Periodic physical inventory counting means that counting for the entire range of items is carried out at one time, or for whole groups of items at one time. This type of physical inventory counting normally takes place once a year in conjunction with the inventory valuation for accounting purposes. It may take place more often, such as once a month, if the aim is to correct stock on hand quantities for material planning reasons. Quite often production is stopped when this type of physical inventory counting is carried out and the permanent production and stores personnel take part in the counting process.

Cycle counting

Cycle counting is a method of counting stock which takes place more or less continuously during the year. Instead of counting the entire range or a large part of the range on one occasion, smaller groups of items are counted cyclically. Companies often try to adapt the number of occasions per year to the volume value of each item so that items with high volume values

Volume value class	Number of items	Counting frequency	Number of counts
A	780	12 times/year	9360
B	1920	4 times/year	7680
C	4560	1 time/year	4560
Total number of counts			21 600
Number of counts per day			96

FIGURE 18.5 Example of choice of counting frequency for different volume value classes

are counted more frequently, while items with lower volume values are counted less frequently. One example of a choice of counting frequency for items in different volume value classes is shown in Figure 18.5. As described in the figure, A items are counted 12 times a year, B items 4 times a year and C items once a year. If it is assumed that counting work can be carried out on 225 days per year, the number of counts per day will be equal to 96. Most ERP systems support the generation of daily cycle counting requests so that the desired frequency for each item is ensured.

Cycle counting is generally carried out during ongoing production by specialised counting personnel. In this way it is possible to allow more frequent counting of stock on hand. The aim of this type of physical inventory is to cost-effectively ensure high stock on hand quality for material planning rather than producing information for financial inventory valuations.

Direct physical inventory counting

Direct physical inventory counting is used to count individual items selectively when certain events occur. This type of physical inventory counting is often used as a complement to the other two types. Examples of events which may necessitate this type of direct physical inventory counting are store transactions on inbound delivery of a new order to stock, and stock transactions that have caused the stock on hand reported to be zero or negative.

Using direct physical inventory counting in conjunction with inbound deliveries or when the reporting stock on hand quantity is zero, the quantities which must be counted are minimised and counting work will therefore be less time-consuming. The risk of making errors will also be reduced. If negative stock on hand quantities occur, this is generally an indication of an error in stock reporting, and there is therefore good reason to directly check and adjust the stock on hand figures.

18.4 Summary

This chapter has described principles and strategies for inventory accounting at the company, i.e. a process that keeps track of stock on hand.

There exist two different methods of recognising current stock on hand: transactional inventory accounting and periodic inventory accounting. In transactional inventory accounting, stock on hand is updated as each transaction affecting stock takes place. Inbound deliveries are added and withdrawals are subtracted. In periodic inventory accounting the balance is only updated when there is a need to physically check how large remaining stocks are, and possibly to enter this quantity in the inventory accounting system. It is thus considered as a form of physical inventory counting. Only transactional inventory accounting was treated in this chapter. Different aspects of inventory accounting for inbound delivery transactions and withdrawal transactions were described. Physical inventory counting was also taken up as a method of correcting inaccuracies that have occurred. Inventory accounting was only discussed with respect to quantities in stock. Different methods of pricing items and valuation of inventories were not included.

🔑 Key concepts

Backflushing 395	Material planning 392
Batch accounting 394	Periodic physical inventory counting 399
Cycle counting 399	Physical inventory counting 392
Direct physical inventory counting 399	Purchase order 392
Fixed position stock location system 394	Random stock location system 394
Inventory accounting 392	Stock on hand 392
Kanban 392	Synchroflushing 397
Manufacturing order 392	Withdrawal 392

Discussion tasks

1 If a delivery from a purchase order takes place and the quantity on the order is put away in stock, the quantity on the order must be reduced at the same time and with the same quantity as the quantity is added to stock on hand. What will the consequences be if that is not the case?

2 Backflushing may be used as a way to reduce the administrative work involved when reporting withdrawals manually. If this method is used, what will the consequences for work in process be if the backflushing takes place when the order starts and when the order is finished respectively?

3 When companies apply cycle counting they often count items with high volume values more frequently than items with low volume values. What is the argument for doing so?

Further reading

Ballou, R. (2004) *Business logistics/supply chain management.* Prentice-Hall, Upper Saddle River (NJ).

Bernard, P. (1999) *Integrated inventory management.* Oliver Wight Publications, Essex Junction (VT).

Brooks, R. and Wilson, L. (1993) *Inventory record accuracy.* Oliver Wight Publications, Essex Junction (VT).

Frazelle, E. (2001) *World class warehousing and materials management.* Logistics resource international, Atlanta (GA).

Lambert, D., Stock, J. and Ellram, L. (1998) *Fundamentals of logistics management.* McGraw-Hill, Irwin/McGraw-Hill, New York (NY).

Silver, E., Pyke, D. and Peterson, R. (1998) *Inventory management and production planning and scheduling.* John Wiley & Sons, New York (NY).

Tompkins, J. and Smith, J. (1988) *The warehouse management handbook.* McGraw-Hill, New York (NY).

Vollmann, T., Berry, W., Whybark, C. and Jacobs, R. (2005) *Manufacturing planning and control for supply chain management.* Irwin/McGraw-Hill, New York (NY).

Appendix A: Solutions to Problems

Problem 4.1

a) The inventory turnover rate is equal to the annual demand divided by the average tied up capital, i.e. $520 / 80 = 6.5$ times per year.

b) The inventory days is calculated as the number of days per year divided by the inventory turnover rate, i.e. $250 / 6.5 = 38.5$ days.

c) The amount of tied up capital at an inventory turnover rate (ITR) of 10 is $520 / 10 =$ €52 million. The tied up capital consequently has to be decreased by $80 - 52 =$ €28 million in order to achieve an ITR of 10.

Problem 4.2

a) Items in part A of the graph have run out times longer than one year. The majority of these items are or may become obsolete.

b) Items in part B of the graph have negative run out times, i.e. the number of back-logged items is larger than the quantity in stock.

c) The ideal graph should start at the origin if no shortages and backlogs should occur. It should end in the upper right corner of the graph. The run-out time in this part of the graph should correspond to half the economic order quantity plus safety stock for the item with longest run-out time.

Problem 4.3

a) The tied up capital before outsourcing was $80 + 70 + 60 + 40 =$ €250 million. This corresponds to an inventory turnover rate of $800 / 250 = 3.2$.

b) The total tied up capital at the main plant after the outsourcing decision is $80 + 60 =$ €140 million, corresponding to an ITR of $800 / 140 = 5.7$. The new plant gets a tied up capital of $70 + 40 =$ €110 million, corresponding to an ITR of $450 / 110 = 4.1$. Consequently, both the main plant and the new plant get higher ITRs than that of the main plant before the outsourcing.

Problem 4.4

a) Only order lines 240-1, 246-3, 249-1, 250-1 and 254-1 have been delivered on time according to the promised delivery date. This corresponds to an order line fill rate of $5 / 11 = 45\%$. If split deliveries are accepted but requiring deliveries on promised delivery date, order rows 240-2, 246-2, 250-2 and 254-2 are also defined as on time. The order line fill rate would then be $9 / 11 = 82\%$.

b) Only order 249 is completely delivered on time, i.e. the order fill rate is 1 / 5 = 20%. If split orders on order row level are accepted, then orders 250 and 254 are also defined as being on time. The order fill rate is then 3 / 5 = 60%.

Problem 5.1

a) Indented bill of material with quantities per item for table EUF:

Table EUF

Level		Item	Quantity per item
1		Table top	1
	2	Chipboard	0.2
1		Leg	4
	2	Wood pole	0.25
1		Console	4
1		Thumbscrew	16

b) If all manufacturing and assembling is conducted in the same work centre, and table tops and legs do not have to be made in order to cover demand from other products, then it is not necessary to consider them in the planning as individual items. Instead, they should be possible to make in quantities equivalent to the demand of a batch of 100 tables. The indented bill of material would then get the following shape:

Table EUF

Level	Item	Quantities per item
1	Chipboard	0.2
1	Wood pole	0.25
1	Console	4
1	Thumbscrew	16

Problem 5.2

a) The bill of material for the mouse trap has the following shape:

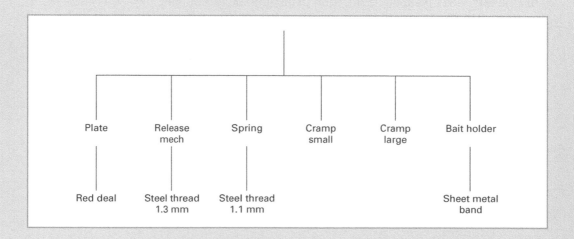

b) Corresponding indented bill of material gets the following shape:

Mouse trap standard

Level	Item	Quantity per item
1	Plate	1
2	Red deal 2m	0.05
1	Release mech.	1
2	Steel thread 1.3 mm	0.0014
1	Spring	1
2	Steel thread 1.1 mm	0.005
1	Cramp, small	3
1	Cramp, large	1
1	Bait holder	1
2	Sheet metal band	0.001

Problem 5.3

a) A single-level where-used analysis for item A:

Level	Item	Quantity per item
2	I	4
0	P1	4
0	P2	4

b) An indented where-used analysis for item E:

Level	Item	Quantity per item
1	B	4
0	P1	4
1	N	2
0	P2	4

c) Each product P1 requires $1 \cdot 4 \cdot 2 = 8$ pieces and each product P2 requires $2 \cdot 2 \cdot 2 = 8$ pieces of item K. Making 100 pieces of K1 and 50 pieces of P2 consequently requires $100 \cdot 8 + 50 \cdot 8 = 1200$ pieces of item K.

Problem 5.4

a) Because the stock balance is not enough for the planned manufacturing batch starting 20 April, and traceability between components and manufacturing batches is required, the projected available stock balance on 25 April will be 150 pieces.

b) Because component A1 is used as a spare part, the stock can partly supply the spare part sales. If 150 pieces is considered too large for the expected spare part demand, then some of the extra quantity should be scrapped.

c) If instead A1 and A2 are fully exchangeable (i.e. the product function and reliability are not affected when switching from A1 to A2), then of the batch of 500 pieces, 150 can be made with component A1 and 350 with component A2. The stock will then be zero on 25 April.

Problem 5.5

a) Lead-time analysis for the product:

Level	Item	Lead time	Cumulative lead time
1	A	4 weeks	4 weeks
1	B	1 week	7 weeks
2	E	2 weeks	6 weeks
3	G	1 week	1 week
3	H	4 weeks	4 weeks
2	F	3 weeks	3 weeks
1	C	8 weeks	8 weeks
1	D	2 weeks	11 weeks
2	I	1 week	5 weeks
3	L	4 weeks	4 weeks
2	J	2 weeks	2 weeks
2	K	5 weeks	9 weeks
3	M	4 weeks	4 weeks

b) The cumulative lead time for the product, including 2 weeks of assembling, is 13 weeks.

c) To decrease the cumulative lead time for the product, the lead times for items P, D, K and M should first be decreased because they are along the product's critical path.

Problem 6.1

Costs for negotiation and supplier selection, freight transport and quality control (because all items are controlled) are not dependent on the number of orders. The total quantity (number of orders) dependent ordering costs are therefore €78 000. With 3000 purchasing orders a year, the ordering cost per item will be 78 000 / 3000 = €26.

Problem 6.2

Storage costs, depreciation costs and materials handling costs should not be included in the inventory carrying factor calculation because the warehouse is not fully utilised and the decision concerns having more or less in stock. These costs occur, no matter size of the stock. The same is true for general administrative and data costs. The costs to include in the incremental inventory carrying costs and calculation of the inventory carrying factor are, thus, the insurance costs of €22 000, costs for wastage of €13 000, inventory counting costs of €45 000, i.e. a total of €80 000 for purchasing and manufacturing items together. For purchase items obsolescence costs of €52 500 are added.

The common inventory carrying costs in relation to the amount of tied up capital are (22 000 + 13 000 + 45 000) / 1 500 000 = 5.3% and the obsolescence costs for purchase items in relation to the tied up capital are 52 500 / 750 000 = 7%. The total inventory carrying factor is then 15 + 5.3 = 20.3% for manufacturing items and 15 + 5.3 + 7 = 27.3% for purchasing items.

Problem 6.3

There are 11 items in the assortment and about 20 per cent of them should be A items. Two to three of the items should then be A items. The table shows that three items have volume values which are considerably higher than those of the other eight items. Items 104, 107 and 111 should therefore be A items.

Problem 6.4

a) Sorting based on turnover (volume value) gives the following ascending order of the items: A, E, D, F, B, I, M, N, C, K, O, G, H, J, L. Highest service level should be given to items with lowest volume value in order to minimise the tied up capital. An example of an ABC-based service level strategy would be to give L, J, H low service level, G, O, K, C, N, M medium service level and I, B, F, D, E, A high service level.

b) Sorting based on number of picking events gives the following ascending order: O, D, G, M, K, H, N, I, J, F, A, L, C, E. The items should be located in the warehouse so that items with most picking events are given the shortest transport/walking distance and the most rational picking. An example of an ABC-based location strategy could be to place O, D, G in the closest storage zone, M, K, H, N in a zone requiring somewhat more transportation, and I, J, F, A, L, C, E in the storage zone resulting in longest transport distance.

c) Sorting based on annual contribution margin gives the following ascending order: L, J, H, G, O, C, K, N, I, M, B, F, D, E, A. To lose as little contribution margin as possible when shortage, the highest service level should be given the items with highest contribution margin. An example of an ABC-based service level strategy would be to give highest service level to L, J, H, medium service level to G, O, C, K, N, I and the lowest service level to M, B, F, D, E, A.

Problem 7.1

a) BR203: few periods (e.g. 6 months)
DB34: many periods (e.g. 12 months)

b) BR203: 111 units (110.8)
DB34: 94 units (93.8)

Problem 7.2

a) STSP01 has constant demand and random variations. Chose low alpha (e.g. 0.1)
STVP07 has clear trend. Choose larger alpha (e.g. 0.3)

b) STSP01: 190 units (189.7) with starting forecast of 187.8 (using 19 periods)
STSP07: 207 units (206.8) with starting forecast of 198.5 (using 6 periods)

Problem 7.3

Using the formula "Base forecast + (n − 1) · Trend" results in the following forecasts (corrected for a trend of 8 units per month):

Month	Jan	Feb	Mar	Apr	May	Jun
Forecast	150	158	166	174	182	190

Problem 7.4

Mean demand per month for product GMT65 is 187 units. Seasonal index per month:

Month	Seasonal index	Month	Seasonal index
1	0.98	7	0.96
2	1.15	8	0.84
3	1.26	9	0.80
4	1.17	10	0.81
5	1.16	11	0.88
6	0.98	12	1.03

For example, seasonal index for month 1 is 183 / 187 = 0.98.

Problem 7.5

a) Forecasts with alpha = 0.3:

Month	Table 1 Demand	Table 1 Forecast	Table 2 Demand	Table 2 Forecast	Table 3 Demand	Table 3 Forecast
1	160		110		450	
2	195	160.0	120	110.0	375	450.0
3	170	170.5	150	113.0	440	427.5
4	210	170.4	140	124.1	380	431.3
5	185	182.2	155	128.9	365	415.9
6	155	183.1	130	136.7	410	400.6
7	167	174.7	140	134.7	440	403.4
8	195	172.4	155	136.3	450	414.4
9	210	179.1	110	141.9	395	425.1
10	220	188.4	120	132.3	385	416.1
11	195	197.9	160	128.6	350	406.7
12	240	197.0	150	138.0	420	389.7
13		209.9		141.6		398.8

b) Forecast for the product group:

Month	Demand	Forecast
1	720	
2	690	720.0
3	760	711.0
4	730	725.7
5	705	727.0
6	695	720.4
7	747	712.8
8	800	723.0
9	715	746.1
10	725	736.8
11	705	733.3
12	810	724.8
13		750.3

c) Forecast for quarters 2 to 5:

	Table 1		Table 2		Table 3	
Quarter	Demand	Forecast	Demand	Forecast	Demand	Forecast
1	525		380		1265	
2	550	525.0	425	380.0	1155	1265.0
3	572	532.5	405	393.5	1285	1232.0
4	655	544.4	430	397.0	1155	1247.9
5		577.5		406.9		1220.0

d) Forecasts for product groups during quarters 2 to 5:

Quarter	Demand	Forecast
1	2170	
2	2130	2170.0
3	2262	2158.0
4	2240	2189.2
5		2204.4

e) MAD for the respective forecast is estimated as the sum of absolute deviations between actual and forecasted demand for the respective month, divided by the number of months (here 11 months). Mean absolute percentage error (MAPE) is determined as MAD divided by average actual demand.

	Table 1			Table 2			Table 3			Product group		
Month	Demand	Forecast	Abs. Dev.	Demand	Forecast	Abs. Dev.	Demand	Forecast	Abs. Dev.	Demand	Forecast	Abs. Dev.
1	160											
2	195	160.0	35.0	120	110.0	10.0	375	450.0	75.0	690	720.0	30.0
3	170	170.5	0.5	150	113.0	37.0	440	427.5	12.5	760	711.0	49.0
4	210	170.4	39.7	140	124.1	15.9	380	431.3	51.3	730	725.7	4.3
5	185	182.2	2.8	155	128.9	26.1	365	415.9	50.9	705	727.0	22.0
6	155	183.1	28.1	130	136.7	6.7	410	400.6	9.4	695	720.4	25.4
7	167	174.7	7.7	140	134.7	5.3	440	403.4	36.6	747	712.8	34.2
8	195	172.4	22.6	155	136.3	18.7	450	414.4	35.6	800	723.0	77.0
9	210	179.1	30.9	110	141.9	31.9	395	425.1	30.1	715	746.1	31.1
10	220	188.4	31.6	120	132.3	12.3	385	416.1	31.1	725	736.8	11.8
11	195	197.9	2.9	160	128.6	31.4	350	406.7	56.7	705	733.3	28.3
12	240	197.0	43.0	150	138.0	12.0	420	389.7	30.3	810	724.8	85.2
		MAD: 22.2			MAD: 18.8			MAD: 38.1			MAD: 36.2	
		MAPE: 11.6%			MAPE: 14.8%			MAPE: 10.4%			MAPE: 5.4%	

	Table 1			Table 2			Table 3			Product group		
Quarter	Demand	Forecast	Abs. Dev.	Demand	Forecast	Abs. Dev.	Demand	Forecast	Abs. Dev.	Demand	Forecast	Abs. Dev.
1	525			380			1265			2170		
2	550	525.0	25.0	425	380.0	45.0	1155	1265.0	110.0	2130	2170.0	40.0
3	572	532.5	39.5	405	393.5	11.5	1285	1232.0	53.0	2262	2158.0	104.0
4	655	544.4	110.7	430	397.0	33.1	1155	1247.9	92.9	2240	2189.2	50.8
		MAD: 58.4			MAD: 29.9			MAD: 85.3			MAD: 64.9	
		MAPE: 9.9%			MAPE: 7.1%			MAPE: 7.1%			MAPE: 2.9%	

The relative forecast error is lower for product groups compared to products and for quarterly forecasts compared to monthly.

f) The forecast accuracy increases when the period length is increased. When grouping products into product groups a similar smoothing effect is achieved as when using longer planning periods.

Problem 7.6

a) Forecast for period 4 calculated as the mean during periods 1 to 3, i.e. $(143 + 142 + 153) / 3 = 146$. Forecast for period 5 as mean during periods 2 to 4, i.e. $(142 + 153 + 157) / 3 = 151$. And so on. Forecasts for periods 5 to 25: 151, 150, 151, 159, 172, 171, 170, 171, 178, 181, 187, 194, 201, 211, 214, 219, 218, 229, 234, 241 and 254.

b) Forecast for period 13 calculated as the mean of the demand during the previous 12 periods, i.e. $(143 + 142 + 153 + \ldots + 182 + 171) / 12 = 161$. And so on. Forecasts for periods 13 to 25: 161, 165, 170, 173, 178, 185, 189, 193, 197, 204, 209, 214 and 223.

c) If assuming that the forecast for period 1 is the same as the actual demand for period 1, i.e. 143 units, then the forecast for period 2 is $0.1 \cdot 143 + (1 - 0.1) \cdot 143 = 143$. The forecast for period 3 is $0.1 \cdot 142 + (1 - 0.9) \cdot 143 = 143$ and the forecast for period 4 is $0.1 \cdot 153 + (1 - 0.9) \cdot 143 = 144$. And so on. Forecasts for periods 2 to 25: 143, 143, 144, 145, 145, 146, 149, 152, 152, 155, 158, 159, 162, 166, 169, 173, 179, 181, 185, 190, 194, 198, 204 and 210.

d) If assuming that the forecast for period 1 is the same as the actual demand for period 1, i.e. 143 units, then the forecast for period 2 is $0.4 \cdot 143 + (1 - 0.4) \cdot 143 = 143$. The forecast for period 3 is $0.4 \cdot 142 + (1 - 0.4) \cdot 143 = 143$ and the forecast for period 4 is $0.4 \cdot 153 + (1 - 0.4) \cdot 143 = 147$. And so on. Forecasts for periods 2 to 25: 143, 143, 147, 151, 146, 150, 162, 169, 162, 169, 174, 173, 180, 188, 190, 198, 211, 207, 214, 220, 225, 231, 239 and 251.

e) If assuming that the forecast for period 1 is the same as the actual demand for period 1, i.e. 143 units, then the forecast for period 2 is $0.8 \cdot 143 + (1 - 0.8) \cdot 143 = 143$. The forecast for period 3 is $0.8 \cdot 142 + (1 - 0.8) \cdot 143 = 142$ and the forecast for period 4 is $0.8 \cdot 153 + (1 - 0.8) \cdot 142 = 151$. And so on. Forecasts for periods 2 to 25: 143, 142, 151, 156, 142, 153, 175, 179, 157, 175, 181, 173, 187, 198, 193, 207, 226, 205, 221, 228, 231, 238, 248 and 266.

f) Few periods and high alpha values result in forecasts that are almost equal to the last period's actual demand while many periods and low alpha values consider demand further back in the history. The number of periods and alpha value consequently affect the forecast stability against random variations and the adjustment to systematic demand variations.

g) If assuming that the forecast for period 1 was the same as the actual demand during period 1, i.e.143 units, and no trend existed during period 1, i.e. the trend was 0 units, then the forecast for period 2 would be:

Basic forecast: $BF(t + 1) = \alpha \cdot D(t) + (1 - \alpha) \cdot (F(t) + T(t))$
Trend: $T(t + 1) = \beta \cdot (BF(t + 1) - F(t)) + (1 - \beta) \cdot T(t)$
Forecast with trend (F): Basic forecast + Trend

The forecast for next period is calculated in two steps. First, a base forecast without trend correction is calculated. The factor $(BF(t) + T(t))$ is the trend correcting forecast during the last period and is equivalent with $F(t)$ in the formula for exponential smoothing without trend correction.

Using the above formula for calculating the forecast with trend for period 2:

Basic forecast: $BF(2) = 0.1 \cdot 143 + (1 - 0.1) \cdot (143 + 0) = 143$
Trend: $T(2) = 0.4 \cdot (143 - 143) + (1 - 0.4) \cdot 0 = 0$
Forecast with trend: $143 + 0 = 143$

Forecast with trend for period 3:
Basic forecast: $BF(3) = 0.1 \cdot 142 + (1 - 0.1) \cdot (143 + 0) = 142.9$
Trend: $T(3) = 0.4 \cdot (143 - 143) + (1 - 0.4) \cdot 0 = 0$
Forecast with trend: $142.9 + 0 = 143$

Forecast with trend for period 4:
Basic forecast: $BF(4) = 0.1 \cdot 153 + (1 - 0.1) \cdot (142.9 + 0) = 143.9$
Trend: $T(4) = 0.4 \cdot (143.9 - 142.9) + (1 - 0.4) \cdot 0 = 0.4$
Forecast with trend: $143.9 + 0.4 = 144$

Forecast with trend for period 5:
Basic forecast: $BF(5) = 0.1 \cdot 157 + (1 - 0.1) \cdot (143.9 + 0) = 145.5$
Trend: $T(5) = 0.4 \cdot (145.5 - 143.9) + (1 - 0.4) \cdot 0.4 = 0.9$
Forecast with trend: $145.5 + 0.9 = 146$

Forecast with trend for period 6:
Basic forecast: $BF(6) = 0.1 \cdot 139 + (1 - 0.1) \cdot (145.5 + 0.9) = 145.7$
Trend: $T(6) = 0.4 \cdot (145.7 - 145.5) + (1 - 0.4) \cdot 0.9 = 0.6$
Forecast with trend: $145.7 + 0.6 = 146$

Calculated in the same way the forecasts for periods 2 to 30 are: 143, 143, 144, 146, 146, 148, 154, 160, 162, 167, 173, 177, 183, 190, 196, 204, 213, 219, 226, 234, 240, 247, 255 and 264.

h) Calculated in the same way as in **g)** the forecasts for periods 2 to 30 are: 143, 143, 144, 147, 147, 150, 157, 164, 167, 174, 181, 185, 191, 198, 203, 210, 220, 224, 231, 237, 242, 248, 254 and 263.

i) Calculated in the same way as in **g)** the forecasts for periods 2 to 30 are: 143, 143, 147, 152, 149, 153, 166, 177, 173, 179, 185, 184, 189, 198, 201, 209, 224, 222, 229, 235, 240, 245, 253 and 266.

j) Calculated in the same way as in **g)** the forecasts for periods 2 to 30 are: 143, 142, 148, 154, 151, 154, 170, 182, 176, 181, 186, 182, 187, 197, 200, 210, 227, 224, 230, 236, 239, 244, 253 and 268.

k) MAD for the forecasts in **b)** are calculated as the following:

Period	Demand	Forecast	Absolute deviation
13	190	161	29
14	201	165	36
15	192	170	22
16	210	173	37
17	231	178	53
18	200	185	15
19	225	189	36
20	230	193	37
21	232	197	35
22	240	204	36
23	251	209	42
24	270	214	56
		Cumulative absolute deviation:	434
		MAD:	434 / 12 = 36.1

Calculated in the same way the MAD for the forecasts in **a)** to **e)** and **g)** to **j)** are:

a) 13.9
b) 36.1
c) 38.8
d) 16.8
e) 14.7
g) 10.2
h) 8.4
i) 8.9
j) 8.9

When comparing MAD it is clear that methods with few periods and high alpha values perform better than methods with many periods and low alphas. This is because of the clear trend that is better taken into consideration when using few periods or high alphas. When comparing methods with and without trend correction it is clear that all trend correcting methods perform better. The best method is the one with low alpha and high beta values. The alpha value results in a method that smoothes out random variations and the high beta value results in fast adjustment to the clear trend.

Problem 7.7

Assume that MAD for period 1 is 0. MAD after period 2 (for the forecast in **d)**) is then (using alpha = 0.4): $0.4 \cdot |143 - 142| + (1 - 0.4) \cdot 0 = 0.4$. MAD after period 3 is $0.4 \cdot |143 - 153| + (1 - 0.4) \cdot 0.4 = 4.4$. MAD for the following periods are calculated in similar ways, resulting in MAD values for periods 2 to 24 of: 0.4, 4.4, 6.7, 8.8, 9.2, 17.9, 17.8, 18.0, 18.0, 15.9, 10.9, 13.3, 16.5, 11.4, 14.9, 22.2, 17.8, 18.0, 17.2, 15.0, 15.0, 17.0, 22.6.

  MAD values for **h)** for periods 2 to 24 (using alpha = 0.1) are: 0.1, 1.1, 2.2, 2.8, 3.4, 6.2, 7.9, 8.5, 8.9, 8.8, 8.9, 8.5, 8.7, 8.4, 8.2, 9.5, 10.5, 9.5, 8.7, 8.3, 7.7, 7.3, 8.1.

Problem 8.1

Projected inventory balance at the end of the respective week and available-to-promise quantities during the respective weeks:

Week	1	2	3	4	5	6
Customer order reservation	60	45	40	30	20	5
Scheduled receipts	100			100		100
Projected inventory balance	65	20	80	50	30	125
Available to promise	20	20	30	30	30	125

Problem 8.2

An alternative is to promise 20 units for delivery in week 1, 10 units in week 3 and 20 units in week 6. Another alternative that results in fewer deliveries would be to deliver 20 units in week 1 and 30 units in week 6. If a complete delivery is preferred it can be shipped no earlier than in week 6.

Problem 8.3

If the order is accepted the reserved quantity in week 4 will be 70 units. This means that the scheduled receipt in week 3 will not be enough for covering the demand during weeks 3 and 4. There will be a shortage of 10 units, which must be taken from the scheduled receipt in week 1. Available to promise in week 1 will thus be reduced to 35 units. Calculations:

Week	1	2	3	4	5	6
Customer order reservations	80	45	60	70	70	25
Scheduled receipts	120		120		120	
Available to promise	35	35	35	35	60	60

Problem 8.4

When considering the assembly capacity, the following quantities are available to promise:

Day	1	2	3	4	5	6	7	8
Planned customer deliveries	20	20	20	20	18	5	4	0
Assembly plan	20	20	20	20	20	20	20	20
Available to promise	0	0	0	0	2	17	33	53

From the table you can see that 5 PCs could be delivered during day 6, when considering the assembly and testing capacity. Available quantities for each of the delivery time critical components are shown in the following two tables:

Component A:

Day	1	2	3	4	5	6
Reservations	8	2	3	0	0	0
Scheduled receipts		20				20
Available to promise	7	22	22	22	22	42

Component B:

Day	1	2	3	4	5	6
Reservations	12	3	2	0	0	0
Scheduled receipts				20		
Available to promise	0	0	0	20	20	20

From the tables you can see that there are 7 units available of component A, already day 1, but that 5 components B are not available until day 4. Because it takes 4 days to assembly and test, the delivery of 5 PCs can consequently not be promised until day 8. Component B is consequently the critical component for the delivery time determination.

Problem 9.1

a) The total number of production days is $56 + 62 + 42 + 58 = 218$ days and the total delivery volume is $9000 + 12\,000 + 7000 + 9000 = 37\,000$ units. Because they want to decrease the order backlog with 1600 units, the total production need is 38 600 units. The production volume per day is therefore $38\,600 / 218 = 177$ units. This is equivalent to a capacity need of $177 / 4 = 44.2$ operators which require 45 available operators.

b) During quarter 1 and 2 there are $56 + 62 = 118$ production days which results in $8 \cdot 177 = 20\,866$ units of products to be delivered. Because the delivery quantities during the first two quarters are estimated to 21 000, the order backlog will increase to $2200 + (21\,000 - 20\,866) = 2314$ units. If assuming that there are 13 weeks per quarter, the estimated delivery quantity during quarter 3 will be $7000 / 13 = 538$ units per week. This results in a delivery time at the quarter shift 2–3 of $2314 / 538 = 4.3$ weeks.

Problem 9.2

a) The estimated delivery quantity during the first half of the year is $5000 + 5000 = 10\,000$ units. Because the demand is lower during the first half of the year we postpone the plan to decrease the inventory level to the second half of the year. With 120 working days and use of a levelled strategy the planned production rate will be $10\,000 / 120 = 83$ units per day during the first half of the year. During the second half of the year expected deliveries are $6000 + 7000 = 13\,000$ units. Because we want to decrease the inventory level with 500 units during this period, there is a production need of $13\,000 - 500 = 12\,500$ units. This volume is equivalent with a production rate of $12\,500 / 120 = 104$ units per day.

b) Because every operator can make 10 units per day, there will be a need for $83 / 10 = 8.3$ operators during the first half of the year and $104 / 10 = 10.4$ operators during the second half of the year. To round the figures to full operators it could be decided to have 8 operators during the first half of the year and 11 operators during the second half.

c) With 8 operators it is possible to produce $8 \cdot 10 \cdot 60 = 4800$ units during the first quarter of the year. By fully utilising the existing storage room they can deliver $4800 + 1000 = 5800$ units. This means that a demand that exceeds the forecast by 1000 will result in shortage during the first quarter. In addition, the estimated production rate for the rest of the year will not be enough for having 500 units in stock in the end of the year.

Problem 9.3

a) Because the total delivery volume during the year is estimated to €120 million, it is necessary to produce for €10 million per month on average. With a levelled strategy the following production plan and projected inventory balance can be developed:

Month	Jan	Feb	Mar	Apr	May	Jun	Jul	Aug	Sep	Oct	Nov	Dec
Delivery	8	10	7	11	12	7	8	10	9	11	12	15
Production	10	10	10	10	10	10	10	10	10	10	10	10
Inventory	2	2	5	4	2	5	7	7	8	7	5	0

With a chase strategy the following production plan and projected inventory can be developed:

Month	Jan	Feb	Mar	Apr	May	Jun	Jul	Aug	Sep	Oct	Nov	Dec
Delivery	8	10	7	11	12	7	8	10	9	11	12	15
Production	8	10	7	11	12	7	8	10	9	11	12	15
Inventory	0	0	0	0	0	0	0	0	0	0	0	0

b) The average tied up capital in inventory is €4 500 000 when applying a levelled strategy and €0 when applying a chase strategy.

c) For the levelled strategy, inventory carrying cost is $0.25 \cdot 4\ 500\ 000 = €1\ 125\ 000$ and for the chase strategy €0. Using the levelled strategy results in capacity changing costs of €100 000, while using the chase strategy results in capacity changing costs of €1 200 000 (12 changes). The total costs are consequently €1 225 000 compared to €1 200 000.

d) A possible production plan based on a hybrid strategy is shown in the table below. It requires 3 capacity changes and results in an average inventory of €2 750 000. The sum of inventory carrying costs and capacity change costs is $3 \cdot 100\ 000 + 0.25 \cdot 2\ 750\ 000 = €987\ 500$.

Month	Jan	Feb	Mar	Apr	May	Jun	Jul	Aug	Sep	Oct	Nov	Dec
Delivery	8	10	7	11	12	7	8	10	9	11	12	15
Production	9	9	10	10	10	10	10	10	10	10	10	12
Inventory	1	0	3	2	0	3	5	5	6	5	3	0

Problem 10.1

a) In order to use the material requirements plan, the master production schedule must be at least as long as the cumulative manufacturing lead time. For the actual industry gate the cumulative manufacturing lead time is $12 + 2 + 2 + 2 = 18$ weeks. The master production schedule consequently has to be at least 18 weeks long.

b) In accordance with the materials plan, the master production schedule only has to be 18 weeks long. But in order to also consider the possibility of making capacity adjustments the master production schedule has to be 20 weeks long.

c) Because the cumulative lead time is 18 weeks and the maximum acceptable delivery time is 10 weeks, all items cannot be planned with material requirements planning. For the actual product, this counts for items G and L. An alternative would be to control these items with re-order point systems, based on forecasts for the respective item.

Problem 10.2

a) The planning horizon for the master production scheduling should be at least as long as the cumulative lead time, i.e. $2 + 2 + 1 + 4 = 9$ weeks.

b) 8 weeks' planned material requirements are necessary for choosing appropriate order quantities. To achieve these 8 weeks of requirements, the master production schedule must be at least $9 + 8 = 17$ weeks long.

c) Because it takes 2 weeks to assemble the product the release time fence should be at least 2 weeks.

d) Because the cumulative lead time for items E, and H are 6 weeks, the planning time fence should be at least 6 weeks.

Problem 10.3

a) Because orders with shorter delivery times than 2 weeks are not accepted, these 2 weeks are appropriate as demand time fence. The forecast time fence could be chosen around 6 weeks because it is uncommon with customer orders with expected delivery time longer than 6 weeks.

b) A possible policy for merging the forecast and order backlog is to only consider customer orders within the demand time fence and to choose the largest of the weekly customer order and forecast figures between the demand and forecast time fences. Such a policy would result in the following delivery schedule for the next 8 weeks:

Week	1	2	3	4	5	6	7	8
Forecasted demand	10	10	10	10	10	10	10	10
Actual customer orders	8	7	12	6	15	13	0	0
Delivery plan	8	7	12	10	15	13	10	10

c) When using the policy "largest of the customer order and forecast figures" during the period between the demand and forecast time fences there is a risk that the delivery plan will systematically be too large. In the actual case, the total expected quantity to deliver according to the delivery plan is 50 pieces, while it is only 40 pieces according to the forecast during the four weeks. To reduce such effects the period during which the largest of the two values is chosen should not be too long. If, for example, choosing the largest value during two weeks, the delivery plan for weeks 3 and 4 would have been 12 and 10 pieces, respectively. But during weeks 5 and 6 it would have been 10 and 12 respectively.

Problem 10.4

The table shows a possible master production schedule for product PTV 2X for weeks 17 to 24:

Week	17	18	19	20	21	22	23	24
Forecasted delivery	60	60	70	70	80	80	70	70
Inventory build-up					50	50	50	50
Projected inventory balance	137	77	157	87	157	227	157	237
Master production schedule	150		150		150	150		150

Problem 10.5

A levelled production would result in production of 2900 / 4 = 725 packages every second week. This would result in a shortage of 200 pieces in week 34. To produce 800 packages in each of weeks 29, 31 and 33 and 500 pieces in week 35 would, on the other hand, result in that some of the pieces put in stock in week 29 would not be withdrawn and delivered until three weeks later.

The table shows a possible master production schedule for the product during the actual circumstances:

Week	29	30	31	32	33	34	35	36
Forecasted delivery	250	250	300	350	650	600	250	250
Projected inventory	450	200	700	350	600	0	250	0
Master production schedule	700		800		900		500	

Problem 11.1

a) The run-out time for SP-1 is 37 / 5 = 7.4 days and for SP-4 it is 68 / 5 = 13.6 days. Corresponding lead times plus safety times for the two products are 7 + 2 = 9 days and 12 + 5 = 17 days, respectively. Because the run-out times for both products are shorter than the lead time plus safety time, manufacturing orders should be released immediately for both products.

b) The difference between lead time plus safety time for SP-1 is 9 − 7.4 = 1.6 days and for SP-4 is 17 − 13.6 = 3.4 days. The margins for avoiding shortages are consequently smaller for SP-4. It should therefore be manufactured before SP-1.

Problem 11.2

A material requirements planning for the vitamin pill is shown in the table:

Day	1	2	3	4	5	6	1	2	3	4	5	6	1	2
Forecast	6	6	2	2	2	1	6	6	2	2	2	1	6	6
Projected inventory balance	3	7	5	3	1	0	4	8	6	4	2	1	5	9
Planned order receipt		10					10	10					10	10
Planned order start	10				10	10					10	10		

A new order start should consequently be planned already in day 1. Because of the short delivery time, the other planned order starts can be considered to be planned orders and used as delivery schedules for the suppliers.

Problem 11.3

a)

Safety stock $= 1.6 \cdot$ lead time demand $= 1.6 \cdot \dfrac{4}{52} \cdot 8000 = 985$ units

Average inventory $=$ Safety stock $+ \dfrac{\text{Purchasing quantity}}{2} = 985 + \dfrac{1000}{2} = 1485$ units

Re-order point $=$ Safety stock $+$ Demand during lead time $= 985 + \dfrac{4}{52} \cdot 8000 = 1600$ units

Average tied-up capital $=$ Average inventory $\cdot$ Item value $= 1485 \cdot 100 = €148\,500$

Inventory turnover rate $= \dfrac{\text{Demand per year}}{\text{Average tied-up capital}} = \dfrac{8000 \cdot 100}{148\,500} = 5.4$ times per year

Run-out time in stock $= \dfrac{52}{\text{Inventory turnover rate}} = \dfrac{52}{5.4} = 9.6$ weeks

Annual inventory carrying cost $=$ Average tied-up capital $\cdot$ Inventory interest rate $= 148\,500 \cdot 0.20 = €29\,700$ per year

b) If only inspecting the inventory every second week, the lead time will increase by on average one week. The calculated figures in a) will change in the following way:

Safety stock $= 1.6 \cdot$ lead time demand $= 1.6 \cdot \dfrac{5}{52} \cdot 8000 = 1231$ units

Average inventory $=$ Safety stock $+ \dfrac{\text{Purchasing quantity}}{2} = 1231 + \dfrac{1000}{2} = 1731$ units

Re-order point $=$ Safety stock $+$ Demand during lead time $= 1231 + \dfrac{5}{52} \cdot 8000 = 2000$ units

Average tied-up capital $=$ Average inventory $\cdot$ Item value $= 1731 \cdot 100 = €173\,100$

Inventory turnover rate $= \dfrac{\text{Demand per year}}{\text{Average tied-up capital}} = \dfrac{8000 \cdot 100}{173\,100} = 4.6$ times per year

Run-out time in stock $= \dfrac{52}{\text{Inventory turnover rate}} = \dfrac{52}{4.6} = 11.3$ weeks

Annual inventory carrying cost $=$ Average tied-up capital $\cdot$ Inventory interest rate $= 173\,100 \cdot 0.20 = €34\,620$ per year

Problem 11.4

When using periodic ordering the safety stock calculation has to be based on the uncertainty during the lead time plus the replenishment interval, i.e. 6 weeks.

Safety stock $= 1.6 \cdot$ (Demand during lead time $+$ replenishment interval) $= 1.6 \cdot \dfrac{6}{52} \cdot 8000 = 1477$ units

Order-up-to level $=$ Demand per period $\cdot$ (Replenishment interval $+$ Lead time) $+$ Safety stock $= \dfrac{8000}{52} \cdot (2 + 4) + 1477 = 2400$ units

Average inventory = Safety stock + $\dfrac{\text{Order up-to-level} - \text{Safety stock}}{2}$ = $1477 + \dfrac{923}{2}$ = 1939 units

Average tied-up capital = Average inventory · Item value = 1939 · 100 = €193 900

Inventory turnover rate = $\dfrac{\text{Demand per year}}{\text{Average tied-up capital}}$ = $\dfrac{8000 \cdot 100}{193\,900}$ = 4.1 times per year

Run-out time in stock = $\dfrac{52}{\text{Inventory turnover rate}}$ = $\dfrac{52}{4.1}$ = 12.7 weeks

Annual inventory carrying cost = Average tied-up capital · Inventory interest rate = 193 900 · 0.20 = €38 780 per year

Order quantity = Order up-to-level − Inventory balance = 2400 − 1600 = 800 units

Problem 11.5

a) The run-out times for the respective variant, if not using the safety stock, are 5 · (100 − 70) / 35 = 4.3 days for variant A, 5 · (120 − 100) / 50 = 2 days for variant B, 5 · (130 − 80) / 40 = 6.3 days for variant C and 5 · (100 − 80) / 40 = 2.5 days for variant D. The lead time (manufacturing lead time) for the variants are 5 · (0.10 + 0.05) · 100 / 80 = 0.9 day, 5 · (0.20 + 0.05) · 150 / 80 = 2.3 days, 5 · (0.25 + 0.05) · 100 / 80 = 1.9 days and 5 · (0.20 + 0.05) · 200 / 80 = 3.1 days for variants A, B, C and D, respectively. For variants B and D, the lead times are longer than the respective run-out times. Consequently, manufacturing orders should be planned for these variants.

b) The manufacturing orders for B and D could be mutually prioritised by calculating priority measures. Here, the difference between lead time and run-out time is used as priority measure. The variant with the largest difference gets the highest priority. The difference for variant B is 0.3 days (2.3 − 2.0) and for variant D it is 0.6 day (3.1 − 2.5). Variant D should therefore be manufactured before variant B, based on the existing inventory balances.

c) The milling operation of one batch of variant B takes 5 · 0.20 · 150 / 80 = 1.9 days and the assembly takes 5 · 0.05 · 150 / 80 = 0.5 days. Corresponding manufacturing times for variant D are 5 · 0.20 · 200 / 80 = 2.5 days, 5 · 0.05 · 200 / 80 = 0.6 days, respectively. If making variant D first, it will be finished after 3.1 days, i.e. there will be a shortage during 0.6 days. The milling operation for variant B can start after 2.5 days and the order can be finished after 2.5 + 1.9 + 0.5 = 4.9 days, resulting in 2.9 days of shortage. If instead choosing to make variant B first, it will be finished after 2.4 days, resulting in 0.4 days of shortage. Milling of variant D can in this situation start after 1.9 days and the order be finished after 1.9 + 2.5 + 0.6 = 5.0 days, i.e. 2.5 days too late. The total days of shortages for both variants are, if first making variant D, 0.6 + 2.9 = 3.5 days, and if first making variant B, 0.4 + 2.5 = 2.9 days. If the shortage cost per day is the same for the two variants, it would thus be beneficial to make variant B first.

Problem 11.6

Bill of material with quantities for Product X:

Level item		Quantity per X
1	A	3 pieces
2	E	2 pieces
2	F	1 piece
1	B	4 pieces

Bill of material with quantities for Product Y:

Level item		Quantity per Y
1	B	3 pieces
1	C	1 piece
2	A	2 pieces
3	E	2 pieces
3	F	1 piece
2	D	1 piece

From the tables you can see that manufacturing orders for A are generated in periods 2, 4 and 7:

Item C						Low-level code 1						
Week	1	2	3	4	5	6	7	8	9	10	11	12
Where-used: Y (1 unit)	200		300	300		200	100	100	200		500	
Gross requirements	200		300	300		200	100	100	200		500	
Scheduled receipts												
Projected inventory balance	200	200	500	200	200	0	500	400	200	200	300	300
Net requirements			100				100				300	
Planned receipt			600				600				600	
Planned order start	600				600				600			

Item A						Low-level code 2						
Week	1	2	3	4	5	6	7	8	9	10	11	12
Where-used:												
C (2 units)	1200				1200				1200			
X (3 units)		1200		1200		1200		1500		900		1500
Gross requirements	1200	1200		1200	1200	1200		1500	1200	900		1500
Scheduled receipts												
Projected inventory balance	1300	100	100	1900	700	2500	2500	1000	2800	1900	1900	400
Net requirements				1100		500			200			
Planned receipt				3000		3000			3000			
Planned order start		3000		3000			3000					

Problem 11.7

a) If the stock on hand is 30 units at the end of the week, the projected inventory balance will be lower than the safety stock in the beginning of week 7. The MRP will therefore plan for receiving an order in week 7 with planned order start in the beginning of week 3, according to the table:

Week	1	2	3	4	5	6	7	8
Gross requirements	50	75	25	50	70	80	75	70
Scheduled receipts		400						
Projected inventory balance	30	355	330	280	210	130	55	−15
Planned receipt							400	
Planned order start			400					

Lot size: 400 units. Lead time: 4 weeks. Safety stock: 100 units.

b) The conducted MRP after the activities are included in the table below. The planned order start of 400 in week 3 is exchanged with planned order starts of 300 in weeks 2 and 5. Because the lead times have decreased from 4 to 3 weeks, it will be possible to finish production without using the safety stock.

Week	2	3	4	5	6	7	8	9
Where-used (X)	25	25	25	140	10	5	20	20
Where-used (Y)	50		25	40	60	60	40	60
Gross requirements	75	25	50	180	70	65	60	80
Scheduled receipts	250							
Projected inventory balance	205	180	130	250	180	115	355	275
Planned receipt				300			300	
Planned order start	300			300				

Order size: 300 units. Lead time: 3 weeks. Safety stock: 75 units.

Problem 11.8

Purchase order proposals for item B are generated for periods 4, 7 and 9 according to the table.

Period	1	2	3	4	5	6	7	8	9	10	11	12
Where-used in P1 − 3	150	240	150		240	60	180	240		180	150	210
Where-used in P2 − 1	110	200		80	140		50	40	80	120	60	50
Gross requirements	260	440	150	80	380	60	230	280	80	300	210	260
Scheduled receipts												
Projected inventory balance	740	300	150	70	290	230	0	320	240	540	330	70
Net requirements					310			280		60		
Planned order receipt					600			600		600		
Planned order start				600			600		600			

Problem 12.1

a) Total incremental cost for item A when ordering 12 times per year:

$$\text{Total incremental cost} = \frac{\frac{3200}{12}}{2} \cdot 30 \cdot 0.15 + 12 \cdot 25 = €900 \text{ per year}$$

Total incremental cost for item A when ordering 20 times per year:

$$\text{Total incremental cost} = \frac{\frac{3200}{20}}{2} \cdot 30 \cdot 0.15 + 20 \cdot 25 = €860 \text{ per year}$$

Total incremental cost for item B when ordering 12 times per year:

$$\text{Total incremental cost} = \frac{\frac{4500}{12}}{2} \cdot 25 \cdot 0.15 + 12 \cdot 35 = €1123 \text{ per year}$$

Total incremental cost for item B when ordering 30 times per year:

$$\text{Total incremental cost} = \frac{\frac{4500}{30}}{2} \cdot 25 \cdot 0.15 + 30 \cdot 35 = €1331 \text{ per year}$$

b) Economic order quantity for item A:

$$EOQ = \sqrt{\frac{2 \cdot 3200 \cdot 25}{30 \cdot 0.15}} = 189 \text{ pieces}$$

$$\text{Number of orders} = \frac{3200}{189} = 17 \text{ times per year}$$

$$\text{Total incremental costs} = \frac{189}{2} \cdot 30 \cdot 0.15 + 17 \cdot 25 = €850 \text{ per year}$$

Economic order quantity for item B:

$$EOQ = \sqrt{\frac{2 \cdot 4500 \cdot 35}{25 \cdot 0.15}} = 290 \text{ pieces}$$

$$\text{Number of orders} = \frac{4500}{290} = 16 \text{ times per year}$$

$$\text{Total incremental costs} = \frac{290}{2} \cdot 25 \cdot 0.15 + 16 \cdot 35 = €1104 \text{ per year}$$

c) The total incremental cost for inventory carrying and ordering in batches of 1000 pieces of item A and 500 pieces of item B minus the quantity discount is €1466 (see calculations below). It is €488 lower compared to the total incremental cost of €1954 per year from problem **b)**. Consequently, the discount to buy full pallets should be used.

Total incremental cost for item A minus discount:

$$\frac{1000}{2} \cdot 30 \cdot 0.99 \cdot 0.15 + \frac{3200}{1000} \cdot 25 - 0.01 \cdot 3200 \cdot 30 = €1348 \text{ per year}$$

Total incremental cost for item B minus discount:

$$\frac{500}{2} \cdot 25 \cdot 0.99 \cdot 0.15 + \frac{4500}{500} \cdot 35 - 0.01 \cdot 4500 \cdot 25 = €118 \text{ per year}$$

d) In order to include both A and B in the same EOQ calculation, the demand and quantity must be expressed in monetary units. With 17 ordering occasions for both items, the total incremental cost is €1770 per year, which is €184 lower compared to problem **b)**. The optimum number of orders per year is 17.7 which results in an additional €2 lower costs.

Incremental cost when 17 orders $= \dfrac{\frac{3200}{17}}{2} \cdot 30 \cdot 0.15 + \dfrac{\frac{4500}{17}}{2} \cdot 25 \cdot 0.15 + 17 \cdot 50$
$$= €1770 \text{ per year}$$

EOQ when ordering equally often $= \sqrt{\dfrac{2 \cdot \left(3200 \cdot 30 + 4500 \cdot 25\right) \cdot 50}{0.15}} = €11\,790$

Optimum number of orders $= \dfrac{3200 \cdot 30 + 4500 \cdot 25}{11790} = 17.7$ times per year

Incremental cost when 17.7 orders per year $= \dfrac{\frac{3200}{17.7}}{2} \cdot 30 \cdot 0.15 + \dfrac{\frac{4500}{17.7}}{2} \cdot 25 \cdot 0.15 +$
$17.7 \cdot 50 = €1768$ per year

Problem 12.2

a) $EOQ = \sqrt{\dfrac{2 \cdot 10\,000 \cdot 50}{50 \cdot 0.15}} = 365$ pieces

b) The solvency restriction is the strongest restriction and maximises the order quantity to 150 pieces, see below.

Volume restriction: $\dfrac{5500}{30} = 183$ pieces

Shelf life restriction: $\dfrac{20}{365} \cdot 10\,000 = 548$ pieces

Solvency restriction: $\dfrac{7500}{50} = 150$ pieces

Problem 12.3

Annual demand $= 450 \cdot \dfrac{52}{9} = 2600$ pieces

$EOQ = \sqrt{\dfrac{2 \cdot 2600 \cdot 50}{10}} = 161$ pieces

$ERT = \dfrac{161}{2600} \cdot 52 = 3.2$ weeks

The table shows projected available in stock when EOQ = 161 pieces and ERT = 3 weeks, respectively.

	1	2	3	4	5	6	7	8	9
Demand (forecast)	10	45	55	80	60	100	50	20	30
Stock on hand (EOQ)	151	106	51	132	72	133	83	63	33
Stock on hand (ERT)	100	55	0	160	100	0	50	30	0

Calculation of optimum lot sizes using the Silver-Meal method is shown below:

1 *Lot size when replenishing in week 1*:

Incremental cost per week if one week of demand: €50
Incremental cost per week if two weeks of demand: $(50 + 45 \cdot 10 / 52) / 2 = €29$
Incremental cost per week if three weeks of demand: $(50 + (45 + 2 \cdot 55) \cdot 10 / 52) / 3 = €27$
Incremental cost per week if four weeks of demand: $(50 + (45 + 2 \cdot 55 + 3 \cdot 80) \cdot 10 / 52) / 4 = €31$

Three weeks result in the lowest incremental cost per week and is therefore chosen.

2 *Lot size when replenishing in week 4*:

Incremental cost per week if one week of demand: €50
Incremental cost per week if two weeks of demand: $(50 + 60 \cdot 10 / 52) / 2 = €31$
Incremental cost per week if three weeks of demand: $(50 + (60 + 2 \cdot 100) \cdot 10 / 52) / 3 = €33$

Two weeks result in the lowest incremental cost per week and is therefore chosen.

3 *Lot size when replenishing in week 6*:

Incremental cost per week if one week of demand: €50
Incremental cost per week if two weeks of demand: $(50 + 50 \cdot 10 / 52) / 2 = €30$
Incremental cost per week if three weeks of demand: $(50 + (50 + 2 \cdot 20) \cdot 10 / 52) / 3 = €22$
Incremental cost per week if four weeks of demand: $(50 + (50 + 2 \cdot 20 + 3 \cdot 30) \cdot 10 / 52) / 4 = €21$

Four weeks result in the lowest incremental cost per week and is therefore chosen.

The table shows the projected stock on hand for weeks 1 to 9 when using the Silver-Meal method:

	1	2	3	4	5	6	7	8	9
Demand (forecast)	10	45	55	80	60	100	50	20	30
Stock on hand	100	55	0	60	0	100	50	30	0

Average stock on hand when EOQ = 161 is $(151 + 106 + 51 + 132 + 72 + 133 + 83 + 63 + 33) / 9 = 91.6$ pieces. This results in an annual inventory carrying cost of $91.6 \cdot 10 = €916$. There are 16.1 orders per year which results in annual ordering costs of €805, and a total cost of €1721.

Average stock on hand when ERT = 3.2 is $(100 + 55 + 0 + 160 + 100 + 0 + 50 + 30 + 0) / 9 = 55.0$ pieces. This results in an annual inventory carrying cost of $55.0 \cdot 10 = €550$. There are 17.3 orders per year which results in annual ordering costs of €865, and a total cost of €1415.

Average stock on hand when using the Silver-Meal method is $(100 + 55 + 0 + 60 + 0 + 100 + 50 + 30 + 0) / 9 = 43.9$ pieces. This results in an annual inventory carrying cost of $43.9 \cdot 10 = €439$. There are 17.3 orders per year which results in annual ordering costs of €865, and a total cost of €1304.

Economic order quantity results in higher total cost compared to the two more dynamic methods (ERT and Silver-Meal) because of the fluctuating demand. The Silver-Meal method results in somewhat lower total cost compared to ERT.

Problem 12.4

a) Incremental ordering costs:

Install, remove, clean: $(3 + 0.5) \cdot 2 \cdot 30 + (3 + 0.5) \cdot 50 = €385$
Fine tuning: $1 \cdot 30 + 1 \cdot 50 + 30 \cdot 50 = €1580$
Quality control: $0.5 \cdot 30 = €15$

Total incremental ordering cost: $385 + 1580 + 15 = €1980$

$$EOQ = \sqrt{\frac{2 \cdot 90\,000 \cdot 1980}{100 \cdot 0.15}} = 4874 \text{ pieces}$$

b) Incremental ordering costs:

Install, remove, clean: $(0.5 + 0.5) \cdot 2 \cdot 30 + (0.5 + 0.5) \cdot 50 = €110$
Fine tuning: $1 \cdot 30 + 1 \cdot 50 + 30 \cdot 50 = €1580$
Quality control: $0.5 \cdot 30 = €15$

Total incremental ordering cost: $110 + 1580 + 15 = €1705$

$$EOQ = \sqrt{\frac{2 \cdot 90\,000 \cdot 1705}{100 \cdot 0.15}} = 4523 \text{ pieces}$$

c) Incremental ordering costs:

Install, remove, clean: $(3 + 0.5) \cdot 2 \cdot 30 = €210$
Fine tuning: $1 \cdot 30 + 30 \cdot 50 = €1530$
Quality control: $0.5 \cdot 30 = €15$

Total incremental ordering cost: $210 + 1530 + 15 = €1755$

$$EOQ = \sqrt{\frac{2 \cdot 90\,000 \cdot 1755}{100 \cdot 0.15}} = 4589 \text{ pieces}$$

d) Incremental inventory carrying cost:

Calculated interest: 10%
Inventory carrying cost: $(50 \cdot 12) / (100 \cdot 100) = 6\%$
Insurance cost: $1 / 100 = 1\%$
Obsolescence: 0 or 3%

Total incremental inventory carrying cost: 17 or 20%

$$EOQ\ (17\%) = \sqrt{\frac{2 \cdot 90\,000 \cdot 1980}{100 \cdot 0.17}} = 4579 \text{ pieces}$$

$$\text{Average run-out time} = \frac{\dfrac{4579}{2}}{90\,000} \cdot 12$$

$$= 0.31 \text{ months}$$

The average run-out time is less than 1 month. Therefore, 0% of the products are expected to be obsolete.

Problem 12.5

a) Optimum order quantity according to $EOQ = \sqrt{\dfrac{2 \cdot 1200 \cdot 35}{0.25 \cdot 100}} = 58$ Because the packaging size contains 50 pieces, the order quantity is rounded off to 50 pieces.

b) Rounding off to 50 pieces results in somewhat less inventory carrying costs and somewhat higher ordering costs compared to when using the economic order quantity.

Problem 12.6

Three days' run-out time results in the following order quantities:

Order quantity 1 = 4 + 6 + 10 = 20 pieces
Order quantity 2 = 12 + 15 + 20 = 47 pieces
Order quantity 3 = 4 + 6 + 10 = 20 pieces

Problem 12.7

a) The run-out time for the two items:

Glass vase Model Antique 2 $\dfrac{12}{\frac{96}{48}} = 6.0$ weeks rounded off to 6 weeks

Glass vase Nature Experience $\dfrac{13}{\frac{150}{48}} = 4.16$ weeks rounded off to 4 weeks

b) An alternative would be to coordinate ordering of the two vases by ordering the two vases every fifth week, and ordering a quantity equivalent to five weeks of demand every time. Another alternative would be to order vase Nature every third week and vase Antique 2 every sixth week, i.e. at every second ordering.

Problem 13.1

a) From the Poisson table you can see that a safety stock of 4 units is necessary for 90% service level when the average demand is 8 units.

b) Standard deviation of the demand during the lead time $= \sqrt{4 \cdot 2^2 + 1^2 \cdot 4^2} = 5.66$ units
Cycle service = 90% $\rightarrow k = 1.28$
Safety stock = $1.28 \cdot 5.66 = 7.2 \rightarrow 8$ units

c) $k = 1.28 \rightarrow E(z) = 0.0475$

Fill rate $= 1 - \dfrac{5.66 \cdot 0.0475}{20} = 98.7\%$

Problem 13.2

Average lead time $= \dfrac{2 + 3 + 1 + 5 + 3 + 5 + 4 + 2 + 8 + 3}{10} = 3.6$ weeks

$\sigma_{LT} = 1.25 \cdot \dfrac{\begin{array}{c}|3.6 - 2| + |3.6 - 3| + |3.6 - 1| + |3.6 - 5| + |3.6 - 3| + |3.6 - 5| \\ + |3.6 - 4| + |3.6 - 2| + |3.6 - 8| + |3.6 - 3|\end{array}}{10} = 1.9$ weeks

Average demand during lead time $= 2500 \cdot 3.6 = 9000$ units

Standard deviation of demand during lead

time, $\sigma_{DDLT} = \sqrt{3.6 \cdot 500^2 + 1.9^2 \cdot 2500^2} = 4844$ units

a) Cycle service $\geq 0.95 \rightarrow k = 1.65 \rightarrow$ Safety stock $= 1.65 \cdot 4844 = 7993$ units

b) $EOQ = \sqrt{\dfrac{2 \cdot 52 \cdot 2500 \cdot 80}{10}} = 1442$ units

Fill rate $= 99\% \rightarrow E(z) = \dfrac{1442 \cdot (1 - 0.99)}{4844} = 0.0030 \rightarrow z = 2.37$

Safety stock $= 2.37 \cdot 4844 = 11\,480$ units

Fill rate $= 95\% \rightarrow E(z) = \dfrac{1442 \cdot (1 - 0.95)}{4844} = 0.0149 \rightarrow z = 1.78$

Safety stock $= 1.78 \cdot 4844 = 8622$ units

Fill rate $= 85\% \rightarrow E(z) = \dfrac{1442 \cdot (1 - 0.85)}{4844} = 0.0447 \rightarrow z = 1.31$

Safety stock $= 1.31 \cdot 4844 = 6346$ units

c) Cycle service $= 95\% \rightarrow k = 1.65$

$k = 1.65 \rightarrow E(z) = 0.02064 = \dfrac{1442 \cdot (1 - Fill\ rate)}{4844} \rightarrow$ Fill rate $= 93.1\%$

Problem 13.3

a)

Standard deviation of demand (per week) $= \sqrt{\dfrac{25}{4}} = 2.5$ units

Average demand per week $= \dfrac{120}{48} = 2.5$ units

Standard deviation of demand during lead time $= \sqrt{2 \cdot 2.5^2 + 2.5^2 \cdot 1^2} = 4.33$ units

$E(z) = \dfrac{(1 - 0.95) \cdot 20}{4.33} = 0.2309 \rightarrow k = 0.40$

Safety stock $= 0.40 \cdot 4.33 = 1.7 \rightarrow 2$ units

Re-order point $= 2 + 2.5 \cdot 2 = 7$ units

b) The safety stock would increase to 6 units and the re-order point to 16 units, according to the following calculations:

Standard deviation of demand during lead time = $\sqrt{4 \cdot 2.5^2 + 2.5^2 \cdot 2^2} = 7.07$ units

$E(z) = \dfrac{(1 - 0.95) \cdot 20}{7.07} = 0.1414 \rightarrow k = 0.71$

Safety stock = $0.71 \cdot 7.07 = 5.02 \rightarrow 6$ units

Re-order point = $6 + 2.5 \cdot 4 = 16$ units

c) The safety stock would decrease to 4 units, the re-order point to 14 units, the average tied-up capital increase from €80 000 to €95 000 and the inventory turnover rate decrease from 15.6 to 16.2 times per year, according to the following calculations:

$E(z) = \dfrac{(1 - 0.95) \cdot 30}{7.07} = 0.2122 \rightarrow k = 0.45$

Safety stock = $0.45 \cdot 7.07 = 3.18 \rightarrow 4$ units

Re-order point = $4 + 2.5 \cdot 4 = 14$ units

Average inventory at order quantity of 30 units = $\dfrac{30}{2} + 4 = 19$ units

Average inventory at order quantity of 20 units = $\dfrac{20}{2} + 6 = 16$ units

Average tied-up capital at order quantity of 30 units = $19 \cdot 50 = €950$

Average tied-up capital at order quantity of 20 units = $16 \cdot 50 = €800$

Inventory turnover rate at order quantity of 30 units = $\dfrac{120}{19} = 6.3$ times per year

Inventory turnover rate at order quantity of 20 units = $\dfrac{120}{16} = 7.5$ times per year

Problem 13.4

a) Periodic ordering system with inspection interval of 3 weeks, safety stock for product A of 350 units and for product B of 124 units, and target levels of 1250 units and 349 units, respectively for products A and B. Calculations below:

Economic order quantity (in purchasing value):

$$\sqrt{\dfrac{2 \cdot (200 \cdot 52 \cdot 25 + 50 \cdot 52 \cdot 20) \cdot 100}{0.15}} = €20\ 396$$

Optimal inspection interval is $\dfrac{20\ 396}{200 \cdot 52 \cdot 25 + 50 \cdot 52 \cdot 20} \cdot 52 = 3.4$ weeks $\rightarrow 3$ weeks

Inspection interval + Lead time = $3 + 1.5 = 4.5$ weeks

Standard deviation of demand during lead time + Inspection interval for product A: $100 \cdot \sqrt{4.5} = 212$ units

Standard deviation of demand during lead time + Inspection interval for product B: $70 \cdot \sqrt{4.5} = 148$ units

Safety stock for product A: Cycle service = 95% $\rightarrow k = 1.65 \rightarrow$ Safety stock = $1.65 \cdot 212 = 350$ units

Safety stock for product B: Cycle service = 80% $\rightarrow k = 0.84 \rightarrow$ Safety stock = $0.84 \cdot 148 = 124$ units

Target level for product A: $350 + 200 \cdot 4.5 = 1250$ units

Target level for product B: $124 + 50 \cdot 4.5 = 349$ units

b) Average inventories for products A and B are 800 and 236 units, respectively. Calculations below:

Average inventory for product A: $350 + \dfrac{1250 - 350}{2} = 800$ units

Average inventory for product B: $124 + \dfrac{349 - 124}{2} = 236$ units

Problem 13.5

a)

Warehouse	Demand (weeks)	Lead time (weeks)	Lot size (units)	Std.dev. of demand during lead time	$E(z)$	z	Safety stock	Average inventory	Tied-up capital (€)
R1	N(30, 20)	4	250	40	0.125	0.82	33	158	158 000
R2	N(50, 50)	4	322	100	0.06	1.15	115	276	276 000
R3	N(25, 20)	4	227	40	0.11	0.85	34	148	148 000
R4	N(100, 40)	4	455	80	0.11	0.85	68	296	296 000
							Total:	878 units	878 000

Calculating regional warehouse 1 (R1):

$$\sigma_{DDLT} = \sqrt{4 \cdot 20^2} = 40 \text{ units per four weeks}$$

$$E(z) = \frac{250 \cdot (1 - 0.98)}{40} = 0.125 \rightarrow z = 0.82$$

Safety stock = $0.82 \cdot 40 = 33$ units

Average inventory = $\dfrac{250}{2} + 33 = 158$ units

Tied-up capital = $158 \cdot 1000 = €158\,000$

b)

Demand = $N(30 + 50 + 25 + 100, \sqrt{20^2 + 50^2 + 20^2 + 40^2}) = N(205, 70)$

$\sigma_{DDLT} = \sqrt{4 \cdot 70^2} = 140$ units per four weeks

EOQ (R1) = $250 = \sqrt{\dfrac{2 \cdot 30 \cdot 52 \cdot \text{Ordering cost}}{\text{Inventory carry cost}}} \Rightarrow \dfrac{\text{Ordering cost}}{\text{Inventory carrying cost}} = 20$

$$EOQ = \sqrt{2 \cdot 205 \cdot 52 \cdot 20} = 653 \text{ units}$$

$$E(z) = \frac{653 \cdot (1 - 0.98)}{140} = 0.09 \rightarrow z = 0.9$$

Safety stock $= 0.9 \cdot 140 = 135$ units

Average inventory $= \dfrac{653}{2} + 135 = 462$ units

Tied-up capital $= 462 \cdot 1000 = €462\,000$ (i.e. 47% reduction)

c)

$$\text{Demand} = N(30 + 50 + 25 + 100, \sqrt{20^2 + 50^2 + 20^2 + 40^2}) = N(205, 70)$$

$$\sigma_{DDLT} = \sqrt{2 \cdot 70^2} = 99 \text{ units per week}$$

$$EOQ = \sqrt{2 \cdot 205 \cdot 52 \cdot 20} = 653 \text{ units}$$

$$E(z) = \frac{653 \cdot (1 - 0.98)}{99} = 0.13 \rightarrow z = 0.75$$

Safety stock $= 0.75 \cdot 99 = 75$ units

Average inventory $= \dfrac{653}{2} + 75 = 402$ units

Tied-up capital $= 402 \cdot 1000 = €402\,000$ (i.e. 13% reduction compared to "b", 54% reduction compared to "a")

Problem 13.6

a) The safety factor is 1.28 for a service level of 90%, 1.65 for a service level of 95% and 2.33 for a service level of 99%. Safety stocks (SS) for the different service levels are consequently:

$SS(90\%) = 1.28 \cdot 4.46 = 5.71$
$SS(95\%) = 1.65 \cdot 4.46 = 7.36$
$SS(90\%) = 2.33 \cdot 4.46 = 10.39$

b) Corresponding re-order points (ROP) are:

$ROP(90\%) = 5 \cdot 2 + 5.71 = 15.71$ rounded off to 16
$ROP(95\%) = 5 \cdot 2 + 7.36 = 17.36$ rounded off to 18
$ROP(90\%) = 5 \cdot 2 + 10.39 = 20.39$ rounded off to 21

Problem 14.1

Since the resource requirements per piece are equal for all three products, we can use overall factors as the planning unit. If it is assumed that the workforce remains unchanged, available capacity will be $122 + 84 + 96 = 302$ pieces per month. If capacity requirements for the three products are added together, the total capacity requirements for next year will be as shown in the table.

Month	1	2	3	4	5	6	7	8	9	10	11	12
Capacity requirements (pcs)	300	300	305	310	300	280	280	300	320	325	325	330
Available capacity (pcs)	302	302	302	302	302	302	302	302	302	302	302	302
Difference	2	2	−3	−8	2	22	22	2	−18	−23	−23	−28

At the end of the year the maximum difference is 28 pcs. The 42 operators employed manage to produce 302 pcs per month, which means that 28 pcs are equivalent to $42 \cdot 28 / 302 = 3.9$ operators or approximately 9 per cent overload. This could be relieved temporarily by using overtime, but since the trend for demand is on the increase, it may be necessary to employ more operators.

Problem 14.2

a) The following planning factors may be used for year 3:

	Year 1	Year 2	Year 3
Man-hours/piece, model X (h/piece)	3.17[1]	3.22	3.19[2]
Man-hours/piece, model Y (h/piece)	1.29	1.35	1.32
Machine-hours/piece for X (h/piece)	1.08	0.96	1.02
Machine-hours/piece for Y (h/piece)	2.33	2.40	2.37
Proportion of man-hours in WC 100			30%
Proportion of man-hours in WC 200			70%
Proportion machine-hours in WC 100			100%
Proportion machine-hours in WC 200			0%

[1] $3.17 = 3800 / 1200$
[2] $3.19 = (3.17 + 3.22) / 2$

b) Capacity requirements in work centres WC 100 and WC 200 for year 3 are shown in the tables:

Quarter	Capacity requirements (h) year 3 – man-hours				
	1	2	3	4	Total
WC 100	462[1]	538	462	510	1973
WC 200	1079[2]	1256	1079	1190	4603
Total	1541	1794	1541	1700	6576

[1] $462 = 0.30 \cdot (3.19 \cdot 400 + 1.32 \cdot 200)$
[2] $1079 = 0.70 \cdot (3.19 \cdot 400 + 1.32 \cdot 200)$

Quarter	Capacity requirements (h) year 3 – machine-hours				
	1	2	3	4	Total
WC 100	881[1]	865	881	932	3559
WC 200	0	0	0	0	0
Total	881	865	881	932	3559

[1] $881 = 1.00 \cdot (1.02 \cdot 400 + 2.37 \cdot 200)$

Problem 14.3

The capacity planning factors will be 0.3 h/piece in WC 980 and 0.4 h/piece in WC 990. Capacity is 140 h/week in WC 980 and 200 h/week in WC 990.

Capacity requirements and a comparison of available capacity for both work centres in weeks 11–20 are shown in the table:

Week	11	12	13	14	15	16	17	18	19	20
Production plan (pcs)	400	500	450	550	450	400	400	450	500	450
WC 980:										
Capacity requirements (h)	120[1]	150	135	165	135	120	120	135	150	135
Available capacity (h)	140	140	140	140	140	140	140	140	140	140
Deviation	20	–10	5	–25	5	20	20	5	–10	5
WC 990:										
Capacity requirements (h)	160[2]	200	180	220	180	160	160	180	200	180
Available capacity (h)	200	200	200	200	200	200	200	200	200	200
Deviation	40	0	20	–20	20	40	40	20	0	20

[1] $120 = 0.3 \cdot 400$
[2] $160 = 0.4 \cdot 400$

Capacity in WC 980 is insufficient in weeks 12, 14 and 19 and in WC 990 in week 14.

Problem 14.4

a) The capacity bill for product 111 is shown in the table:

Work centre	Capacity requirement (h/piece)
M 1	$(1 / 5 + 0.20) + (1 / 10 + 0.10) + (2 / 20 + 0.25) = 0.95$
M 2	$1 / 10 + 0.15 = 0.25$

b) Calculated nominal capacity and capacity in the work centres are shown in the table:

Work centre	Number of machines	Shifts per day	Hours per shift	Days per week	Nominal capacity	Utilisation factor	Net capacity
M 1	4	2	7.5	5	300[1]	0.85	255[2]
M 2	1	2	7.5	5	75	0.85	64

[1] $300 = 4 \cdot 2 \cdot 7.5 \cdot 5$
[2] $255 = 0.85 \cdot 300$

c) Capacity requirements and available capacity in hours for the work centres M 1 and M 2 are shown in the table:

Week	21	22	23	24	25	26	Total
M 1:							
Capacity requirements	209[1]	237.5	237.5	285	209	237.5	1415.5
Available capacity	255	255	255	255	255	255	1530
Deviation	46	17.5	17.5	–30	46	17.5	114.5
M 2:							
Capacity requirements	55[2]	62.5	62.5	75	55	62.5	372.5
Available capacity	64	64	64	64	64	64	384
Deviation	9	1.5	1.5	–11	9	1.5	11.5

[1] $209 = 0.95 \cdot 220$, where 220 is the requirement in week 21 according to the production plan.
[2] $55 = 0.25 \cdot 220$, where 220 is the requirement in week 21 according to the production plan.

Both work centres are overloaded in week 24, but there is sufficient total capacity over the six-week period.

Problem 14.5

a) The table shows the resource profile (i.e. a capacity bill with lead time offset capacity requirements) in hours per piece for product C:

Work centre	Period relative to requirement week		
	–2	–1	0
WC 1	0.43[1]	0.50	2.30
WC 2	0.56	0.76[2]	0

[1] 0.43 = 0.9 / 30 + 0.4 (operation 10 for item Z)
[2] 0.76 = 2.4 / 60 + 0.2 + 0.6 / 30 + 0.5 (operation 10 for item Y and operation 20 for item Z)

b) The results of the capacity requirements calculations for product C are as shown in the table. The capacity requirement is stated in hours.

Work centre	Week				
	14	15	16	17	18
WC 1	80.4	71.8	94.4	85.4	96.9
WC 2	32.0	36.8	35.8	39.6	39.6

The next table shows in more detail how capacity requirements per hour were calculated:

Work centre	Week					Reason for capacity requirements
	14	15	16	17	18	
WC 1					12.9[1]	30 pcs C in week 20
				12.9[1]	15.0[2]	30 pcs C in week 19
			12.9[1]	15.0[2]	69.0[3]	30 pcs C in week 18
		10.8	12.5	57.5		25 pcs C in week 17
	12.9[1]	15.0[2]	69.0[3]			30 pcs C in week 16
	10.0	46.0				20 pcs C in week 15
	57.5					25 pcs C in week 14
WC 1 total	80.4	71.8	94.4	85.4	96.9	
WC 2					16.8	30 pcs C in week 20
				16.8	22.8	30 pcs C in week 19
			16.8	22.8	0.0	30 pcs C in week 18
		14.0[5]	19.0[4]	0.0		25 pcs C in week 17
	16.8	22.8	0.0			30 pcs C in week 16
	15.2	0.0				20 pcs C in week 15
	0.0					25 pcs C in week 14
WC 2 total	32.0	36.8	35.8	39.6	39.6	

[1] 12.9 = 30 · 0.43
[2] 15.0 = 30 · 0.50
[3] 69.0 = 30 · 2.30
[4] 19.0 = 25 · 0.76
[5] 14.0 = 25 · 0.56

Problem 14.6

The table shows the resource profile in hours per piece for product K:

Work centre	Period relative to requirement week			
	−3	−2	−1	0
WC 200			1.2^2	1.9^1
WC 300	2.1	0.6	1.3^3	

[1] $1.9 = 4.0 / 10 + 1.5$ (item K)   [2] $1.2 = 2.0 / 10 + 1.0$ (item M)   [3] $1.3 = 3.0 / 10 + 1.0$ (item L)

The next table shows the calculation of capacity requirements in hours for product K in both work centres:

Work centre	Week								Reason for capacity requirements
	8	9	10	11	12	13	14	15	
WC 200							24.0^2	38.0^1	20 pcs K in period 15
						0.0	0.0		0 pcs K in period 14
					12.0	19.0			10 pcs K in period 13
				24.0	38.0				20 pcs K in period 12
			12.0	19.0					10 pcs K in period 11
WC 200, total			12.0	43.0	50.0	19.0	24.0	38.0	
WC 300					42.0	12.0	26.0^3		20 pcs K in period 15
				0.0	0.0	0.0			0 pcs K in period 14
			21.0	6.0	13.0				10 pcs K in period 13
		42.0	12.0	26.0					20 pcs K in period 12
	21.0	6.0	13.0						10 pcs K in period 11
WC 300, total	21.0	48.0	46.0	32.0	55.0	12.0	26.0	0.0	

[1] $38 = 20 \cdot 1.9$ [2] $24 = 20 \cdot 1.2$ [3] $26 = 20 \cdot 1.3$

Problem 14.7

a) Material requirements planning is first carried out as in the table:

Item			Week					
			1	2	3	4	5	6
Q	Gross requirements		40	40	40	30	40	30
	Scheduled receipts							
	Stock on hand	50	10	0	0	0	0	0
	Net requirements			30	40	30	40	30
	Planned delivery in			30	40	30	40	30
	Planned order start		30	40	30	40	30	
R	Gross requirements		30	40	30	40	30	
	Scheduled receipts		50					
	Stock on hand	10	30	40	10	20	40	40
	Net requirements			10		30	10	
	Planned delivery in			50		50	50	
	Planned order start		50		50	50		
S	Gross requirements		90	120	90	120	90	
	Scheduled receipts		130					
	Stock on hand	0	40	70	130	10	70	70
	Net requirements			80	20		80	
	Planned delivery in			150	150		150	
	Planned order start		150	150		150		

Every item is manufactured in one single operation. Since the lead time is one week for all operations and the logic of material requirements planning dictates that a planned order start in a certain week means that the order starts at the beginning of the week, all capacity requirements for an order (operation) come in the same week as the planned order start. All planned orders are summarised in the table:

Item	Week	1	2	3	4	5	6
Q	Planned delivery in		30	40	30	40	30
	Planned order start	30	40	30	40	30	
R	Planned delivery in		50		50	50	
	Planned order start	50		50	50		
S	Planned delivery in		150	150		150	
	Planned order start	150	150		150		

Capacity requirements per piece calculated from stated operation data are shown in the table:

Item	Work centre	Capacity requirements/piece (h)
Q	A 10	1.70[1]
R	WC 30	1.06[2]
S	WC 70	1.01

[1] 1.70 = piece time for product Q (set-up time is zero).
[2] 1.06 = 3.0 / 50 + 1.0

With the aid of data from the above tables, capacity requirements in hours in the three work centres for five weeks ahead is calculated as in the next table:

Work centre	Week				
	1	2	3	4	5
M 10	51[1]	68	51	68	51
WC 30	53	0	53	53	0
WC 70	152	152	0	152	0

[1] $51 = 30 \cdot 1.70$

b) Since net capacity in work centre A 10 is 60 hours, the work centre will be over-loaded in weeks 2 and 4. Possible solutions are:

- smooth out the master production schedule (35 pcs in weeks 3–6), which will result in a maximum of 59.5 hours capacity requirements per week.
- use overtime in weeks 2 and 4, but this will mean increased costs.
- make a temporary transfer of personnel from other work centres in weeks 2 and 4.

Problem 14.8

Using the operation data for items and work centre data, lead time templates (backward scheduling) for the two items can be produced. These are shown in the tables.

Lead time template for item 880:

Lead time component	Lead time (h)	Cumulative lead time (h)
Transportation	1	1
Op 20: op time, WC 740	13.5[1]	14.5
Op 20: set-up time, WC 740	2	16.5
Queue time WC 740	10	26.5
Transportation	1	27.5
Op 10: op time, WC 730	4.5	32
Op 10: set-up time, WC 730	1	33
Queue time WC 730	6	39

[1] $13.5 = 15 \cdot 0.9$

Lead time template for item 990:

Lead time component	Lead time (h)	Cumulative lead time (h)
Transportation	1	1
Op 20: op time, WC 750	27	28
Op 20: set-up time, WC 750	1.5	29.5
Queue time WC 750	15	44.5
Transportation	1	45.5
Op 10: op time, WC 740	18	63.5
Op 10: set-up time, WC 740	2	65.5
Queue time WC 740	10	75.5

With the aid of lead time templates and information on the planned start time and finish time of the order, operations and thus capacity requirements can be scheduled using backward scheduling. The table shows how this is carried out for the planned order of 90 pcs of item 990 with delivery in week 43:

Work centre	Planning period (week)									
	41					42				
	M	T	W	Th	F	M	T	W	Th	F
WC 750								Op 20, 1.5 h + 27 h		
WC 740			Op 10, 2 h + 18 h							

Operations are spread out for other orders in the same way. Capacity requirements in hours will then be allocated as in the table:

Planned and ongoing orders	WC 730 Week				WC 740 Week				WC 750 Week			
	40	41	42	43	40	41	42	43	40	41	42	43
Planned delivery: 15 pcs of item 880 in week 42		5.5				15.5						
Planned delivery: 15 pcs of item 880 in week 44				5.5				15.5				
Planned delivery: 90 pcs of item 990 in week 43						20[1]					28.5[1]	
Planned delivery: 90 pcs of item 990 in week 45								20				
Released order: 90 pcs of item 990, delivery in week 41									28.5			
Total capacity requirement	0	5.5	0	5.5	0	35.5	0	35.5	28.5	0	28.5	0

[1] Capacity requirement as in the figure above

The total capacity requirement in the work centres (referred to in the table below) is obtained by adding capacity requirements for items 880 and 990 according to the table above and the capacity requirement for other items at the company.

Week	40	41	42	43
WC 730:				
Cap. requirement (h)	65	65.5	68	69.5
Capacity (h)	70	70	70	70
Deviation	5	4.5	2	0.5
WC 740:				
Capacity requirement (h)	58	69.5	67	85.5
Capacity (h)	70	70	70	70
Deviation	12	0.5	3	−15.5
WC 750:				
Cap. requirement (h)	34.5	34	40.5	33
Capacity (h)	35	35	35	35
Deviation	0.5	1	−5.5	2

Problem 15.1

Demand per hour = 1000 / 40 = 25 pieces

$$n = \frac{25 \cdot 5 \cdot (1 + 0.2)}{10} = 15 \text{ kanbans}$$

Problem 15.2

a)

Demand per hour = 240 / 8 = 30

Total lead time $= \dfrac{30}{60} + \dfrac{20}{60} + \dfrac{20}{60} + 4 = 5$ hours

Number of kanbans, $n = \dfrac{30 \cdot 5 \cdot (1 + \alpha)}{30} = 8$

Safety factor, $\alpha = \dfrac{8 \cdot 30}{30 \cdot 5} - 1 = 0.60$

Safety stock $= D \cdot L \cdot \alpha = 30.5 \cdot 0.60 = 90$ pieces

b)

x = cycle time

$$7 = \frac{30 \cdot x \cdot (1 + 0.60)}{30}$$

$$x = \frac{7 \cdot 30}{30 \cdot (1 + 0.60)} = 4.375 \text{ hrs} = 4 \text{ hours and } 23 \text{ minutes}$$

The cycle time needs to be reduced by 37 minutes (from 5 hours to 4 hours and 23 minutes)

c)

$$7 = \frac{37.5 \cdot 5 \cdot (1 + \alpha)}{30}$$

Safety factor, $\alpha = \frac{7 \cdot 30}{37.5 \cdot 5} - 1 = 0.12$

Safety stock $= 37.5 \cdot 5 \cdot 0.12 = 22.5$ pieces

Problem 15.3

Lead time $= 4 \cdot 60 + 10 \cdot 6 + 6 = 306$ minutes

Number of kanbans $= 10$

Pallet capacity $= 10$ pieces

Demand per minute $= \dfrac{10 \cdot 10}{306 \cdot (1 + 0.10)} = 0.297$ pieces per minute

If 8 working hours per day, then the daily demand could be $0.297 \cdot 60 \cdot 8 = 143$ pieces

Problem 16.1

The table shows dispatch lists generated from the three priority rules:

Priority order	(a) Shortest operation time		(b) Earliest start time for operation		(c) Critical ratio	
	Order	Operation time	Order	Start time for operation	Order	Critical ratio
1	107	0.5	109	95	109	0.63[1]
2	101	1.0	107	100	107	1.00
3	110	1.5	110	101	103	1.04
4	109	2.0	103	102	110	1.05[2]
5	105	2.5	101	103	101	1.33
6	103	3.0	105	108	105	2.25

[1] $0.63 = (105 - 100) / 8$
[2] $1.05 = (110 - 100) / 9.5$

Problem 16.2

The table shows a backward planning of order TO4321 from the end of week 5. The delivery time for the order is on Monday week 6.

Operation	Work centre	Start time	Op.tid (hrs)	Finish time	Slack time (hrs)	Capacity requirement
10	WC 100	W3 Hour 31	9	W3 Hour 40	16	9 hrs in week 3
20	WC 200	W4 Hour 16	5	W4 Hour 21	16	5 hrs in week 4
30	WC 300	W4 Hour 37	14	W5 Hour 11	16	3 hrs in week 4
						11 hrs in week 5
40	WC 400	W5 Hour 27	5	W5 Hour 32	8	5 hrs in week 5

The capacity requirements in the different work centres are received by summing the capacity requirements for order TO4321 and all released orders. The result is shown in the table below. The available capacity is shown in hours per week.

Work centre	Available capacity	Capacity requirement (hours) Week						
		1	2	3	4	5	6	7
WC 100	70	35	26	18 + 9	9	4	0	0
WC 200	35	35	35	26	35 + 5	13	9	5
WC 300	70	70	70	70	70 + 3	61 + 11	52	70
WC 400	35	35	35	35	35	31 + 5	30	26

It is not possible to deliver TO4321 on Monday week 6 because the capacity requirement exceeds the available capacity in WC 200 in week 4, in WC 300 in weeks 4 and 5 and in WC 400 in week 5. The earliest delivery date for the order is on Monday in week 7, which gives the following capacity requirement:

Work centre	Available capacity	Capacity requirement (hours) Week						
		1	2	3	4	5	6	7
WC 100	70	35	26	18	9 + 9	4	0	0
WC 200	35	35	35	26	35	13 + 5	9	5
WC 300	70	70	70	70	70	61 + 3	52 + 11	70
WC 400	35	35	35	35	35	31	30 + 5	26

Problem 16.3

a) The throughput time is $1 + 40 \cdot 1 + 1 + 2 + 40 \cdot 1.5 + 1 + 2 + 40 \cdot 1.0 = 147$ hours, which corresponds to 3.7 weeks. An alternative is to prepare for set-up in WC 2 and WC 3 so that the machines are ready for production when orders arrive. However, this would only reduce the throughput times by 4 hours (to 143 hours, which corresponds to 3.6 weeks). It is consequently not possible to deliver the order on time.

b) By splitting the order into two batches, each of 20 pieces, and producing the two batches in parallel in WC 2 and WC 3 the throughput time would be reduced to $1 + 40 \cdot 1 + 1 + 2 + 20 \cdot 1.5 + 1 + 2 + 20 \cdot 1.0 = 97$ hours, which corresponds to 2.4 weeks. By preparing for set-up in WC 2 and WC 3 the throughput time can be further reduced by 4 hours to 93 hours, which corresponds to 2.4 weeks. The order can consequently be delivered on time. However, this requires that the order gets highest priority in the factory, i.e. no queue time at the work centres.

c) By dividing the order into several batches and applying overlapping, the throughput time can be reduced without using both machines WC 2 and WC 3 in parallel. Also consider that the set-up in WC 2 and WC 3 is prepared and finished when material is received from the previous WC.

First, the maximum batch sizes (X pieces) are determined. Because WC 2 is the constraining machine, the batches will queue up before this work centre. In WC 3, the batches will be processed immediately when arriving. The following table is generated for the first batch up to WC 2 and for the last batch from WC 2:

Part of throughput time	Time (hours)
Making the first batch in WC 1	$1 + X \cdot 1$
Moving first batch to WC 2	1
Making all batches in WC 2	$40 \cdot 1.5$
Moving last batch to WC 3	1
Making last batch in WC 3	$X \cdot 1$
Total throughput time	$1 + X \cdot 1 + 1 + 40 \cdot 1.5 + 1 + X \cdot 1 = 63 + 2X$

If considering that the throughput time cannot be longer than 3 weeks minus 1 day, i.e. 112 hours, we get $63 + 2X = 112$ which results in $X = 24.5$ pieces. This means that the maximum batch size should be 24 pieces. It would be practical to choose batch sizes of 20 pieces which is half the order quantity.

Appendix B: Safety Stock Tables

Normal distribution

Safety factor	Service level %	Safety factor	Service level %	Safety factor	Service level %	Safety factor	Service level %
0.00	50.0	0.72	76.4	1.44	92.5	2.16	98.5
0.02	50.8	0.74	77.0	1.46	92.8	2.18	98.5
0.04	51.6	0.76	77.6	1.48	93.1	2.20	98.6
0.06	52.4	0.78	78.2	1.50	93.3	2.22	98.7
0.08	53.2	0.80	78.8	1.52	93.6	2.24	98.7
0.10	54.0	0.82	79.4	1.54	93.8	2.26	98.8
0.12	54.8	0.84	80.0	1.56	94.1	2.28	98.9
0.14	55.6	0.86	80.5	1.58	94.3	2.30	98.9
0.16	56.4	0.88	81.0	1.60	94.5	2.32	99.0
0.18	57.1	0.90	81.6	1.62	94.7	2.34	99.0
0.20	57.9	0.92	82.1	1.64	94.9	2.36	99.1
0.22	58.7	0.94	82.6	1.66	95.2	2.38	99.1
0.24	59.5	0.96	83.1	1.68	95.4	2.40	99.2
0.26	60.3	0.98	83.6	1.70	95.5	2.42	99.2
0.28	61.0	1.00	84.1	1.72	95.7	2.44	99.3
0.30	61.8	1.02	84.6	1.74	95.9	2.46	99.3
0.32	62.6	1.04	85.1	1.76	96.1	2.48	99.3
0.34	63.3	1.06	85.5	1.78	96.2	2.50	99.4
0.36	64.1	1.08	86.0	1.80	96.4	2.52	99.4
0.38	64.8	1.10	86.4	1.82	96.6	2.54	99.4
0.40	65.5	1.12	86.9	1.84	96.7	2.56	99.5
0.42	66.3	1.14	87.3	1.86	96.9	2.58	99.5
0.44	67.0	1.16	87.7	1.88	97.0	2.60	99.5
0.46	67.7	1.18	88.1	1.90	97.1	2.62	99.6
0.48	68.4	1.20	88.5	1.92	97.3	2.64	99.6
0.50	69.1	1.22	88.9	1.94	97.4	2.66	99.6
0.52	69.8	1.24	89.3	1.96	97.5	2.68	99.6
0.54	70.5	1.26	89.6	1.98	97.6	2.70	99.7
0.56	71.2	1.28	90.0	2.00	97.7	2.72	99.7
0.58	71.9	1.30	90.3	2.02	97.8	2.74	99.7
0.60	72.6	1.32	90.7	2.04	97.9	2.76	99.7
0.62	73.2	1.34	91.0	2.06	98.0	2.78	99.7
0.64	73.9	1.36	91.3	2.08	98.1	2.80	99.7
0.66	74.5	1.38	91.6	2.10	98.2	2.82	99.8
0.68	75.2	1.40	91.9	2.12	98.3	2.84	99.8
0.70	75.8	1.42	92.2	2.14	98.4	2.86	99.8

Service loss function E(z)

Safety factor	Service function	Safety factor	Service function	Safety factor	Service function	Safety factor	Service function
0.00	0.3989	0.72	0.1381	1.44	0.0336	2.16	0.0055
0.02	0.3890	0.74	0.1334	1.46	0.0321	2.18	0.0052
0.04	0.3793	0.76	0.1289	1.48	0.0307	2.20	0.0049
0.06	0.3699	0.78	0.1245	1.50	0.0293	2.22	0.0046
0.08	0.3602	0.80	0.1202	1.52	0.0280	2.24	0.0044
0.10	0.3509	0.82	0.1160	1.54	0.0267	2.26	0.0041
0.12	0.3418	0.84	0.1120	1.56	0.0255	2.28	0.0039
0.14	0.3328	0.86	0.1080	1.58	0.0244	2.30	0.0037
0.16	0.3240	0.88	0.1042	1.60	0.0232	2.32	0.0035
0.18	0.3154	0.90	0.1004	1.62	0.0222	2.34	0.0033
0.20	0.3069	0.92	0.0968	1.64	0.0211	2.36	0.0031
0.22	0.2986	0.94	0.0933	1.66	0.0201	2.38	0.0029
0.24	0.2904	0.96	0.0899	1.68	0.0192	2.40	0.0027
0.26	0.2824	0.98	0.0865	1.70	0.0183	2.42	0.0026
0.28	0.2745	1.00	0.0833	1.72	0.0174	2.44	0.0024
0.30	0.2668	1.02	0.0802	1.74	0.0166	2.46	0.0023
0.32	0.2592	1.04	0.0772	1.76	0.0158	2.48	0.0021
0.34	0.2518	1.06	0.0742	1.78	0.0150	2.50	0.0020
0.36	0.2445	1.08	0.0714	1.80	0.0143	2.52	0.0019
0.38	0.2374	1.10	0.0686	1.82	0.0136	2.54	0.0018
0.40	0.2304	1.12	0.0660	1.84	0.0129	2.56	0.0017
0.42	0.2236	1.14	0.0634	1.86	0.0123	2.58	0.0016
0.44	0.2169	1.16	0.0609	1.88	0.0116	2.60	0.0015
0.46	0.2104	1.18	0.0584	1.90	0.0111	2.62	0.0014
0.48	0.2040	1.20	0.0561	1.92	0.0105	2.64	0.0013
0.50	0.1978	1.22	0.0538	1.94	0.0100	2.66	0.0012
0.52	0.1917	1.24	0.0517	1.96	0.0094	2.68	0.0011
0.54	0.1857	1.26	0.0495	1.98	0.0090	2.70	0.0011
0.56	0.1799	1.28	0.0475	2.00	0.0085	2.72	0.0010
0.58	0.1742	1.30	0.0455	2.02	0.0080	2.74	0.0009
0.60	0.1687	1.32	0.0437	2.04	0.0076	2.76	0.0009
0.62	0.1633	1.34	0.0418	2.06	0.0072	2.78	0.0008
0.64	0.1580	1.36	0.0400	2.08	0.0068	2.80	0.0008
0.66	0.1528	1.38	0.0383	2.10	0.0065	2.82	0.0007
0.68	0.1478	1.40	0.0367	2.12	0.0061	2.84	0.0007
0.70	0.1429	1.42	0.0351	2.14	0.0058	2.86	0.0006

Glossary

ABC analysis*   A method for dividing items, customers, suppliers and other objects into different classes based on specific criteria, e.g. volume value per item, profit contribution per product, turnover per customer, purchase value per supplier. ABC analysis is an application of the minority principle, meaning that in every group of objects there is a small number which accounts for a large proportion of the value. ABC analysis is also called the 80/20 rule.

Accumulated production lead time   The time from ordering of raw materials and other start-up materials for manufacturing until the point in time when the product can be delivered.

Activity-based costing*   A cost accounting system that accumulates costs based on activities performed and then uses cost drivers to allocate these costs to products or other bases, such as customers, markets or projects. It is an attempt to allocate overhead costs on a more realistic basis than direct labour or machine-hours.

Advance shipment notice   A notification sent by the supplier to notify the customer that a delivery is on the way.

Advanced planning and scheduling (APS) system*   Techniques that deal with analysis and planning of logistics and manufacturing over the short-, intermediate- and long-term time periods. APS describes any computer program that uses advanced mathematical algorithms or logic to perform optimisation or simulation on finite capacity scheduling, sourcing, capital planning, resource planning, forecasting, demand management and others. The five main components of APS systems are demand planning, production planning, production scheduling, distribution planning and transportation planning.

Allocation*   The process of designating stock for a specific order or schedule. Also called reservation.

Assemble to order (ATO)*   A production environment where a good or service can be assembled after receipt of a customer's order.

Available capacity*   The capability of a system or resource to produce a quantity of output in a particular time period.

Available to promise (ATP)*   The uncommitted portion of a company's inventory and planned production, maintained in the master schedule to support customer order promising.

Backflushing*   A method of inventory bookkeeping where the book inventory of components is automatically reduced by the computer after completion of activity on the component's upper-level parent item based on what should have been used as specified on the bill of material and allocation records.

Backward scheduling*   A technique for calculating operation start dates and due dates. The schedule is computed starting with the due date for the order and working backward to determine the required start date and/or due dates for each operation.

Basic data   Basic data means fundamental information about a company's products, which items they consist of, how they are manufactured and information about the company's manufacturing resources.

Batch accounting   Inventory accounting carried out for every batch delivered. It requires that every batch delivered is kept separate from other delivered batches.

Bill of material (BOM)*   A listing of all sub-assemblies, intermediates, parts and raw materials that go into a parent assembly showing the quantity of each required to make an assembly. Also called product structure.

Bull-whip effect   Variations in demand which are amplified upstream in supply chains with storage points.

Call-off   The part of the delivery plan furthest in time. It is normally considered as a clear order from the customer.

Capacity*   The capability of a worker, machine, work centre, plant or organisation to produce output per time period.

Capacity bill   A capacity bill expresses the total capacity requirements per piece of a product in terms of the resources required for its manufacture.

Capacity planning*   The process of determining the amount of capacity required to produce in the future.

Capacity requirements planning (CRP)*   The function of establishing, measuring and adjusting limits or levels of capacity. The term capacity requirements planning in this context refers to the process of determining in detail the amount of labour and machine resources required to accomplish the tasks of production. Open shop orders and planned orders in the MRP system are input to CRP, which through the use of parts routings and time standards translates these orders into hours of work by work centre by time period.

Capacity utilisation*   The extent to which the available capacity (e.g. work centre, machine, truck, etc.) is used.

Carrying cost*   The cost of holding inventory, usually defined as a percentage of the euro value of inventory per unit of time (generally one year). Carrying cost depends mainly on the cost of capital invested as well as such costs of maintaining the inventory as taxes and insurance, obsolescence, spoilage, and space occupied. Such costs vary from 10 per cent to 35 per cent annually, depending on type of industry. Carrying cost is ultimately a policy variable reflecting the opportunity cost of alternative uses for funds invested in inventory.

Chase strategy*   A production planning method that maintains a stable inventory level while varying production to meet demand. Companies may combine chase and level production schedule methods.

Collaborative planning forecasting and replenishment (CPFR)**   A collaboration process whereby supply chain trading partners can jointly plan key supply chain activities from production and delivery of raw materials to production and delivery of final products to end customers.

Configurator*   Software system that creates, uses, and maintains product models that allow complete definition of all possible product options and variations with a minimum of data entries.

Constraint*   Any element or factor that prevents a system from achieving a higher level of performance with respect to its goal. Constraints can be physical, such as a machine centre or lack of material, but they can also be managerial, such as a policy or procedure.

Constraint-based material requirements planning   A technique similar to material requirements planning to calculate requirements for materials and to plan replenishment orders. The difference is that constraint-based material requirements planning considers capacity constraints in production and delays in deliveries from external suppliers. It is based on the theory of constraints (TOC).

Continuous line process   Continuous line process is a linear process in which material flow is continuous. If there is discrete manufacturing, this means that every unit of a manufactured item is handed over to the next manufacturing step as it is produced, and not after a whole batch is produced.

Control stock   A type of inventory that arises accidentally. Its size is related to imperfections in control and caused by items having to wait for other items to achieve simultaneous availability. A common example is items required for an assembly but which must wait in stock for starting the assembly operation due to another item being out of stock.

CONWIP   CONWIP stands for constant work in process. It is a pull-based system similar to the generic kanban method in the sense that it generates and sends a signal upstream when it is time to produce items but it does not specify what to produce. It is thus an alternative method to apply in a mixed-model production environment. In a CONWIP system, the last unit in a process is the only unit sending a signal and the first unit in the process is the only unit receiving a signal, no matter how many individual operations are included in the process. This procedure secures a constant level of work in process in the system at the same time as it allows for mix variations over time.

Co-ordination stock   A co-ordination stock is a type of inventory caused by deliberate co-ordination of material flows. Two examples are the simultaneous inbound delivery of several items from the same supplier to decrease transport costs, and the simultaneous ordering of manufacturing to decrease set-up costs in production resources, even though the items have somewhat different requirements dates.

Cover time   Synonym: run-out time.

Cover time planning   Also known as Run-out time planning, is the time for which available stocks, i.e. physically present stocks plus planned inbound deliveries, are expected to last.

Customer-managed ordering   A customer/supplier ordering process in which the customer is responsible for entering customer orders directly into the supplier's ERP system or web-shop.

Customer order*   An order from a customer for a particular product or a number of products. It is often referred to as an actual demand to distinguish it from a forecasted demand.

Customer order decoupling point   The point in time at which a product becomes earmarked for a particular customer. Downstream this point, the material flow is driven by customer orders in contrast to upstream flows which are driven by forecasts.

Customer order point   The position in a product's bill of material from which the product has customer-specific appearance and characteristics.

Customer order process   One of the two processes in a company that is related to demand management for the company's products. It covers the known part of demand and the forecast process covers that part of demand which must be estimated in different ways.

Cycle counting*   An inventory accuracy audit technique where inventory is counted on a cyclic schedule rather than once a year. A cycle inventory count is usually taken on a regular, defined basis (often more frequently for high-value or fast-moving items and less frequently for low-value or slow-moving items).

Cycle service   A measure of the service level in stock defined as the probability that no shortages occur in an inventory cycle.

Cycle stock*   One of the two main conceptual components of any item inventory, the cycle stock is the most active component, i.e. that which depletes gradually as customer orders are received and is replenished cyclically when suppliers' orders are received. The other conceptual component of the item inventory is the safety stock, which is a cushion of protection against uncertainty in the demand or in the replenishment lead time.

Cycle time*   (1) In industrial engineering, the time between completion of two discrete units of production. For example, the cycle time for motors assembled at a rate of 120 per hour would be 30 seconds. (2) In materials management, it refers to the length of time from when material enters a production facility until it exists.

Damping functions   Parameters used to suppress unimportant rescheduling action massages in material requirements planning.

Delivery agreement   A delivery agreement may be drawn up in many different ways, but in general it is an agreement to purchase a certain minimum quantity of a specific item or group of items at a certain price during a certain time period, normally six months or one year.

Delivery flexibility   In general terms flexibility can be defined as the ability to quickly and efficiently react to changing conditions. Specifically, delivery flexibility refers to this ability concerning changing open orders, quantity wise as well as time wise.

Delivery lead time*   The time from receipt of a customer order to the delivery of the product.

Delivery plan   A delivery plan refers to a plan on the sales- and operations-planning level regarding volumes to deliver over the next year or so. Sales plan is a commonly used alternative name.

Delivery precision   Delivery precision refers to the extent to which deliveries take place at the delivery dates agreed with the customer. While the efficiency measurement of stock service-level is intended to state the delivery capacity from stock, the measurement of delivery precision is intended as an expression of delivery capacity for companies with assemble-to-order, make-to-order and engineer-to-order type of operations.

Delivery reliability   A delivery service measure meaning the extent to which the right products are delivered in the right quantities.

Delivery schedule*   The required or agreed time or rate of delivery of goods or services purchased for a future period.

Demand*   A need of a particular product or component.

Demand time fence   (1) That point in time inside of which the forecast is no longer included in total demand and projected available inventory calculations; inside this point, only customer orders are considered. Beyond this point, total demand is a combination of actual orders and forecasts, depending on the forecast consumption technique chosen. (2) In some contexts, the demand time fence may correspond to that point in the future inside which changes to the master schedule must be approved by an authority higher than the master scheduler. Note, however, that customer orders may still be promised inside the demand time fence without higher authority approval if there are quantities available to promise (ATP). Beyond the demand time fence, the master scheduler may change the MPS within the limits of established rescheduling rules without the approval of higher authority.

Dependent demand* Demand that is directly related to or derived from the bill of material structure for other items or end products. Such demands are therefore calculated and need not and should not be forecast. A given inventory item may have both dependent and non dependent demand at any given time. For example, a part may simultaneously be the component of an assembly and sold as a service part.

Design bill A design bill means a bill of material for a product which is formed to describe how a product is composed from a design and functional point of view. In a design bill, the product's structure is generally considered from a functional perspective.

Direct physical inventory counting Direct physical inventory counting is used to count individual items selectively when certain events occur. This type of physical inventory counting is often used as a complement to other types of physical inventory counting. An example of event which may necessitate this type of physical inventory counting are stock transactions that have caused the stock on hand in the inventory accounting system to be negative.

Dispatch list* A listing of manufacturing orders in priority sequence. The dispatch list, which is usually communicated to the manufacturing floor via paper or electronic media, contains detailed information on priority, location, quantity and the capacity requirements of the manufacturing order by operation. Dispatch lists are normally generated daily and oriented by work centre.

Drum–buffer–rope (DBR)* In the theory of constraints, the generalised process used to manage resources to maximise throughput. The drum is the rate or pace of production set by the system's constraint. The buffers establish protection against uncertainty. The rope is the communication process from the constraint.

Economic order quantity (EOQ)* A type of fixed order quantity model that determines the amount of an item to be purchased or manufactured at one time. The intent is to minimise the combined costs of acquiring and carrying inventory.

Economic run-out time Economic run-out time is a run-out time calculated as a trade-off between ordering costs and inventory carrying costs that arise so that the total costs associated with storage are minimised.

Efficiency factor A measurement of the actual output to the standard output expected. It measures how well operations are performing in relation to existing standards. The efficiency factor is often expressed in percent.

Electronic data interchange (EDI)* The paperless (electronic) exchange of trading documents, such as purchase orders, shipment authorisations, advanced shipment notices and invoices, using standardised document formats.

Engineer to order (ETO)* Products whose customer specifications require unique engineering design, significant customisation or new purchased materials. Each customer order results in a unique set of part numbers, bills of material and routings.

Engineering change* A revision of drawing or design released by engineering to modify or correct a part. The request for the change can be from a customer or from production, quality control, another department or a supplier.

Enterprise resource planning (ERP)* Framework for organising, defining and standardising the business processes necessary to effectively plan and control an organisation so the organisation can use its internal knowledge to seek external advantage.

Execution and control A planning level used for more detailed planning of manufacturing orders. Planning objects are the operations, i.e. manufacturing steps, belonging to the manufacturing order planned to be carried out. The level of execution and control covers planning of new order releases to the shop floor, including material availability checks and sequencing in which order the various operations should be carried in the available manufacturing resources.

Explosion* The process of calculating the demand for the components of a parent item by multiplying the parent item requirements by the component usage quantity specified in the bill of material. Also called requirements explosion.

Exponential smoothing forecast* A type of weighted moving average forecasting technique in which past observations are geometrically discounted according to their age.

Fill rate* A measure of delivery performance of finished goods, usually expressed as a percentage. In a make-to-stock company, this percentage usually represents the number of items or euros (on one or more customer orders) that were shipped on schedule for a specific time period, compared with the total that were supposed to be shipped in that time period.

Fill rate service   A measure of the service level in stock defined as the percentage of demand that can be delivered directly from stock.

Final assembly schedule (FAS)*   A schedule of end items to finish the product for specific customers' orders in a make-to-order or assemble-to-order environment. It is also referred to as the finishing schedule because it may involve operations other than just the final assembly; also, it may not involve assembly, but simply final mixing, cutting, packaging, etc. The FAS is prepared after receipt of a customer order as constrained by the availability of material and capacity, and it schedules the operations required to complete the product from the level where it is stocked (or master scheduled) to the end-item level.

Finite capacity scheduling   A scheduling methodology where work is scheduled on work centres in a way such that no work centre capacity requirement exceeds the capacity available for that work centre.

Firm planned order*   A planned order that can be frozen in quantity and time. The computer is not allowed to change it automatically; this is the responsibility of the planner in charge of the item that is being planned. This technique can aid planners working with MRP systems to respond to material and capacity problems by firming up selected planned orders. In addition, firm planned orders are the normal method of stating the master production schedule.

Fixed position stock location system*   A method of storage in which a relatively permanent location is assigned for the storage of each item in a storeroom or warehouse. Although more space is needed to store parts than in a random-location storage system, fixed locations become familiar, and therefore a locator file may not be needed.

Flexibility*   (1) The ability of the manufacturing system to respond quickly, in terms of range and time, to external or internal changes. (2) The ability of a supply chain to mitigate, or neutralise, the risks of demand forecast variability, supply continuity variability, cycle time plus lead-time uncertainty, and transit time plus customs-clearance time uncertainty during periods of increasing or diminishing volume.

Focus forecasting*   A system that allows the user to simulate the effectiveness of numerous forecasting techniques, enabling selection of the most effective one.

Forecast*   An estimate of future demand. A forecast can be constructed using quantitative methods, qualitative methods, or a combination of methods, and it can be based on extrinsic (external) or intrinsic (internal) factors.

Forecast consumption*   The process of reducing the forecast by customer orders or other types of actual demands as they are received. The adjustments yield the value of the remaining forecast for each period.

Forecast error*   The difference between actual demand and forecast demand, stated as an absolute value or as a percentage.

Forecast horizon*   The period of time into future for which a forecast is prepared.

Forecast monitoring   Activity aiming at automatically detecting and signalling when forecasts are systematically too high or too low, i.e. being able to verify that forecasts are close to average.

Forecast time fence   The forecast time fence used in master production scheduling is equivalent to the point in time when the order backlog is normally so small as to be negligible compared with the forecast.

Forward scheduling*   A scheduling technique where the scheduler proceeds from a known start date and computes the completion date for an order, usually proceeding from the first operation to the last. Dates generated by this technique are generally the earliest start dates for operations.

Generic kanban   A generic kanban system works basically in the same way as the traditional kanban system, but with one exception. The generic kanban is an authorisation to start producing or moving an item but it doesn't specify exactly what item to produce or to move. It may specify a product family but not the exact model, variant or option.

Goods reception   Goods reception is the activity and function of receiving goods.

Grassroots approach   Forecasting where all salesmen and other personnel in direct contact with the market make their own forecast assessments separately and work out their own proposals for forecasts.

Gross requirement*   The total of independent and dependent demand for a component before the netting of on-hand inventory and scheduled receipts.

Harris formula   Synonym: Economic order quantity.

Hedging*   (1) An action taken in an attempt to shield the company from an uncertain event such as a strike, price increase, or currency reevaluation. (2) In master scheduling, a scheduled quantity to protect against uncertainty in demand or supply. The hedge is similar to safety stock, except that a hedge has the dimension of timing as well as amount. (3) In purchasing, any purchase or sale transaction having as its purpose the elimination of the negative aspects of price fluctuations.

Heijunka*   In the just-in-time philosophy, an approach to level production throughout the supply chain to match the planned rate of end product sales.

Indented bill – indented bill of material*   A form of multi-level bill of material. It exhibits the highest level parents closest to the left margin, and all the components going into these parents are shown indented toward the right. All subsequent levels of components are indented farther to the right. If a component is used in more than one parent within a given product structure, it will appear more than once, under every sub-assembly in which it is used.

Independent demand*   The demand for an item that is unrelated to the demand for other items. Demand for finished goods, parts required for destructive testing, and service parts requirements are examples of independent demand.

Input/output control*   A technique for capacity control where planned and actual inputs and planned and actual outputs of a work centre are monitored. Planned inputs and outputs for each work centre are developed by capacity requirements planning and approved by manufacturing management. Actual input is compared to planned input to identify when work centre output might vary from the plan because work is not available at the work centre. Actual output is also compared to planned output to identify problems within the work centre.

Intermittent line process   A line process in which entire batch quantities of manufactured items are completed in one manufacturing step before they are transported to the next step.

Internet*   A worldwide network of computers belonging to businesses, governments and universities that enables users to share information in the form of files and to send electronic messages and have access to a tremendous store of information.

Inventory accounting*   The branch of accounting dealing with valuing inventory. Inventory may be recorded or valued using either a perpetual or a periodic system. A perpetual inventory record is updated frequently or in real time, while a periodic inventory record is counted or measured at fixed time intervals, e.g. every second week or monthly. Inventory valuation methods of LIFO, FIFO, or average costs are used with either recording system.

Inventory carrying costs   The costs for keeping goods in stock. It is made up of a financial fraction, a physical fraction and an uncertainty fraction.

Inventory carrying factor   A percentage expressing the inventory carrying cost in relation to the average stock value.

Inventory turnover rate*   The number of times that an inventory cycles, or "turns over", during the year. A frequently used method to compute inventory turnover is to divide the average inventory level into the annual cost of sales. For example, an average inventory of €3 million divided into an annual cost of sales of €21 million means that inventory turned over seven times.

Job reporting   Process of describing how the planned orders have been executed in practice. There are three purposes of job reporting; (1) follow-up released orders, (2) updating resource availability, (3) supplying information on real resource consumption.

Jobbing or job shop process   A process type characterised by production resources being organised by manufacturing function, and the flow of materials during the process of production is adapted to this organisation and production layout.

Just-in-time*   Any unique manufactured or purchased part, material, intermediate, sub-assembly or product.

Kanban*   A method of just-in-time production that uses standard containers or lot sizes with a card attached to each. It is a pull system in which work centres signal with a card that they wish to withdraw parts from feeding operations or suppliers. The Japanese word kanban, loosely translated, means card, billboard or sign but other signalling devices such as coloured golf balls have also been used. The term is often used synonymously for the specific scheduling system developed and used by the Toyota Corporation in Japan.

Labour card   Part of the shop packet. A labour card contains detailed information on the operation, including set-up and run times. It is used to authorise start of operation, but also for reporting actual labour time.

Labour reporting   Labour reporting from the workshop monitors how the execution of manufacturing orders is progressing. It is also required for taking corrective actions, follow-up of real operation times, checking of workloads and follow-up of capacity utilisation in work centres.

Lag strategy   Lag strategy is a reactive strategy and means that investments in new capacity are not made until a change in the size of demand is stated and real. This strategy limits volume flexibility with rising demand and puts greater demands on being able to utilise changes in stock or changes in delivery times to avoid losing sales and market shares.

Lead strategy   A lead strategy is a capacity planning strategy meaning that capacity is increased or decreased before demand rises or falls. It is a proactive strategy. When demand is increasing, this strategy provides volume flexibility which enables a company to gain market share and thereby expand its operations.

Lead time*   The span of time required to perform a process or series of operations. In a logistics context, the time between recognition of the need to order and the receipt of goods.

Lean production*   A philosophy of production that emphasises the minimisation of the amount of all the resources (including time) used in the various activities of the enterprise. It involves identifying and eliminating non-value-adding activities in design, production, supply chain management, and dealing with customers. Lean producers employ teams of multiskilled workers at all levels of the organisation and use highly flexible, increasingly automated machines to produce volumes of products in potentially enormous variety. It contains a set of principles and practices to reduce cost through the relentless removal of waste and through the simplification of all manufacturing and support processes.

Least total cost method*   A dynamic lot sizing technique that calculates the order quantity by comparing the setup (or ordering) costs and carrying cost for various lot sizes and selects the lot size where these costs are most nearly equal.

Least unit cost method*   A dynamic lot sizing technique that adds ordering cost and inventory carrying cost for each trial lot size and divides by the number of units in the lot size, picking the lot size with the lowest unit cost.

Level strategy*   A production planning method that maintains a stable production rate while varying inventory levels to meet demand.

Levelled production*   A production plan that maintains a stable production rate while varying inventory levels to meet demand.

Levelling stock   A stock used to decouple rate of production from rate of consumption. It is used when producing goods at a relatively even rate to achieve a smooth utilisation of capacity and thereby decrease capacity costs.

Lot for lot*   A lot sizing technique that generates planned orders in quantities equal to the net requirements in each period.

Lot sizing*   The process of, or techniques used in, determining lot size.

Make to delivery schedule   Refers to manufacturing of completely known and specified products similar to the make-to-stock operation. The difference is that manufacturing is not initiated by replenishment orders released based on availability in stock but by delivery schedules received from the customer.

Make to order (MTO)*   A production environment where a good or service can be made after receipt of customer's order.

Make to stock (MTS)*   A production environment where products can be and usually are finished before receipt of a customer order.

Manufacturing bill   A manufacturing bill refers to a bill of material for a product which reflects how it is intended to manufacture the product, the structure of material flows, which assembly and sub-assembly steps are required.

Manufacturing lead time*   The total time required to manufacture an item, exclusive of lower level purchasing lead time.

Manufacturing order* A document, group of documents, or schedule conveying authority for the manufacture of specified parts or products in specified quantities.

Manufacturing resource planning (MRP II)* A method for the effective planning of all resources of a manufacturing company. Ideally, it addresses operational planning units, financial planning in euros, and has a simulation capability to answer what-if questions. It is made up of a variety of processes, each linked together: business planning, production planning (sales and operations planning), master production scheduling, material requirements planning, capacity requirements planning, and the execution support systems for capacity and material. Output from these systems is integrated with financial reports such as the business plan, purchase commitment report, shipping budget, and inventory projections in euros. Manufacturing resource planning is a direct outgrowth and extension of closed-loop MRP.

Master production schedule (MPS)* The master production schedule is a line on the master schedule grid that reflects the anticipated build schedule for those items assigned to the master scheduler. The master scheduler maintains this schedule, and in turn, it becomes a set of planning numbers that drives material requirements planning. It represents what the company plans to produce expressed in specific configurations, quantities and dates.

Master production scheduling The planning level where delivery plans and production schedules for the company's products are drawn up on the basis of current customer orders and/or forecasts with adjustments for any stock on hand.

Material availability checking A manufacturing order cannot start unless starting materials are available. The process of checking the availability of necessary starting materials for orders is called materials availability checking.

Material planning* The activities and techniques of determining the desired levels of items, whether raw materials, work in process, or finished products including order quantities and safety stock levels.

Material requirements planning (MRP)* A set of techniques that uses bill of material data, inventory data and the master production schedule to calculate requirements for material. It makes recommendations to release replenishment orders for material. Further, because it is time phased, it makes recommendations to reschedule open orders when due dates and need dates are not in phase. Time-phased MRP begins with the items listed on the MPS and determines (1) the quantity of all components and materials required to make those items and (2) the date that the components and material are required. Time-phased MRP is accomplished by exploding the bill of material, adjusting for inventory quantities on hand and offsetting the net requirements by the appropriate lead times.

Mean absolute deviation (MAD)* The average of the absolute values of the deviations of observed values from some expected value.

Mean absolute percentage error (MAPE) The mean absolute percentage error is a relative measure of MAD, i.e. the relationship between MAD and demand.

Mean error (ME) Mean error means the average forecast error over a number of forecasting periods.

Mean square error (MSE) Mean square error is an estimate of the variance of demand when the mean demand is 0.

Mixed model production* Making several different parts or products in varying lot sizes so that a factory produces close to the same mix of products that will be sold that day. The mixed-model schedule governs the making and the delivery of component parts, including those provided by outside suppliers. The goal is to build every model every day, according to daily demand.

Move kanban* In a just-in-time context, a card or other signal indicating that a specific number of units of a particular item are to be taken from a source (usually an outbound stockpoint) and taken to a point of use (usually an inbound stockpoint). It authorises the movement of one part number between a single pair of work centres. The kanban card circulates between the outbound and the inbound stockpoint of the using work centre.

Moving average* An arithmetic average of a certain number (n) of the most recent observations. As each new observation is added, the oldest observation is dropped. The value of n (the number of periods to use for the average) reflects responsiveness versus stability in the same way that the choice of smoothing constant does in exponential smoothing.

Multi-site master production scheduling   In multi-site master production scheduling, master plans for more than one production site are developed at the same time. The aim is to balance demand and supply and synchronise the material flow along a supply chain by developing master plans which are optimal for the supply chain as a whole. It not only balances demand with available capacities but also assigns demands (production and distribution amounts) to sites in order to avoid bottlenecks, wherefore it has to cover one full seasonal cycle, or at least 12 months in terms of weekly or monthly time buckets. The optimisation models used are normally defined as linear programming (LP) or mixed integer programming (MILP) models considering the total supply chain costs, forecasted demand and sales prices and defined constraints and restrictions of the entities included. These models are included in advanced planning and scheduling systems (APS) software.

Net capacity   The capacity calculated as available for performing planned manufacturing activities.

Net requirement*   In MRP, the net requirements for a part or an assembly are derived as a result of applying gross requirements and allocations against inventory on hand, scheduled receipts and safety stock.

Normal distribution*   A particular statistical distribution where most of the observations fall fairly close to one mean, and a deviation from the mean is as likely to be plus as it is to be minus. When graphed, the normal distribution takes the form of a bell-shaped curve.

Obsolescence stock*   Stock items that have met the obsolescence criteria established by the organisation. For example, stock that has been superseded by a new model or otherwise made obsolescent. Obsolete stock will never be used or sold at full value. Disposing of the stock may reduce a company's profit.

Operation*   (1) A job or task, consisting of one or more work elements, usually done essentially in one location. (2) The performance of any planned work or method associated with an individual, machine, process, department or inspection. (3) One or more elements that involve one of the following: the intentional changing of an object in any of its physical or chemical characteristics; the assembly or disassembly of parts or objects; the preparation of an object for another operation, transportation, inspection or storage; planning, calculating or giving or receiving information.

Operation reporting*   The recording and reporting of every manufacturing operation occurrence on an operation-to-operation basis.

Order acknowledgement   A message from a supplier to advise the customer that a purchase order has been received and accepted.

Order backlog*   All the customer orders received but not yet shipped. Sometimes referred to as open orders or the order board.

Order confirmation   A note by which the supplier confirms that he undertakes to deliver a certain quantity of a certain product at a certain point in time.

Order line   Each unique item on an order represents an order line.

Order planning   The planning level related to materials supply, i.e. to ensure that all raw materials, purchased components, parts and other semi-finished items are purchased or manufactured internally in such quantities and at such times that the master production schedules drawn up under the master production scheduling process can be carried out.

Order release   The activity of releasing materials to a production process to support a manufacturing order.

Order release control   The control of order release involves determining which manufacturing orders will be released to the workshop and executed. An important starting point is the start time and due time established at the material planning level, but current workload must also be taken into consideration.

Order splitting   Dividing an order into two or more batches and simultaneously performing the same operation on each batch using two or more machines.

Ordering costs*   Used in calculating order quantities, the costs that increase as the number of orders placed increases. It includes costs related to the clerical work of preparing, releasing, monitoring, receiving orders, the physical handling of goods, inspections, and setup costs, as applicable.

Order-to-delivery process   From customer order received to invoiced dispatch.

Order-up-to level*   The order-up-to level is used in a period review system and equals the order point plus a variable order quantity. It is sometimes called target inventory level.

Overlapping   The overlapping of successive operations means that a portion of the order quantity finished at one work centre is moved to be processed at the succeeding work centre while the remaining quantity still is processed at the proceeding work centre.

Overtime*   Work beyond normal established working hours that usually requires that a premium be paid to the workers.

Pegging*   In MRP and MPS, the capability to identify for a given item the sources of its gross requirements and/or allocations. Pegging can be thought of as active where-used information.

Periodic ordering system   A periodic ordering system is an inventory control system where new replenishment orders are released at predetermined regular intervals irrespective of current stock on hand. The ordered quantity is set equal to the difference between an order-up-to level and stock on hand at the time of ordering.

Periodic physical inventory counting   A periodic ordering system is an inventory control system where new replenishment orders are released at predetermined regular intervals irrespective of current stock on hand. The ordered quantity is set equal to the difference between an order-up-to level and stock on hand at the time of ordering.

Phantom items   An item type in an ERP system representing administrative phenomena. The use of phantom items varies somewhat, but there are two general areas of use: (1) when adding and maintaining bills of materials; (2) for creating more rational planning procedures and material flows.

Physical inventory counting   The determination of actual quantity in inventory. The purpose is to adjust the quantity in the inventory accounting system if differences occur.

Picking*   The process of withdrawing from stock the components to make assemblies or finished goods. In distribution, the process of withdrawing goods from stock to ship to a distribution warehouse or to a customer.

Planned order*   A suggested order quantity, release date, and due date created by the planning system's logic when it encounters net requirements in processing MRP.

Planning horizon*   The amount of time a plan extends into the future.

Planning time fence*   A point in time denoted in the planning horizon of the master scheduling process that marks a boundary inside of which changes to the schedule may adversely affect component schedules, capacity plans, customer deliveries and cost. Outside the planning time fence, customer orders may be booked and changes to the master schedule can be made within the constraints of the production plan. Changes inside the planning time fence must be made manually by the master scheduler.

Priority control*   The process of communicating start and completion dates of manufacturing departments in order to execute a plan. The dispatch list is the tool normally used to provide these dates and priorities based on the current plan and status of all open orders.

Priority rule   A priority rule is a rule used to sequence orders and operations available to be executed at a work centre. First in, first out, shortest operation time and earliest due date are examples of commonly used priority rules.

Procurement*   The business functions of procurement planning, purchasing, inventory control, traffic, receiving, incoming inspection and salvage operations.

Product mix flexibility   The ability within the existing capacity to rapidly adapt production and material supplies to shifts in demand between existing products and product variants.

Product model   A group of product variants which represents all the different possible combinations of modules.

Product type   Product variants are often based on a number of different basic designs, called product types.

Production activity control*   The function of routing and dispatching the work to be accomplished through the production facility and of performing supplier control.

Production kanban*   In a just-in-time context, a card or other signal for indicating that items should be made for use or to replace some items removed from pipeline stock.

Production plan*   The agreed-upon plan that comes from the production planning (sales and operations planning) process, especially the overall level of manufacturing output planned to be produced, usually stated as a monthly rate for each product family (group of products, items, options, features and so on). Various units of measurement can be used to express the plan: units, tonnage, standard hours, number of workers and so on. The production plan is management's authorisation for the master scheduler to convert it into a more detailed plan, that is, the master production schedule.

Production rate   Expresses production volume per time unit.

Project process   A manufacturing process characterised by basically no product flow at all; instead, production resources are organised around the product as it evolves.

Pull   A characteristic of materials management, meaning that the manufacturing and the movement of materials only take place on the initiative of and authorised by the consuming unit in the flow of materials.

Pull system*   (1) In production, the production of items only as demanded for use or to replace those taken for use. (2) In material control, the withdrawal of inventory as demanded by the using operations. Material is not issued until a signal comes from the user. (3) In distribution, a system for replenishing field warehouse inventories where replenishment decisions are made at the field warehouse itself, not at the central warehouse or plant.

Purchase item*   An item sourced from a supplier.

Purchase order*   The purchaser's authorisation used to formalise a purchase transaction with a supplier. A purchase order, when given to a supplier, should contain statements of the name, part number, quantity, description and price of the goods or services ordered; agreed-to terms as to payment, discounts, date of performance, and transportation; and all other agreements pertinent to the purchase and its execution by the supplier.

Purchase requisition*   An authorisation to the purchasing department to purchase specified materials in specified quantities within a specified time.

Push   A characteristic of materials management, meaning that manufacturing and materials movement takes place without the consuming unit authorising the activities, i.e. they have been initiated by the supplying unit itself or by a central planning unit in the form of plans or direct orders.

Pyramid forecasting*   A forecasting technique that enables management to review and adjust forecasts made at an aggregate level and to keep lower level forecasts in balance. The procedure begins with the roll up (aggregation) of item forecasts into forecasts by product group. The management team establishes a (new) forecast for the product group. The value is then forced down (disaggregation) to individual item forecasts so that they are consistent with the aggregate plan. The approach combines the stability of aggregate forecasts and the application of management judgement with the need to forecast many end items within the constraints of an aggregate forecast or sales plan.

Quantity discount*   A price reduction allowance determined by the quantity or value of a purchase.

Queue time*   The amount of time a job waits at a work centre before set-up or work is performed on the job. Queue time is one element of total manufacturing lead time. Increases in queue time result in direct increases to manufacturing lead time and work-in-process inventories.

Quotation*   A statement of price, terms of sale, and description of goods or services offered by a supplier to a prospective purchaser; a bid. When given in response to an enquiry, it is usually considered an offer to sell.

Random stock location system*   A storage technique in which parts are placed in any space that is empty when they arrive at the storeroom. Although this random method requires the use of a locator file to identify part locations, it often requires less storage space than a fixed-location storage method.

Rate-based scheduling*   A method for scheduling and producing based on a periodic rate, e.g. daily, weekly, or monthly. This method has traditionally been applied to high-volume and process industries. The concept has also been applied within job shops using cellular layouts and mixed-model level schedules where the production rate is matched to the selling rate.

Regulated order release   When orders are released through scheduled start times, they are released entirely on the basis of timing by material planning in which only material requirements are considered, without considering available capacity. Regulated order release, on the other hand, means that the release of orders to the shop floor is adapted to the capacity available through checks made on short-term workload.

Release time fence   The point in time when manufacturing and purchase orders at the latest must be released is called the release time fence.

Released order*   An open order that has an assigned due date. Also called open order or scheduled receipt.

Re-order point system*   The inventory method that replaces an order for a lot whenever the quantity on hand is reduced to a predetermined level known as the order point.

Report card   Part of the shop packet. For every operation there is at least one report card. The aim of the report card is, among other things, to report that operations have been started and completed.

Request for quotation*   A document used to solicit vendor responses when a product has been selected and price quotations are needed from several vendors.

Rescheduling*   The process of changing order or operation due dates, usually as a result of their being out of phase with when they are needed.

Resource planning*   Capacity planning conducted at the business plan level. The process of establishing, measuring and adjusting limits or levels of long-range capacity. Resource planning is normally based on the production plan but may be driven by higher level plans beyond the time horizon for the production plan, e.g. the business plan. It addresses those resources that take long periods of time to acquire. Resource planning decisions always require top management approval.

Rough-cut capacity planning (RCCP)*   The process of converting the master production schedule into requirements for key resources, often including labour, machinery, warehouse space, suppliers' capabilities and, in some cases, money. Comparison to available or demonstrated capacity is usually done for each key resource. This comparison assists the master scheduler in establishing a feasible master production schedule. Three approaches to performing RCCP are the bill of labour (resources, capacity) approach, the capacity planning using overall factors approach, and the resource profile approach.

Routing*   Information detailing the method of manufacture of a particular item. It includes the operations to be performed, their sequence, the various work centres involved and the standards for set-up and run. In some companies, the routing also includes information on tooling, operator skill levels, inspection operations, instruction sheet, manufacturing data sheet, operation chart, operation list, operation sheet, route sheet and routing sheet.

Run time*   The time required to process a piece or lot at a specific operation. Run time does not include set-up time.

Run-out time   Run-out time refers to the time that available stock on hand, i.e. current stock on hand plus planned deliveries within lead time, is expected to last. It is calculated by dividing stock on hand plus planned deliveries by expected demand per time unit.

Run-out time planning   An inventory control method similar to the re-order point system. Instead of comparing a re-order point with current stock on hand to decide whether to order or not, the run-out time is compared with the replenishment lead time. If the run-out time is less than the lead time an order is released.

Safety factor   The numerical value used when calculating safety stocks to provide a given level of customer service.

Safety lead time*   An element of time added to normal lead time to protect against fluctuations in lead time so that an order can be completed before its real need date. When used, the MRP system, in offsetting for lead time, will plan both order release and order completion for earlier dates than it would otherwise.

Safety stock*   (1) In general, a quantity of stock planned to be in inventory to protect against fluctuations in demand or supply. (2) In the context of master production scheduling, the additional inventory and capacity planned as protection against forecast errors and short-term changes in backlog. Overplanning can be used to create safety stock.

Sales and operations planning (SOP)**   A strategic process that reconciles conflicting business objectives and plans future actions. SOP planning usually involves various business functions such as sales, operations and finance working together to agree on a single plan/forecast that can be used to drive the entire business.

Sales management approach   Forecasting where the management personnel are gathered to one or more meetings to discuss and determine forecasts of future sales or related issues. Sales statistics and other types of assessment data are compiled, processed and distributed in preparation for these meetings.

Sales plan   Also called delivery plan. A time-phased statement of expected customer orders anticipated to be received (incoming sales, not outgoing shipments) for each major product family or item. It represents sales and marketing management's commitment to take all reasonable steps necessary to achieve this level of actual customer orders. The sales plan is a necessary input to the production planning process (or sales and operations planning process). It is expressed in units identical to those used for the production plan (as well as in sales euros).

Scheduled receipt*   An open order that has an assigned due date.

Seasonal variation*   A repetitive pattern of demand from year to year (or other repeating time interval) with some periods considerably higher than others.

Service level*   A measure (usually expressed as a percentage) of satisfying demand through inventory or by the current production schedule in time to satisfy the customer's requested delivery dates and quantities.

Set-up costs*   Costs such as scrap costs, calibration costs, downtime costs and lost sales associated with preparing the resource for the next product.

Set-up time*   The time required for a specific machine, resource, work centre, process or line to convert from the production of the last good piece of item A to the first good piece of item B.

Shop calendar*   A calendar used in inventory and production planning functions that consecutively numbers only the working days so that the component and work order scheduling may be done based on the actual number of workdays available.

Shop packet*   A package of documents used to plan and control the shop floor movement of an order. The packet may include a manufacturing order, operation sheets, engineering, blueprints, picking lists, move tickets, inspection tickets and time tickets.

Shortage costs*   The marginal profit that is lost when a customer orders an item that is not immediately available in stock.

Signal kanban   A signal kanban is a special kanban card tagged onto the edge of the pallet or container positioned as the re-order point in the store room.

Silver-Meals method   Is a dynamic lot sizing method, also known as the least period cost method.

Single minute exchange of die (SMED)*   The concept of set-up times of less than 10 minutes, developed by Shigeo Shingo in 1970 at Toyota.

Speculation stock   A type of cycle stock in which the stock replenishment is completely decoupled from expected short-term consumption. The motivation for building up a speculation stock is expected price increases in the future.

Square root formula   Synonym: Economic order quantity.

Standard deviation*   A measurement of dispersion of data or of a variable. The standard deviation is computed by finding the differences between the average and actual observations, squaring each difference, adding the squared differences, dividing by $n - 1$ (for a sample), and taking the square root of the result.

Standard time*   The length of time that should be required to (1) set up a given machine or operation and (2) run one batch or one or more parts, assemblies, or end products through that operation. This time is used in determining machine requirements and labour requirements. Standard time assumes an average worker following prescribed methods and allows time for personal rest to overcome fatigue and unavoidable delays. It is also frequently used as a basis for incentive pay systems and as a basis of allocating overhead in cost accounting systems.

Stock on hand   The quantity shown in the inventory accounting system as being physically available is called stock on hand.

Stock point*   A designated location in an active area of operation into which material is placed and from which it is taken. Not necessarily a stockroom isolated from activity, it is a way of tracking and controlling active material.

Stock service level   Stock service level is a part of the overall customer service and means the extent to which products can be delivered to the customer directly from stock. It may refer to either an individual product, a product group or to the overall performance level for the entire product range in stock.

Subcontracting*   Sending production work outside to another manufacturer.

Supplier kanban   A move kanban is called supplier kanban when the supplying unit is an external supplier.

Supply*   (1) The quantity of goods available for use. (2) The actual or planned replenishment of a product or component. The replenishment quantities are created in response to a demand for the product or component or in anticipation of such a demand.

Supply chain planning (SCP)*   The determination of a set of policies and procedures that govern the operation of a supply chain. Planning includes the determination of marketing channels, promotions, respective quantities and timing, inventory and replenishment policies and production policies. Planning establishes the parameters within which the supply chain will operate.

Synchroflushing   A variant of backflushing when manufacturing orders and allocations are used. This lies somewhere between start and final reporting, and means that quantity reporting takes place at each operation, or at selected operations during the process of manufacture.

Takt time*   Sets the pace of production to match the rate of customer demand and becomes the heartbeat of any lean production system. It is computed as the available production time divided by the rate of customer demand.

Theory of constraints (TOC)*   A management philosophy developed by Dr Eliyahu M. Goldratt that can be viewed as three separate but interrelated areas – logistics, performance measurement and logic thinking.

Throughput capacity   Throughput capacity is a measurement of how many hours per time period can be set aside to perform a certain manufacturing operation within one resource unit.

Throughput time*   Synonym: Cycle time.

Tied-up capital   The capital tied up in the flow of materials, i.e. materials that are held in raw material and component stocks, in production, in finished stocks or distribution stocks and in transport.

Time fence*   A policy or guideline established to note where various restrictions or changes in operating procedures take place. For example, changes to the master production schedule can be accomplished easily beyond the cumulative lead time, while changes inside the cumulative lead time become increasingly more difficult to a point where changes should be resisted. Time fences can be used to define these points.

Time hedging   Uncertainty hedging where a delivery is made earlier than the expected time of need.

Time series*   A set of data that is distributed over time, such as demand data in monthly time periods. Various patterns of demand must be considered in time series analysis: seasonal, trend, cyclical and random.

Time-phased order point (TPOP)*   MRP-like time planning logic for independent demand items, where gross requirements come from a forecast, not via explosion. This technique can be used to plan distribution centre inventories as well as used to plan for service (repair) parts, because MRP logic can readily handle items with dependent demand, independent demand, or a combination of both. Time-phased order point is an approach that uses time periods, thus allowing for lumpy withdrawals instead of average demand. When used in distribution environments, the planned order releases are input to the master schedule dependent demands.

Total quality management (TQM)*   A management approach to long-term success through continuous improvement and customer satisfaction.

Trend*   General upward or downward movement of a variable over time, e.g. demand, process attribute.

Two-bin system*   A type of fixed-order system in which inventory is carried in two bins. A replenishment quantity is ordered when the first bin (working) is empty. During the replenishment lead time, material is used from the second bin (which contains a quantity to cover demand during lead time plus some safety stock). When material is received, the second bin is refilled and the excess is put into the working bin. At this time, stock is drawn from the first bin until it is again exhausted. This term is also used loosely to describe a fixed-order system even when physical "bins" do not exist.

Utilisation rate   The percentage of time a machine or work centre is being used for production is called the utilisation rate. It compares actual time used to available time.

Vendor-managed inventory (VMI)*   A means of optimising supply chain performance in which the supplier has access to customer's inventory data and is responsible for maintaining the inventory level required by the customer. This activity is accomplished by a process in which resupply is done by the vendor through regularly scheduled reviews of the on-site inventory. The on-site inventory is counted, damaged or outdated goods are removed, and the inventory is restocked to predefined levels. The vendor obtains a receipt for the restocked inventory and accordingly invoices the customer.

Wilson's formula   Synonym: Economic order quantity.

Withdrawal* (1) Removal of material from stores. (2) A transaction issuing material to a specific location, run, or schedule.

Volume flexibility* The ability of the transformation process to quickly accomodate large variations in production levels.

Work centre* A specific production area, consisting of one or more people and/or machines with similar capabilities, that can be considered as one unit for purposes of capacity requirements planning and detailed scheduling.

Work centre where-used* A listing (constructed from a routing file) of every manufactured item that is routed (primary or secondary) to a given work centre.

Work in process (WIP)* A good or goods in various stages of completion throughout the plant, including all material from raw material that has been released for initial processing up to completely processed material awaiting final inspection and acceptance as finished goods inventory. Many accounting systems also include the value of semi-finished stock and components in this category.

Notes:
*Adapted from Blackstone, J. and Cox, J. (eds) (2005) *APICS dictionary*, 11th edition, APICS, Fall Church.
**Adapted from CSCMP (2006) *Supply chain and logistics terms and glossary* (updated October 2006), Council of Supply Chain Management Professionals (CSCMP), www.cscmp.org.

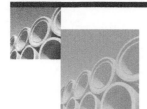

Index